W9-BCG-415

Mathematics, science and epistemology

Mathematics, science and epistemology

Philosophical Papers

Volume 2

IMRE LAKATOS

EDITED BY

JOHN WORRALL AND GREGORY CURRIE

CAMBRIDGE UNIVERSITY PRESS

CAMBRIDGE

LONDON · NEW YORK · MELBOURNE

Published by the Syndics of the Cambridge University Press
The Pitt Building, Trumpington Street, Cambridge CB2 1RP
Bentley House, 200 Euston Road, London NW1 2DB
32 East 57th Street, New York, NY 10022, USA
296 Beaconsfield Parade, Middle Park, Melbourne 3206, Australia

First published 1978

Printed in Great Britain
at the University Press, Cambridge

Library of Congress Cataloguing in Publication Data

Lakatos, Imre.
Mathematics, Science and Epistemology.

(His Philosophical papers; v. 2)
Bibliography: p.
Includes index.
1. Mathematics – Philosophy – Collected works.
2. Science – Philosophy – Collected works. I. Title.
Q175.L195 vol. 2 [QA8.6] 501s [510'.1] 77-71415
ISBN 0 521 21769 5

Contents

Editors' introduction

When Imre Lakatos died in 1974, many friends and colleagues expressed the hope that his unpublished papers would be made available. Some were also interested in seeing his contributions to journals and conference proceedings collected together in a book. At the request of the managing committee of the Imre Lakatos Appeal Fund we have prepared two volumes of selected papers which we hope will meet these demands.

None of the papers published here for the first time was regarded by Lakatos as entirely satisfactory. Some are early drafts, while others seem not to have been intended for publication. We have pursued a fairly liberal policy, including papers which, at least in their present form, Lakatos would not have allowed to go to print. As for previously published papers, we have included them all except for the two papers, 'The Role of Crucial Experiments in Science' and 'Criticism and the Methodology of Scientific Research Programmes', which would have introduced undue repetition, and *Proofs and Refutations*, which recently appeared in book form.

Information about the history of the material published here is included as introductory footnotes to each paper. These and other editorial footnotes are indicated by asterisks. (We have aimed to minimize the number of editorial footnotes, particularly in the case of papers which Lakatos had himself published.)

Offprints of some of the published papers found in Lakatos's library contained handwritten corrections and we have incorporated these wherever possible. In preparing the previously unpublished papers for the press, we have taken the liberty of introducing some presentational alterations where the original text was incomplete, or seemed likely to be misleading, or where minor alterations seemed to produce major increases in readability. We felt justified in making these changes because Lakatos always took great care over the presentation of any of his material which was to be published and, prior to publication, he always had such material widely circulated among colleagues and friends for criticism and suggested improvements. The newly published papers would undoubtedly have undergone this treatment and the resulting changes would have been much more far-reaching than those we have dared to introduce. Wherever the device of enclosing our alterations within square brackets worked easily and smoothly we have adopted it. (However, square brackets within quotations from other authors enclose Lakatos's own insertions.)

Where Lakatos mentioned a paper reprinted in either of the present volumes, we have altered the style of reference. So, for example, 'Lakatos [1970*a*]' becomes 'volume 1, chapter 1', and 'Lakatos [1968*b*]' becomes 'this volume, chapter 8'.

Chapter 1 is reprinted from the *Aristotelian Society Supplementary Volumes*, **36**, 1962, by kind permission of the Editor. Chapter 8 is reprinted with the permission of the Noth-Holland Publishing Company. The editing of 'Cauchy and the Continuum' (this volume, chapter 3) was kindly undertaken by Dr J. P. Cleave of the University of Bristol. The editorial footnotes to that chapter (marked 'J.P.C.') were written by him. (For Cleave's own interesting development of the problem which Lakatos deals with in that chapter, see his [1971].)

A generous grant from the Fritz Thyssen Stiftung made possible the creation of an archive of Lakatos's papers – an essential preliminary to the publication of these volumes. We should like to thank Nicholas Krasso and Professors Kneale, Pearce Williams and Szabo for helping us to supply some missing references, and Dr Cleave for his work on chapter 3. We are grateful to Allison Quick who constructed the indexes to this volume. Once again our thanks are due to Sandra Mitchell for her help in preparing these volumes, to John Watkins for helpful advice and to Gillian Page for her generous co-operation in making Lakatos's papers available to us.

J.W.
G.C.

Part 1

Philosophy of Mathematics

I

Infinite regress and foundations of mathematics*

INTRODUCTION

[Sceptical philosophy has been teaching for more than two thousand years that it is impossible to achieve either the aim of conclusively establishing meaning or the aim of conclusively establishing truth. But to establish the meaning and truth of mathematics is precisely the aim of 'foundations'.]

The classical sceptical argument is based on the infinite regress. One can try to pin down the meaning of a term either by defining it in other terms – this leads to infinite regress – or by defining it in 'perfectly well-known terms'. But are the four terms of the expression 'perfectly well-known terms' really all perfectly well-known terms? One sees that the abyss of infinite regress opens up again. How then could mathematical philosophy still claim that in mathematics we have or we should have exact concepts? How does it hope to avoid the sceptics' strictures? How can it claim that it has offered foundations of mathematics – logicist, meta-mathematical or intuitionist? But even allowing for 'exact' concepts, how can we prove that a proposition is true? How can we avoid the infinite regress in proofs, even if we could avoid the infinite regress in definitions? Meaning and truth can only be transferred, but not established. But if so, how can we *know*?

This controversy between *dogmatists* – who claim that we can know – and *sceptics* – who claim that we either cannot know or at least cannot

* This paper was first published in the Aristotelian Society Supplementary Volume, 36, 1962. An offprint of that paper in Lakatos's library contained some handwritten corrections some of which we have included. Lakatos's paper was originally delivered as the second paper in a symposium on the foundations of mathematics at the Aristotelian Society–Mind Association Joint Session at the University of Leicester in July 1962. It began with a brief discussion of the first paper in the symposium given by R. L. Goodstein (Goodstein [1962]). This discussion is difficult to understand out of context and so we have omitted it. The burden of the paper is entirely unaffected by this omission. Lakatos's introductory footnote reads: 'The connoisseur will appreciate the impact of Karl Popper's philosophy throughout the paper. It was technically impossible for me to give the proper references to him – I have to assume that the reader will recognise, in what follows, many of the ideas of *Logic of Scientific Discovery* and *Conjectures and Refutations*. I am also indebted to A. Musgrave and Dr T. J. Smiley, who read the first draft, for their many valuable suggestions and criticisms. W. W. Bartley drew my attention to the central role of the sceptic–dogmatist controversy in the history of epistemology. I also gained much from discussing the first two sections with Professor S. Körner and J. C. Shepherdson.' (*Eds.*)

know that we can know and when we can know – is the basic issue in epistemology. In discussing modern efforts to establish foundations for mathematical knowledge one tends to forget that these are but a chapter in the great effort to overcome scepticism by establishing foundations for knowledge in general. *The aim of my contribution is to exhibit modern mathematical philosophy as deeply embedded in general epistemology and as only to be understood in this context.* This is why the first section inevitably contains a potted history of epistemology. Respectable historians sometimes say that the sort of 'rational reconstruction' here attempted is a caricature of real history – of the way things actually did happen – but one might equally well say that both history and the way things actually did happen are just caricatures of the rational reconstruction.

1 STOPPING INFINITE REGRESS IN SCIENCE

The sceptics used infinite regress to show that it is hopeless to find foundations for knowledge. They were – just like their dogmatist opponents – epistemological justificationists, i.e. their main problem was *How do you know?*, and they thought that they had to fall back on a prostrating *I do not know* because there can be no firm foundations for meaning and truth. They concluded that rational effort to obtain knowledge is powerless; science and mathematics are sophistry and illusion. So it has become a vital problem for rationalism to stop these exasperating twin infinite regresses and to provide knowledge with a firm bedrock. Three huge rationalist enterprises tried to achieve this: (1) the *Euclidean programme*, (2) the *Empiricist programme* and (3) the *Inductivist programme.*

All these three set out to organize knowledge in *deductive systems.* The basic definitional characteristic of a (not necessarily formal) deductive system is the *principle of retransmission of falsity* from the 'bottom' to the 'top', from the conclusions to the premises: a counterexample to a conclusion will be a counterexample to at least one of the premises. If the principle of retransmission of falsity applies, so does the *principle of transmission of truth* from the premises to the conclusions. We do not demand, however, from a deductive system that it should transmit falsehood or retransmit truth.

(1) I call a deductive system a '*Euclidean theory*' if the propositions at the top (*axioms*) consist of perfectly well-known terms (*primitive terms*), and if there are *infallible truth-value-injections* at this top of the truth-value *True,* which flows downwards through the deductive channels of truth-transmission (*proofs*) and inundates the whole system. (If the truth-value at the top was *False,* there would of course be no current of truth-value in the system.) Since the Euclidean programme implies that all knowledge can be deduced from a finite set of trivially true propositions consisting only of terms with a trivial meaning-load,

I shall call it also the *Programme of Trivialization of Knowledge*.[1] Since a Euclidean theory contains only indubitably true propositions, it operates neither with conjectures nor with refutations. In a fully-fledged Euclidean theory meaning, like truth, is injected at the top and it flows down safely through meaning-preserving channels of nominal definitions from the primitive terms to the (abbreviatory and therefore theoretically superfluous) defined terms. A Euclidean theory is *eo ipso* consistent, for all the propositions occurring in it are true, and a set of true propositions is certainly consistent.

(2) I call a deductive system an '*empiricist theory*' if the propositions at the bottom (*basic statements*) consist of perfectly well-known terms (*empirical terms*) and there is a possibility of *infallible truth-value-injection* at this bottom which, if the truth-value is *False*, flows upward through the deductive channels (*explanations*) and inundates the whole system. (If the truth-value is *True*, there is, of course, no current of truth-value in the system.) Thus an empiricist theory is either conjectural (except possibly for true statements at the very bottom) or consists of conclusively false propositions.[2] In an empiricist theory there are *theoretical* or '*occult*' *terms* which – like the middle terms of Aristotelian syllogisms – do not figure in any basic statements, and there are no meaning-preserving channels leading upwards to them.

If, in a rationalist zeal to keep out 'metaphysics', we allow, apart from logical meaning-injections, meaning-injections only at the bottom, we have a '*Strictly Empiricist Theory*'. This requirement – devised to demarcate science from gibberish – is however suicidal, since a strictly empiricist theory with theoretical terms is, not counting its bottom level, meaningless.[3] An empiricist theory may be consistent or may be inconsistent. Therefore for an empiricist theory the need for a consistency proof emerges.[4]

The Euclidean programme proposes to build up Euclidean theories with foundations in meaning and truth-value at the top, lit by the *natural light of Reason*, specifically by arithmetical, geometrical, metaphysical, moral, etc. intuition. The Empiricist programme proposes to build up Empiricist theories with foundations in meaning and truth-value at the bottom, lit by the *natural light of Experience*. Both programmes however rely on Reason (specifically on logical intuition) for the safe transmission of meaning and truth-value.

[1] The *locus classicus* for the description of this programme can be found in Pascal [1657–8].

[2] For a most lyrical description of some aspects of an empiricist theory see M. Schlick's [1934], translated in Ayer (*ed.*) [1959], pp. 209–27. For a very lucid and picturesque discussion see R. B. Braithwaite [1953], *passim*, but particularly pp. 350–4.

[3] R. B. Braithwaite showed that a strictly empiricist theory without theoretical terms can be meaningful but is incapable of growth ([1953], p. 76). Strict empiricists – like Schlick and Ramsey – try to get out of the embarrassing meaninglessness of higher level hypotheses by dubbing them 'rules'.

[4] Cf. K. R. Popper [1959], pp. 91–2. I do not know who first suggested that we test respectable scientific theories for consistency.

I should stress the difference between the usual concept of an *empirical* theory and the more general concept of an '*empiricist* theory'. My only condition for an empiricist theory is that truth-value is injected at the bottom, whatever that happens to be – 'factual', 'singular spatio-temporal', 'arithmetical' or what not. The point of this stretching of the concept of basic statement is to make the concepts of the Empiricist and Inductivist programmes applicable to mathematics – or to metaphysics, ethics, etc.

In traditional epistemology the two crucial concepts are not '*Euclidean*' and '*Empiricist*', but on the one hand *a priori* and *a posteriori*, on the other '*analytic*' and '*synthetic*'. These concepts refer to propositions and not to theories; epistemologists were slow to notice the emergence of highly organized knowledge, and the decisive rôle played by the specific patterns of this organization. It makes an immense epistemological difference at what level we inject truth-value in the theory; for this determines the pattern of flow of truth and falsity in the system. From which source the injection comes – from experience, from self-evidence or from anything else – is, however, of secondary importance for the solution of many problems. We can get a long way merely by discussing *how* anything flows in a deductive system without discussing the problem of *what in fact flows* there, infallible truth or only, say, Russellian 'psychologically incorrigible' truth, Braithwaitian 'logically incorrigible' truth, Wittgensteinian 'linguistically incorrigible' truth[1] or Popperian corrigible falsity and 'verisimilitude', Carnapian probability.

The fascinating story of the Euclidean programme and of its breakdown has not yet been written, although it is generally known how in the upper regions of deductive structures modern science led to terms ever more theoretical and to propositions ever more unlikely, instead of to ever more trivial terms and propositions. To switch to the empiricist programme and fix the foundations at the bottom was very difficult; it was indeed one of the most dramatic shocks in the history of human thought, for it implied radical changes in the originally Euclidean rational outlook. If one can inject truth-value only at the bottom, then a theory is either conjectural or false. Thus while a Euclidean theory is *verified*, an empiricist theory is *falsifiable*, but not verifiable. Both programmes offer truths, which are trivial and uninteresting if taken in isolation, but *because of its location*, trivial truth inundates the whole Euclidean theory but not the empiricist one.

A Euclidean never *has* to admit defeat: his programme is irrefutable. One can never refute the pure existential statement that there exists a set of trivial first principles from which all truth follows. Thus science may be haunted for ever by the Euclidean programme as a regulative principle, 'influential metaphysics'.[2] A Euclidean can always

[1] Cf. the [1938] papers by Braithwaite, Russell and Waismann. [2] Watkins [1958].

deny that the Euclidean programme as a whole has broken down when a particular candidate for a Euclidean theory is tottering. In fact rigorous Euclideans themselves constantly reveal that the 'Euclidean' theories of their predecessors were not *really* Euclidean, that the intuition which established the truth of the axioms was inadmissible, misleading, that it was a will-o'-the-wisp, not the truly genuine guiding Light of Reason. They may either make a completely fresh start or claim that in Euclidean fairyland the tortuous path to the sunlit peaks of triviality must unavoidably lead through sombre gorges. One must simply hope and climb further.

Short-sighted or tired Euclideans may be deluded into taking a dark gorge for a sunlit peak. While on the one hand criticism, and, surely, refutation, can detrivialize the most trivial-looking background knowledge – a beautiful example is the Einsteinian criticism of simultaneity – on the other hand authoritarian handling and corroboration can trivialize (push into unquestioned background knowledge) the most sophisticated-looking speculation, an amusing example is the Kantian attitude towards Newtonian mechanics. Refutation makes us learn; corroboration makes us forget. Thus conceited rationalism can – by a sort of 'rubber-Euclideanism' – stretch the boundaries of self-evidence, and it may do this, not only in a period of victory, but also in a period of desperate retreat.[1]

(3) Some dogmatists tried to save Knowledge from the sceptics by a non-Euclidean method. Defeated at the top, reason sought refuge and anchor at the bottom. But truth at the bottom does not have the power of truth at the top. Induction was expected to restore the symmetry. The Inductivist Programme was a desperate effort to build a channel through which truth flows *upwards* from the basic statements, thus establishing an additional logical principle, the *principle of re-transmission of truth*. Such a principle would enable the inductivist to inundate the whole system with truth from below. An 'inductivist theory', just like a Euclidean theory, is of course consistent, for all the propositions occurring in it are true.

An inductive channel did not look so obviously impossible in the seventeenth century as it does today, if one based deduction on Cartesian intuition and disparaged Aristotelian formal logic. If there is a deductive intuition, why not have also an inductive intuition on a par? However, the history of logic (i.e. of the theory of truth-value-channels) from Descartes till today is essentially *the history of*

[1] Rubber-Euclideanism sometimes yields *proofs* with an amusing sham-rigour. Mach calls Euclideanism in science 'the mania for demonstration' (Mach [1882], chapter I, § 5). He gives an impressive enumeration: 'In this manner Archimedes *proves* his law of the lever, Stevinus his law of inclined pressure, Daniel Bernoulli the parallelogram of forces, Lagrange the principle of virtual displacements.' He could have added of course many more names like those of Maupertuis and Euler, whose Euclidean inclinations he discusses in another context in chapter IV, § 2. (But he misses Euler's proofs of Newtonian axioms.)

criticizing and improving the deductive channels and destroying the inductive channels by making logic 'formal'.

If inductivism wants to *prove* dubitable, occult, theoretical propositions from below, from the usually empirical bottom, it must also clarify perfectly the meaning of the theoretical terms – without final concepts no final truths. Thus the inductivist has to define theoretical terms in 'observables'. This cannot be done by explicit definitions so the inductivist tries to do it by contextual, implicit definitions, by 'logical constructs'. When in mathematics one wanted to prove everything from above, one had to redefine, reconstruct everything in the perfectly well-known terms of the top. When in science one wants to prove everything from below, one has first to redefine, reconstruct everything in the perfectly well-known terms of the bottom. (Particularly if one is a 'strict inductivist', for then not only does truth have to flow upwards, but meaning has to do the same, since truth cannot flow into meaningless propositions.) *The problem of inductive proof* and that of the definition of theoretical terms in observables – it could be called *the problem of inductive definition* – are thus twin-problems, and their solubility twin-illusions.[1]

The original version of the Inductivist programme has been destroyed by sceptical criticism. But most people still cannot put up with the empiricist revolution, they still consider it an affront to the dignity of Reason. Some modern ideologists of inductivism – I am now referring to a characteristic brand of logical positivism – created an immense literature in the defence of a new weak version of the old programme: *Probabilistic Inductivism.* Above all, they – rightly – cannot admit that a scientific deductive system should be meaningless, except for the very bottom of the system. In fact they claim that a theory is meaningful if its bottom reaches the level of observation-statements. But while their 'Verification Principles' admit theoretical statements to be meaningful, one is left in the dark as to their actual meaning. Strict empiricists – wrongly – cannot admit meaning injections, other than at the bottom of the theory. But are then theoretical statements meaningful without having any particular meaning? They wriggle out of the dilemma by radically expanding the concept of definition – of meaning-transference – in such a way that it embraces 'reduction', a logical sleight-of-hand devised to retransmit, if not full, at least some partial *ersatz*-meaning upwards, from the observables to the theoretical terms.[2]

Then, since they know and accept formal logic, they have to regard induction as invalid. But now, after having expanded the concept of meaning-transference, they expand the meaning of truth-transference

[1] Russell's method of 'constructionalism' was an attempt to solve the problem of inductive definition and thereby establish a firm conceptual foundation for his inductivism. For an excellent discussion cf. Weitz [1948].

[2] Cf. Carnap [1936–7], and for the recent literature, some articles and references in Feigl, Scriven and Maxwell (*eds.*) [1958].

in such a way that there should be a retransmission upwards, if not of truth, at least of partial, probable truth, some 'degree of confirmation', from the observation-statements to the theoretical statements.[1]

A theory with probabilistic induction is probably consistent. A probabilistic theory of probable consistency can be expected at any time.

When criticizing outdated, inept and pretentious modern inductivism one should not forget its noble origin. Seventeenth and eighteenth century inductivist *credo* had a very important and progressive role. It was the great *lebenslüge* of young speculative science in the dark pre-Popperian age of enlightenment, when mere guessing was despised, a refutation was a solecism, and where to establish an authoritative source of truth was a question of survival. The shift of authority from Revelation to the facts met, of course, the opposition of the Church. Scholastic logicians and 'humanists' kept pointing out the doom of the inductivist venture, and showed – on the basis of formal Aristotelian syllogistic – that there can be no valid inference from effects to causes, and thus scientific theories cannot be true but only instruments of fallible prediction: 'mathematical hypotheses'. They challenged those ideologists of modern science who rejected Aristotelian logic and preached informal, intuitive logic and induction. While defending truth of revelation they subjected truth of reason and experience to devastating criticism. The seventeenth century alliance of Euclideanism and Inductivism defended science from humiliation and fought for its high status.

Empiricists excelled in the criticism of Euclideanism. They criticized the guarantee of the intuitive Euclidean truth-injection: self-evidence. The conclusive empiricist destruction of Inductivism was, however, paradoxically accomplished by a philosopher who carried the epistemological revolution beyond Empiricism: Popper. Popper showed, in his criticism of the probabilistic version of the theory of inductive inference, that there cannot even be a partial transference of meaning and truth upwards. But he then showed that injections of meaning and of truth-value at the bottom level are far from being trivial; that there are no 'empirical' terms, but *only* 'theoretical' ones, and that there is nothing conclusive about the truth-value of basic statements, thus refurbishing the old Greek criticism of sense-experience.

(4) Popperian *critical fallibilism* takes the infinite regress in proofs and definitions seriously, does not have illusions about 'stopping' them, accepts the sceptic criticism of any infallible truth-injection. In this approach there are no Foundations of Knowledge, either at the top or at the bottom of theories, but there can be tentative truth-injections and tentative meaning-injections at any point. An 'empiricist theory' is either false or conjectural. A 'Popperian theory' can only

[1] The idea can be traced back to Leibniz [1678], and to Huyghens ([1690], Preface). Inductive Logic has been replaced by the new, weaker Probability Logic by Keynes, Reichenbach and Carnap. For references and criticism see Popper [1959], chapter x.

be conjectural. We never *know*, we only guess. We can, however, turn our guesses into criticizable ones, and criticize and improve them. In this Critical Programme many of the old problems – like those of probabilistic induction, reduction, justification of *synthetic a priori*, justification of sense-experience, and so on – become pseudo-problems, since they all answer the wrong dogmatist question *How do you know?* Instead of these old problems, however, many new problems emerge. The new central question, *How do you improve your guesses?* will give enough work for philosophers for centuries; and how to live, act, fight, die when one is left with guesses only, will give more than enough work for future political philosophers and educationalists.

The indefatigable sceptic however will ask again: 'How do you *know* that you improve your guesses?' But now the answer is easy: 'I guess.' There is nothing wrong with an infinite regress of guesses.

2 STOPPING INFINITIVE REGRESS BY THE LOGICO-TRIVIALIZATION OF MATHEMATICS

From the seventeenth to the twentieth century Euclideanism has been on a great retreat. The occasional rearguard skirmishes to break through beyond the hypotheses, towards the peaks of *first principles*, all failed. The fallible sophistication of the empiricist programme has won, the infallible triviality of Euclideans has lost. Euclideans could only survive in those underdeveloped subjects where knowledge is still trivial, like ethics, economics etc.[1]

These four hundred years of retreat seems to have by-passed mathematics completely. Euclideans here retained their original stronghold. The mess of eighteenth-century analysis was of course a set-back. But since Cauchy's revolution of rigour they headed slowly but safely towards the peaks. By a – very self-conscious – Euclidization, Cauchy and his successors performed the miracle: they turned the 'tremendous obscurity of analysis'[2] into a crystal-clear Euclidean theory. 'This great school of mathematicians, in virtue of startling definitions, have saved mathematics from the sceptics and provided a rigid demonstration of its propositions.'[3] Mathematics has been trivialized, derived from indubitable, trivial axioms in which only absolutely clear trivial terms figure, and from which truth pours down in clear channels. Concepts like 'continuity', 'limit', etc. were defined in terms of concepts like 'natural number', 'class', 'and', 'or' etc. The 'arithmetization of mathematics' was a most wonderful Euclidean achievement. Even empiricists had to admit that Euclid, the 'evil genius' of science, is to

[1] For ethics, cf. Sidgwick [1874], Book III, Intuitionism, and for recent references M. Warnock [1960]. For economics cf. for example L. C. Robbins [1932], pp. 78–9, and L. von Mises [1960], pp. 12–13.

[2] Abel [1826*b*], p. 263.

[3] Ramsey [1931], p. 56, and, following him, Russell [1959], p. 125, use this phrase to characterize their own purpose and method.

be recognized as the 'good genius' of mathematics.[1] In fact, modern logical empiricists, though far from being radical 'empiricists' in science (most of them are inductivists), are radical Euclideans in mathematics. Hard-core Euclideans however – like the young Russell – never resigned themselves to this restricted kingdom: they worked hard to complete their programme in mathematics, and then hoped to reconquer the lost territory: to Euclidize and trivialize the whole universe of knowledge.

No Euclidean theory, however, can ever stand up to sceptical criticism. And the most incisive sceptical arguments against mathematical dogmatism came from the self-tormenting doubts of the dogmatists themselves: 'Have we *really* reached the primitive terms? Have we *really* reached the axioms? Are our truth-channels *really* safe?' These questions played a decisive role in Frege's and Russell's great enterprise to go back to still more fundamental first principles, beyond the Peano axioms of arithmetic. I shall particularly concentrate on Russell's approach, showing how he failed in his original Euclidean programme, how he finally fell back on Inductivism, how he chose confusion rather than facing and accepting the fact that what is interesting in mathematics is conjectural.

The main problem of Russellian philosophy was always to save Knowledge from the sceptics. 'Scepticism, while logically impeccable, is psychologically impossible, and there is an element of frivolous insincerity in any philosophy which pretends to accept it' (Russell [1948], p. 9). In his youth he hoped to escape scepticism with the help of a vast Euclidean programme. His 'philosophical development' is virtually the piecemeal retreat from Euclideanism, bravely fighting every inch of the way, and rescuing as much certainty as he could.

It is intriguing to recall the optimism of his early plans. Russell thought that before 'extending the sphere of certainty to other sciences' he had first to arrive 'at a perfected mathematics which should leave no room for doubts' (Russell [1959], p. 36). For this he had to 'refute mathematical scepticism' (*ibid.*, p. 209), and thus secure a firm Euclidean bridgehead for the later general attack. Thus the starting-point of Russell's philosophical career was to establish mathematics as the Euclidean bridgehead.

He found mathematical proofs shockingly unreliable. 'A great deal of the argumentation that I had been told to accept was obviously fallacious' (*ibid.*, p. 209). And he was not quite happy about the certainty of the axioms – geometrical or arithmetical. He was aware of the sceptical criticism of intuition: the *leitmotiv* of his first-ever publication was to fight 'the confusion between the psychologically subjective and the logically *a priori*' (Russell [1895], p. 245). How can one know that truth-injections at the top are justified beyond doubt? In pursuing the problem he analysed the axioms of geometry and

[1] R. B. Braithwaite [1953], p. 353.

arithmetic one by one and found that their justification was based on very different sorts of intuition. In his first published paper [1896] Russell analyses the axioms of Euclidean geometry from this point of view and finds that some of the axioms are certainly true, and in particular *a priori* true, for 'their denial would involve logical and philosophical absurdities' (p. 3). He classified for instance the homogeneity of space as *a priori* true, the 'want of homogeneity and passivity is...absurd; no philosopher has ever thrown doubt, so far as I know, on these two properties of empty space; indeed they seem to flow from the maxim that nothing can act on nothing...We must, then, on purely philosophical grounds, admit...the axiom, e.g., of Conguence' (p. 4). On the other hand he classified the axiom about the three-dimensionality of space as empirical, but he claimed that its certainty is 'almost as great' as if it were *a priori* (p. 14); it is, however, not '*logically* inevitable' (my italics) and only 'may be supposed to derive its evidence from intuition' (p. 23).

So Russell tried to establish a hierarchy of *a priori* truths, of 'mathematical beliefs', geometrical or arithmetical. He 'read whatever books he could find that seemed to offer a firmer foundation for them' (Russell [1959], p. 209). This is how he came across Frege. He at once opted for Frege's solution: to derive all mathematics from trivial *logical* principles. Arithmetical intuition was to be scrapped and doomed to follow mechanical and geometrical intuition into the wastepaper basket for old detrivialized trivialities – while logical intuition was to be enthroned, not just as an 'intuition', but as infallible insight, as a super-trivial super-intuition. The arithmetico-trivialization of mathematics was to be dethroned and replaced by its logico-trivialization.

To appreciate this step we have to see the special place of logical intuition. Euclideans excel in dethroning the intuitive sources of truth-injections-at-the-top enthroned by their predecessors. The story of Euclideanism is a story full of such dethronements. Mathematical Euclideanism offers an example. The discovery of irrational numbers led the Greeks to abandon Pythagorean arithmetical for Euclidean geometrical intuition: arithmetic had to be translated into crystal-clear geometry. To accomplish this translation they elaborated their complicated 'theory of proportions'. The nineteenth century, in 'clarifying' the concept of irrational numbers, switched back to arithmetical intuition as the *dominant intuition*. Later on, Cantorian set-theoretical, Russellian logical, Hilbertian 'global', and Brouwerian 'constructivist' intuitions competed for this role. Throughout this battle for the exclusive right to inject truth-values at the top, logical intuition plays a very special role: for, whoever wins the battle for the axioms, logical intuition has to be relied upon to carry truth from the top to the remote parts of the system. Even empiricists, who in science routed all top-level intuitions (while enthroning bottom-level factual intuition), have to rely on a trivially safe logic to carry their refutations upwards. If criticism is meant to be conclusive it must hit with a deadly thrust conveyed by an inexorable logic. The special status of logical intuition explains why even arch-enemies of intuition do not list logical intuition under the head of 'intuition' at all – for they need logical intuition to criticize the others.[1] But if *any* dogmatist programme – Euclidean of any denomination, Inductivist, Empiricist – needs a trivial, truly infallible logical intuition, then to show that all mathematics does not need any other *but* logical intuition will certainly be a huge gain: there will be only one single source of certainty both for the axioms and for the truth-transmission.

Logical intuition however had first to be made autonomous, had to be purged of extraneous intuitions. In classical Euclidean theories each relevant logical step had to be justified by a special axiom. Any statement of the form '*A* entails *B*' or rather '*A* obviously entails *B*' had to be seen to be valid independently. Cartesian logic contains an undetermined infinity of topic-dependent axioms. Russell envisaged a powerful logic consisting of a few specified, trivial, 'topic-neutral'[2] axioms. He had not realized at the start that if logic is to become a super-trivial Euclidean deductive system, it has to contain on the one hand super-trivial *axioms* and on the other hand a super-super-trivial logic of this logic, containing specified *rules* to transfer truth in it: 'All pure mathematics – Arithmetic, Analysis and Geometry – is built up by combinations of the primitive ideas of logic, and its propositions are deduced from the general axioms of logic, such as the syllogism and the other rules of inference' (Russell [1901*b*], p. 76). These 'axioms' will *now* be *really* trivially true, shining beyond doubt in the natural light of *purely logical* reason, 'cornerstones, fastened in an eternal foundation, reachable but not movable by human reason' (Frege [1893], p. xvi). The *terms* occurring in them will be *really* perfectly clear logical terms. The dictionary will essentially consist only of two trivial terms: *relation* and *class*. 'What these ideas mean it is necessary to know if you wish to become an arithmetician.' But nothing is easier than that. 'It must be admitted that what a mathematician has to know to begin with is not much' (Russell [1901*b*], pp. 78–9). In this period – a month or two before the discovery of his Paradox – he thought that the *definitive* Euclidization of mathematics had been provided and scepticism for ever defeated: 'In the whole philosophy of mathematics, which used to be at least as full of doubt as any other part of philosophy, order and certainty have replaced the confusion and hesitation which formerly reigned' (*ibid.*, pp. 79–80). And thus

[1] For example, according to Couturat ([1905], chapter 1): 'self-evidence was not a condition, but an obstacle for logical rigour...self-evidence is fully subjective...and therefore alien to logic'.

[2] The term is G. Ryle's [1953].

against that kind of scepticism which abandons the pursuit of ideals because the road is arduous and the goal not certainly attainable, mathematics, within its own sphere, is a complete answer. Too often it is said that there is no absolute truth, but only opinion and private judgment; that each of us is conditioned, in his view of the world, by his own peculiarities, his own taste and bias; that there is no external kingdom of truth to which, by patience and discipline, we may at last obtain admittance, but only truth for me, for you, for every separate person. By this habit of mind one of the chief ends of human effort is denied, and the supreme virtue of candour, of fearless acknowledgment of what is, disappears from our moral vision. Of such scepticism mathematics is a perpetual reproof; for its edifice of truths stands unshakable and inexpugnable to all the weapons of doubting cynicism [*ibid.*, p. 71].

We all know how the brief Euclidean 'honeymoon' gave place to 'intellectual sorrow' (Russell [1959], p. 73), how the intended logico-trivialization of mathematics degenerated into a sophisticated system, including 'axioms' like that of reducibility, infinity, choice, and also ramified type theory – one of the most complicated conceptual labyrinths a human mind ever invented. '*Class*' and '*membership-relation*' turned out to be obscure, ambiguous, anything but 'perfectly well known'. There even emerged the completely un-Euclidean need for a consistency proof to ensure that the 'trivially true axioms' should not contradict one another. All this and what followed must strike any student of the seventeenth century as *déjà vu*: proof had to give way to explanation, perfectly well known concepts to theoretical concepts, triviality to sophistication, infallibility to fallibility, Euclidean theory to empiricist theory. We also encounter the same refusal to accept the dramatic change: the same rearguard skirmishes, hopes, and *ersatz*-solutions.

Russell's first reactions to his unintended, unwanted and counter-trivial *Principia* follow the same pattern as the classical seventeenth-century attempts to rescue dogmatism. I mentioned two of them: (1) to stick to the original Euclidean programme and either to try to break through the hypotheses to first principles, or to stretch intuition and turn the paradoxical speculation of yesterday into the obvious of today; or, if this does not help, (2) to try, by a justification of induction, to send truth, injected at the bottom, upwards to fill the whole system.

(1) Like Newton hoping to explain the Law of Gravitation by principles of Cartesian push-mechanics, Russell hoped for the trivialization of the reducibility axiom (Russell [1925], pp. 59–60): 'Although it seems very improbable that the axiom should turn out to be false, it is by no means improbable that it should be found to be deducible from some other more fundamental and more evident axiom.' Later he gave up this hope: 'Viewed from this strictly logical point of view, I do not see any reason to believe that the axiom of reducibility is logically necessary...The admission of this axiom into a system of logic is therefore a defect, even if the axiom is empirically true' (Russell [1919], p. 193).

Russell described this standard pattern in respect of the parallel-axiom:

For the Kantian view, it was necessary to maintain that all the axioms are self-evident – a view which honest people found it hard to extend to the axiom of parallels. Hence arose a search for more plausible axioms, which might be declared *a priori* truths. But though many such axioms were suggested, all could sanely be doubted, and the search only led to scepticism (Russell [1903], §353).

Would he have agreed that *his* search for 'plausible' logical axioms 'which might be declared *a priori* truths', only led to scepticism?

In the case of type theory Russell fell back on 'rubber-Euclideanism'. He was convinced that there was a trivial solution of the Russell-paradox. This, of course, had to be a very dim hope since, unlike the case of the sophisticated Burali–Forti paradox, here the most trivial common sense assertions were shown to be inconsistent, so that to improve it one would have to assume that the negation of some common-sense axiom was true. Zermelo's solution – to adopt consciously the negation of the trivially true-looking Principle of Abstraction – was in this line. The Euclidean-minded Russell, however, abhorred such a solution. He never reconciled himself to axiomatic set theory. He thought that we have only to purge our common sense from error by devoted efforts – again a seventeenth-century pattern – and we shall *see*, when new natural light comes to us, that there was *of course* always something manifestly wrong with the argument. As spotting a *lemma* in an argument, and saying that it is not trivially true but trivially false, might make Euclidean self-deceit too difficult, Russell discovered that one can replace this *de facto* detrivializing method by another: the guilty *lemma* is *not trivially false, but trivially meaningless* – only it has not occurred to us until now to look at it from this point of view. So now we have first to see whether a proposition is meaningful or a meaningless monster. If it is meaningless, it cannot be true or false; but if we do not test it for (manifest) meaningfulness, but test it at once for truth, we might be misled into taking it to be trivially true.

This 'monster-barring method' is a standard, though usually barren, Euclidean defence mechanism.[1] It nevertheless became a major principle in logical positivism, as a monstrous generalization of Russellian type theory. The main danger of the monster-barring method consists in hiding vital sophisticated assumptions in the definitions and thus behind the facade of the conceptual framework. In meta-mathematical terminology, type theory is part of the formation rules (about what constitutes a well-formed formula) and not of the axioms. We can see the significance of this move in Kemeny's advocacy of logicism. He says in his semi-popular [1959] (p. 21), that

[1] Cf. my [1961], chapter 1 (*now published in a much revised version as Lakatos [1976*c*], chapter 1 (*Eds.*)).

> Mathematics is shown to be no more than highly developed Logic. In this process two new logical principles turn up, the axioms of infinity and choice, whose somewhat controversial nature need not concern us here. Let it suffice that if we recognize these two as legitimate logical principles – as most logicians do – then all of Mathematics follows and becomes just advanced Logic.

Kemeny does not mention type-theory – which would of course spoil this picture of the infallible triviality of logic which he draws for his readers – and he can justify this omission because type theory belongs to the formation rules and not to the axioms.

Russell of course knew that for his original Euclidean programme the triviality of type-theory is in fact vital. This is why he insisted on the 'vicious-circle principle', on the meaninglessness of self-referential statements, as the basic idea of type theory. He thought that this principle would be recognized as manifest, and thus his elimination of the inconsistency of naive logic would comply with the Euclidean tenet 'that the solution should, *on reflection*, appeal to what may be called "logical common sense" – i.e. that it should seem, in the end, just what one ought to have expected all along' (Russell [1959], pp. 79–80). The – by then obviously hopeless – search for a *trivial* solution thus trapped him into the methodological poverty of monster-barring, into the particularly poor mistake of the anti-self-referential crusade, and into the 'rather sloppy' (Ramsey [1931], p. 24) deduction of type-theory from the principle.[1] Type-theory, when presented as a piece of self-evidence, as 'inherently credible' (Russell [1925], p. 37), is a fine example of rubber-Euclideanism. Russell's search for Euclidean triviality also explains his dismay at Quine's speculative 'logical dexterity' (Russell [1959], p. 80). A rubber-Euclidean tends to discard the trivialities of others as speculations, while insisting that his own speculations are trivialities.

(2) Russell occasionally despairs of Euclidean manifestness and opts for a sort of inductivism:

> That the axiom of reducibility is self-evident is a proposition which can hardly be maintained. But in fact self-evidence is never more than a part of the reason for accepting an axiom, and is never indispensable. The reason for accepting an axiom, as for accepting any other proposition, is always largely inductive, namely, that many propositions which are nearly indubitable can be deduced from it, and that no equally plausible way is known by which these propositions could be true if the axiom were false, and nothing which is probably false can be deduced from it. If the axiom is apparently self-evident, that only means, practically, that it is nearly indubitable; for things have been thought to be self-evident and have yet turned out to be false. And if the axiom itself is nearly indubitable, that merely adds to the inductive evidence derived from the fact that its consequences are nearly indubitable: it does not provide new evidence of a radically different kind. Infallibility is never attainable, and therefore some element of doubt should always attach to every axiom and to all its consequences. In formal logic, the element of doubt is less than in most

[1] Also cf. Wang [1959].

sciences, but it is not absent, as appears from the fact that the paradoxes followed from premisses which were not previously known to require limitations. In the case of the axiom of reducibility, the inductive evidence in its favour is very strong, since the reasonings which it permits and the results to which it leads are all such as appear valid [Russell [1925], p. 59].

Or later:

When pure mathematics is organized as a deductive system – *i.e.*, as the set of all those propositions that can be deduced from an assigned set of premisses – it becomes obvious that, if we are to believe in the truth of pure mathematics, it cannot be solely because we believe in the truth of the set of premisses. Some of the premisses are much less obvious than some of their consequences, and are believed chiefly because of their consequences. This will be found to be always the case when a science is arranged as a deductive system. It is not the logically simplest propositions of the system that are the most obvious, or that provide the chief part of our reasons for believing in the system. With the empirical sciences this is evident. Electro-dynamics, for example, can be concentrated into Maxwell's equations, but these equations are believed because of the observed truth of certain of their logical consequences. Exactly the same thing happens in the pure realm of logic; the logically first principles of logic – at least some of them – are to be believed, not on their own account, but on account of their consequences. The epistemological question: 'Why should I believe this set of propositions?' is quite different from the logical question 'What is the smallest and logically simplest group of propositions from which this set of propositions can be deduced?' Our reasons for believing logic and pure mathematics are, in part, only inductive and probable, in spite of the fact that, in their *logical* order, the propositions of logic and pure mathematics follow from the premisses of logic by pure deduction. I think this point important, since errors are liable to arise from assimilating the logical to the epistemological order, and also, conversely, from assimilating the epistemological to the logical order. The only way in which work on mathematical logic throws light on the truth or falsehood of mathematics is by disproving the supposed antinomies. This shows that mathematics *may* be true. But to show that mathematics *is* true would require other methods and other considerations [Russell [1924], pp. 325–6].

It is intriguing how mathematical logicians who are so squeamish about rigour, and who set out to achieve absolute certainty, can slip into the morass of inductivism. For instance A. Fraenkel, the distinguished logician, dares to state that some axioms of logic receive their 'full weight' from 'the evidence of their consequences' (Fraenkel [1927], p. 61).

Like Newton in celestial mechanics, Russell had to realize the defeat of the Euclidean venture in mathematics. Some of his followers however made a virtue of the defeat without facing its important implications. Thus Rosser:

We wish to make one point clear about our use of the word 'axiom'. Originally the word was used by Euclid to mean a 'self-evident truth'. This use of the word 'axiom' has long been completely obsolete in mathematical circles. For us, the axioms are a set of arbitrarily chosen statements which, together with the rule of *modus ponens*, suffice to derive all the statements which we wish to derive [Rosser [1953], p. 55].

Rosser obviously meant 'all and only all' – since he obviously does not advocate inconsistent axiom systems. But what are the statements which we wish to derive? Those which are self-evidently true? In this case Rosser's statement only shifts the problem of self-evidence from the axioms to the 'statements which we wish to derive'.[1] Russell himself – just like Newton – never made a virtue of his defeat. He despised this sort of 'postulating': 'The method of "postulating" what we want has many advantages; they are the same as the advantages of theft over honest toil' (Russell [1919], p. 71). Postulationists are not necessarily authoritarian – they may be 'liberal' and say that they are interested in 'axiomatizing' any consistent set of arbitrary statements, true or false. This game of course has then nothing to do with truth or truth-transmission. Russell never even considered this possibility. Rejecting postulation, shaken in his Euclidean hopes, he desperately clung to Induction which, he hoped, would keep away the ghost of fallibility, first from mathematics and then from science: 'I do not see any way out of a dogmatic assertion that we *know* the inductive principle, or some equivalent; the only alternative is to throw over almost everything that is regarded as knowledge by science and common sense' (Russell [1944], p. 683).[2] He never considered the possibility that mathematics may be conjectural and that this might not necessarily mean giving up reason altogether.

It would be of only historical interest to follow the minor details of Russell's 'retreat from Pythagoras' (Russell [1959], chapter XVII): 'The splendid certainty which I had always hoped to find in mathematics was lost in a bewildering maze' (*ibid.*, p. 212). He was forced to give up Euclideanism, which would have rested on 'thought emancipated from sense...The hope of finding perfection and finality and certainty has been lost' (*ibid.*). He never really recovered from the confusion into which he was pushed by recalcitrant mathematics. In his [1912] he hesitated in putting forward his views on mathematics. Instead, with a surprising but understandable *volte face*, he gave credit to Kant, who after all was his ally in the big task of justifying science and defeating scepticism (cf. pp. 82–4, 87, 109). He wrote a wary Preface to his [1919], warning that the book is not about mathematical philosophy proper, where 'comparative certainty is not yet attained'; it is only an *introduction*. 'The utmost endeavour has been made to avoid dogmatism on such questions as are still open to serious doubt.' In his [1948], mathematical knowledge – which he earlier thought to be the paradigm of human knowledge – is not discussed at all. The

[1] Or: 'The acceptance of logical principles as canonical need be neither on arbitrary grounds nor on grounds of their allegedly inherent authority, but on the ground that they effectively achieve *certain postulated ends*' (E. Nagel [1944], p. 82, my italics).

[2] Cf. Fries's trilemma (Fries [1831]). Cf. Popper [1959], p. 93ff.

Russell paradox made Frege give up mathematical philosophy immediately.* Russell persisted for a time, but eventually he followed Frege.

Let us now draw some of the conclusions which Russell refused to draw. The infinite regress in proofs and definitions in mathematics cannot be stopped by a Euclidean logic. Logic may *explain* mathematics but cannot *prove* it. It leads to sophisticated speculation which is anything but trivially true. The domain of triviality is limited to the uninteresting decidable kernel of arithmetic and of logic – but even this trivial kernel might some time be overthrown by some detrivializing sceptic criticism.

The logical theory of mathematics is an exciting, sophisticated speculation like any scientific theory. It is an empiricist theory and thus if not shown to be false, will remain conjectural for ever. Dogmatists who despise mere conjectures can choose between hoping for an ultimate trivialization and hoping to justify induction.[1] Sceptics will point out that the established empiricist character of Russellian theory only shows that mathematics does not offer any knowledge, but only sophistry and illusion. Pure-grained sceptics are rare: we find, however, that pessimistic dogmatists are virtually sceptics. These pessimistic dogmatists demand that we should abandon speculation and restrict our attention to some narrow field which they gracefully – but without any real justification – acknowledge to be safe. In modern mathematical philosophy Intuitionism represents this brand of destructive, sceptical dogmatism, 'a treason to our science', as Hilbert put it in his [1926]. Weyl characterizes Russell's work in very similar terms to those of Cardinal Bellarmino, who characterized Galileo's theories as mere 'mathematical hypotheses'. According to Weyl, the *Principia* bases mathematics 'not on logic alone, but on a sort of logician's paradise, a universe endowed with an "ultimate furniture" of rather complex structure...would any realistically-minded man dare say he believes in this transcendental world?' (Weyl [1949], p. 233). Intuitionists are certainly right in claiming that Russellian logic is counter-intuitive, fallible. But for all that it might still be true.

An empiricist theory, however, should be severely tested. How could we test Russellian logic? All true basic statements – the decidable kernel of arithmetic and of logic – are derivable in it, and thus it does not seem to have any potential falsifiers. So the only way of criticizing

* This is false – as Lakatos later realised (see, e.g., volume 1, p. 126, n. 5). (*Eds.*)

[1] Another dogmatist way out is that of the ostrich: pretend not to see. Logical positivists particularly excelled at this. They had vested interests in hiding the defeat of the Russellian venture to justify mathematical certainty, since they claimed to perform the greatest revolution in the history of philosophy with the help of 'the inexorable judgment of the new logic' (Carnap [1930–1]). 'In the new logic is the point at which the old philosophy is to be removed from its hinges' (*ibid.*). No wonder that the paper carefully avoids even hinting at the fact that the 'new logic', this powerful bulwark of their philosophy, may possibly be false. Hempel claimed that logicism has shown that 'the propositions of mathematics have the same unquestionable certainty which is typical of such propositions as "All bachelors are unmarried"' (Hempel [1945*a*], p. 159).

this peculiar empiricist theory is, on the face of it, to test it for consistency.[1] This leads us to the Hilbertian circle of ideas.

3 STOPPING INFINITE REGRESS BY A TRIVIAL META-THEORY

Hilbertian meta-mathematics was 'designed to put an end to scepticism once and for all'.[2] Thus its aim was identical with that of the logicists.

> One has to admit that in the long run the situation in which we find ourselves because of the paradoxes is an unbearable one. Just imagine: in mathematics, in this paradigm of certainty and truth, the most common concept-formations and inferences that are learned, taught and used, lead to absurdities. But if even mathematics fails, where are we to look for certainty and truth? There is however a completely satisfactory method of avoiding paradoxes [Hilbert [1926]].

Hilbert's theory was based on the idea of formal axiomatics. He claimed (*a*) that all arithmetical propositions which are formally proved – the arithmetical theorems – will certainly be true if the formal system is consistent, in the sense that A and $\bar{A}$ are not both theorems, (*b*) that all arithmetical truths can be formally proved, and (*c*) that meta-mathematics, this new branch of mathematics set up to prove the consistency and completeness of formal systems, will be a particular brand of Euclidean theory: a 'finitary' theory, with trivially true axioms, containing only perfectly well known terms, and with trivially safe inferences. 'It is contended that the principles used in the meta-mathematical proof that the axioms of mathematics do not lead to contradiction, are so obviously true that not even sceptics can doubt them'.[3] A meta-mathematical argument will be 'a concatenation of self-evident intuitive (*inhaltlich*) insights' (Neumann [1927], p. 2). Arithmetical truth – and, because of the already accomplished arithmetization of mathematics, all sorts of mathematical truths – will rest on a firm, trivial, 'global' intuition, and thus, on 'absolute truth'.[4]

Gödel's *second theorem* was a decisive blow to this hope for a Euclidean meta-mathematics. The infinite regress in proofs cannot peter out in a 'finitarily' trivial meta-theory: consistency proofs have to contain enough sophistication to render the consistency of the theory in which they are carried out dubitable, and therefore they are bound to be fallible. For instance, Goldbach's conjecture – that every even number is the sum of two primes – might be formally proved to-morrow, but we shall never *know* that it is true. For it would only be true if

[1] In fact there do exist other methods. For instance Rosser and Wang [1950] showed that if Quine's system should be consistent, it is false.

[2] Cf. Ramsey [1926*b*], p. 68

[3] Ramsey, *loc. cit.* p. 69.

[4] Hilbert [1898]. *For a more accurate account of Hilbert's theory see *below*, chapter 2, pp. 31–2.

meta-mathematics, meta-meta-mathematics...*ad infinitum* are consistent. This we shall never *know*. The formalization may well have misfired and our axiomatic system may have *no model* at all.

Gödel's *first theorem* showed a second way in which a formal theory could misfire: if it has a model at all, it has more models than intended. In a consistent formal theory we can prove those and only those propositions which are true in all models;[1] so we cannot formally prove propositions which are, though true in the intended model, false in an unintended one. Gödel's first theorem showed that the selectivity of formal systems embracing arithmetic is irreparably bad, for in no consistent formalization of arithmetic can we 'tune out' unintended models, which are essentially different from the intended one.[2] Consequently in any consistent formalization there will be formally unprovable arithmetical truths. If the Goldbach conjecture is true in its intended interpretation, but false in an unintended one, there will be no formal proof leading to it in any formalization.

Gödel's *discovery of ω-inconsistent systems* was still worse. It turned out that 'the consistency of the system will not prevent the possibility of structural falsehood'(Tarski [1933], p. 295). A formalized arithmetic might be consistent, i.e. have models, but none of the models might be the intended one; every model, if containing all the numbers, might contain some other 'class-alien' elements which might provide counter-examples to propositions which are true in the narrower domain of the intended interpretation. In a consistent, but ω-inconsistent system we might prove the negation of the Goldbach conjecture even if the Goldbach conjecture is true. In a formalization which has misfired in this – or some similar – crooked way, 'provability' and truth part. An inconsistent system of arithmetic or of logic has no model, i.e. is about nothing, and an ω-inconsistent system of arithmetic or of logic does not have the intended model, i.e. is not about arithmetic or logic.

The discovery of ω-inconsistency and related phenomena have put an end to Hilbertian 'formalism', the central idea of which was that after 'formalization' 'there is no longer ambiguity about what constitutes a proof in the theory...The purpose of formalising a theory is to get an explicit definition of what constitutes proof in the theory. Having achieved this, there is no need always to appeal directly to the intuition' (Kleene [1952], pp. 63, 86). That this conjecture has been refuted, is usually expressed by the euphemism that 'the syntactical concept of proof gave way to the semantical idea of proof', a euphemism that hides the defeat of a major dogmatist enterprise to save mathematics from the sceptics.

[1] Henkin [1947].

[2] We are using here Kemeny's terminology: 'Two models are *essentially* different if there are sentences true in one but false in the other. (The requirement of being essentially different is stronger than the requirement that the models be non-isomorphic.)' (Kemeny [1958], p. 164.)

Thus the Hilbertian programme of trivialization on the meta-level collapsed. But soon a powerful campaign started to fill the gaps. Gentzen contributed to the gap-filling by his ingenious *consistency* proof, which Hilbertians contend is in compliance with the minimum standards of Gödelian sophistication and yet does not exceed the limits of triviality. Some of Tarski's results showed a way to fill the gaps in *completeness*:

> The definition of truth, and, more generally, the establishment of semantics, enables us to match some important negative results which have been obtained in the methodology of the deductive sciences with parallel positive results, and thus to fill up *in some measure* [my italics] the gaps thereby revealed in the deductive method and in the edifice of deductive knowledge itself [Tarski [1956], pp. 276–7].

Unfortunately, some logicians tend to ignore Tarski's wary qualification. In a recent textbook we learn that 'the Gödelian "negative" [*sic*] result is matched by Tarski's positive result' (Stegmüller [1957], p. 253). It is right not to put 'positive' in inverted commas, as sceptics would – but why put 'negative' into those belittling commas?

Thus rubber-Euclideanism turns up again, this time as the new party line of post-Hilbertians. One is amazed how sophisticated triviality can be. Self-evidence – if once admitted – is of course stretchable, and testing a proposition for self-evident truth is the same as testing it for truth: to show that it is inconsistent or false. If we refuse to stretch our intuition infinitely, we have to admit that *meta-mathematics does not stop the infinite regress in proofs which now reappears in the infinite hierarchy of ever richer meta-theories.* (Gödel's first theorem is in fact a Principle of Conservation of Sophistication, or a Principle of Conservation of Fallibility.) But this need not make us give in to mathematical scepticism: we only have to admit the fallibility of daring speculation. Gentzen's consistency proof – and Tarski's semantic results – are real and not 'Pyrrhic victories' as Weyl calls them,[1] even if one admits not only the 'substantially lower standard of evidence',[2] but the definitely conjectural character of the new methods. As meta-mathematics grows, its sophisticated triviality grows ever more sophisticated and ever less trivial. Triviality and certainty are *Kinderkrankheiten* of knowledge.

Let us stress again that the Euclidean can always stick to his guns after any defeat: either by hoping to find, higher up, *real* first principles, or by fooling himself into believing, with some logical or epistemological somersault, that what in fact is fallible speculation is obvious truth. For the logicist programme the favourite somersault was induction. The Hilbertian somersault is a weird plea for belief in the new revelation and a sudden, and indeed surprising, enthronement of meta-mathematical rubber-intuition, which first was just finitary-

[1] Weyl [1949], p. 220.
[2] *Ibid.*

Brouwerian, then transfinite-Gentzenian, and even semantic-Tarskian.[1] We read in one of the most competent books written on the subject that 'the ultimate [*sic*] test whether a method is admissible in meta-mathematics must of course [*sic*] be whether it is intuitively convincing' (Kleene [1952], p. 63). But why then don't we stop a step earlier, why not say that 'the ultimate test whether a method is admissible in *arithmetic* must of course be whether it is intuitively convincing', and omit meta-mathematics altogether, as is actually done by Bourbaki?[2] Meta-mathematics – like Russellian logic – has its origin in the *criticism* of intuition; now meta-mathematicians – as did the logicists – ask us to accept *their* intuition as the 'ultimate' test; thereby both fall back on the same subjectivistic psychologism which they once attacked. But why on earth have '*ultimate*' tests, or 'final' authorities?[3] Why foundations, if they are admittedly subjective? Why not honestly admit mathematical fallibility, and try to defend the dignity of *fallible* knowledge from cynical scepticism, rather than delude ourselves that we shall be able to mend invisibly the latest tear in the fabric of our 'ultimate' intuitions?

[1] Of course, by 'postulating' one can dissolve any problem. If one gives up intuition, despairs of certainty, *and* equates knowledge with certainty, then one may turn one's back on truth and play around with formal systems 'unhampered by the striving after "correctness"' and by outmoded Russello-Hilbertian ideas such as 'a new language-form must be proved to be "correct" and to constitute a faithful rendering of "the true logic"' (Carnap [1937], Foreword). It is sad how many 'logicians' followed this advice and soon forgot that logic is about truth-transmission and not about strings of symbols – even after Carnap started to realize his mistake. In their work the *technique* of logic overpowered its subject and started a perverted life of its own.

[2] Bourbaki [1949*a*], p. 8.

[3] The mathematician 'should not forget that his intuition is the final authority' (Rosser [1953], p. 11).

2

A renaissance of empiricism in the recent philosophy of mathematics?*

INTRODUCTION

[According to logical empiricist orthodoxy, while science is *a posteriori*, contentful and (at least in principle) fallible, mathematics is *a priori*, tautologous and infallible.[1]] It may therefore come as a surprise for the historian of ideas to find statements by some of the best contemporary experts in foundational studies that seem to herald a renaissance of Mill's radical assimilation of mathematics to science. In the next section I present a rather long list of such statements. I then go on (in section 2) to explain the motivation and rationale of these statements. I then argue (in section 3) for what I call the 'quasi-empirical' nature of mathematics, as a whole. This presents a problem – namely what kind of statements may play the role of potential falsifiers in mathematics. I investigate this problem in section 4. Finally, in section 5, I examine briefly periods of stagnation in the growth of 'quasi-empirical' theories.

* This paper developed out of some remarks made by Lakatos at a Colloquium in the Philosophy of Science in London, 1965. These remarks were in the form of a reply to Professor Kalmár's paper (Kalmár [1967]) and were published in Lakatos (*ed.*) [1967*a*], under the same title as the present paper.

Lakatos expanded these remarks into a longer paper which he completed in 1967. However, he withheld it from publication, intending to improve it further. Other interests prevented him from returning to the paper, and it appears here essentially as he left it in 1967. We have made a few minor presentational changes, and deleted some introductory sentences which relate only to the discussion of Kalmár's paper. (*Eds.*)

[1] This empiricist position (and one of its central difficulties) is very clearly described by Ayer in his [1936]: 'Whereas a scientific generalisation is readily admitted to be fallible, the truths of mathematics and logic appear to everyone to be necessary and certain. But if empiricism is correct no proposition which has a factual content can be necessary or certain. Accordingly the empiricist must deal with the truths of logic and mathematics in one of the two following ways: he must say either that they are not necessary truths, in which case he must account for the universal conviction that they are; or he must say that they have no factual content, and then he must explain how a proposition which is empty of all factual content can be true and useful and surprising' (pp. 72–3).

1 EMPIRICISM AND INDUCTION: THE NEW VOGUE IN MATHEMATICAL PHILOSOPHY?

Russell was probably the first modern logician to claim that the evidence for mathematics and logic may be 'inductive'. He, who in 1901 had claimed that the 'edifice of mathematical truths stands unshakable and inexpugnable to all the weapons of doubting cynicism',[1] in 1924 thought that logic (and mathematics) is exactly like Maxwell's equations of electro-dynamics: both 'are believed because of the observed truth of certain of their logical consequences'.[2]

Fraenkel claimed in 1927 that 'the intuitive or logical self-evidence of the principles chosen as axioms [of set theory] naturally plays a certain but not decisive role; some axioms receive their full weight rather from the self-evidence of the consequences which could not be derived without them'.[3] And he compared the situation of set theory in 1927 with the situation of the infinitesimal calculus in the eighteenth century, recalling d'Alembert's '*Allez en avant, et la foi vous viendra*'.[4]

Carnap, who at the 1930 conference in Königsberg still thought that 'any uncertainty in the foundations of the "most certain of all the sciences" is extremely disconcerting',[5] [had decided by] 1958 that there is an analogy – if only a distant one – between physics and mathematics: 'the impossibility of absolute certainty'.[6]

Curry drew similar conclusions in 1963:

> The search for absolute certainty was evidently a principal motivation for both Brouwer and Hilbert. But does mathematics need absolute certainty for its justification? In particular, why do we need to be sure that a theory is consistent, or that it can be derived by an absolutely certain intuition of pure time, before we use it? In no other science do we make such demands. In physics all theorems are hypothetical; we adopt a theory so long as it makes useful predictions and modify or discard it as soon as it does not. This is what has happened to mathematical theories in the past, where the discovery of contradictions had led to modifications in the mathematical doctrines accepted up to the time of that discovery. Why should we not do the same in the future? Using formalistic conceptions to explain what a theory is, we accept a theory as long as it is useful, satisfies such conditions of naturalness and simplicity as are reasonable at that time, and is not known to lead us into error. We must keep our theories under surveillance to see that these conditions are fulfilled

[1] Russell [1901*a*], p. 57.

[2] Russell [1924], pp. 325–6. He obviously hesitated between the view that one can put up with this state of affairs (and work out some sort of inductive logic for the *Principia*), [and the view that] one has to go on with the search for self-evident axioms. In the Introduction to the second edition of the *Principia*, he says that one *cannot* rest content with an axiom that has mere inductive evidence (p. xiv), while on p. 59 he devotes a little chapter to the (inductive) 'Reasons for Accepting the Axiom of Reducibility' (although still not giving up the hope of deducing it from some self-evident truth).

[3] Fraenkel [1927], p. 61.

[4] *Ibid.*

[5] Carnap [1931], p. 31. English translation in Benacerraf and Putnam (*eds.*) [1964].

[6] Carnap [1958], p. 240.

and to get all the presumptive evidence of adequacy that we can. The Gödel theorem suggests that this is all we can do; an empirical philosophy of science suggests it is all we should do.[1]

To quote Quine:

We may more reasonably view set theory, and mathematics generally, in much the way in which we view theoretical portions of the natural sciences themselves; as comprising truths or hypotheses which are to be vindicated less by the pure light of reason than by the indirect systematic contribution which they make to the organizing of empirical data in the natural sciences.[2]

And later he said:

To say that mathematics in general has been reduced to logic hints at some new firming up of mathematics at its foundations. This is misleading. Set theory is less settled and more conjectural than the classical mathematical superstructure than can be founded upon it.[3]

Rosser too belongs to the new fallibilist camp:

According to a theorem of Gödel...if a system of logic is adequate for even a reasonable facsimile of present-day mathematics, then there can be no adequate assurance that it is free from contradiction. Failure to derive the known paradoxes is very negative assurance at best and may merely indicate lack of skill on our part.[4]

Church, in 1939 thought that: 'there is no convincing basis for a belief in the consistency either of Russell's or of Zermelo's system, even as probable'.[5]

Gödel in 1944 stressed that under the influence of modern criticism of its foundations, mathematics has already lost a good deal of its 'absolute certainty' and that in the future, by the appearance of further axioms of set theory, it will be increasingly fallible.[6]

In 1947, developing this idea, he explained that for some such new axiom,

even in case it had no intrinsic necessity at all, a (probable) decision about its truth is possible also in another way, namely, inductively by studying its 'success', that is, its fruitfulness in consequences demonstrable without the new axiom, whose proofs by means of the new axiom, however, are considerably simpler and easier to discover, and make it possible to condense into one proof many different proofs. The axioms for the system of real numbers, rejected by the intuitionists, have in this sense been verified to some extent owing to the fact that analytical number theory frequently allows us to prove number theoretical theorems which can subsequently be verified by elementary methods. A much higher degree of verification than that, however, is conceivable. There might exist axioms so abundant in their verifiable consequences, shedding so much light upon a whole discipline, and furnishing such powerful methods for solving given problems (and even solving them, as far as that is

[1] Curry [1963], p. 16. See also his [1951], p. 61.
[2] Quine [1958], p. 4.
[3] Quine [1965], p. 125.
[4] Rosser [1953], p. 207.
[5] Church [1939].
[6] Gödel [1944], p. 213.

possible, in a constructivistic way) that quite irrespective of their intrinsic necessity they would have to be assumed at least in the same sense as any well established physical theory.[1]

Also, he is reported to have said a few years later that:

the role of the alleged 'foundations' is rather comparable to the function discharged, in physical theory, by explanatory hypotheses...The so-called logical or set-theoretical 'foundation' for number-theory or of any other well established mathematical theory, is explanatory, rather than really foundational, exactly as in physics where the actual function of axioms is to *explain* the phenomena described by the theorems of this system rather than to provide a genuine 'foundation' for such theorems.[2]

Weyl says that non-intuitionistic mathematics can be tested, but not proved:

No Hilbert will be able to assure us of consistency forever; we must be content if a simple axiomatic system of mathematics has met the test of our elaborate mathematical experiments so far...A truly realistic mathematics should be conceived, in line with physics, as a branch of the theoretical construction of the one real world, and should adopt the same sober and cautious attitude toward hypothetic extensions of its foundations as is exhibited by physics.[3]

Von Neumann, in 1947, concluded that

After all, classical mathematics, even though one could never again be absolutely certain of its reliability...stood on at least as sound a foundation as, for example, the existence of the electron. Hence, if one was willing to accept the sciences, one might as well accept the classical system of mathematics.[4]

Bernays argues very similarly: It is of course surprising and puzzling that the more content and power mathematical methods have, the less is their self-evidence. But 'this will not be so surprising if we consider that there are similar conditions in theoretical physics'.[5]

According to Mostowski mathematics is just one of the natural sciences:

[Gödel's] and other negative results confirm the assertion of materialistic philosophy that mathematics is in the last resort a natural science, that its notions and methods are rooted in experience and that attempts at establishing the foundations of mathematics without taking into account its originating in the natural sciences are bound to fail.[6]

[And Kalmár agrees:] 'the consistency of most of our formal systems is an empirical fact...Why do we not confess that mathematics, like other sciences, is ultimately based upon, and has to be tested in, practice?'[7]

These statements describe a genuine revolutionary turn in the philosophy of mathematics. Some describe their individual *volte-face*

[1] Gödel [1947], p. 521. The word 'probable' was inserted in the reprinted version, Gödel [1964], p. 265.

[2] Mehlberg [1962], p. 86.

[3] Weyl [1949], p. 235.

[4] Neumann [1947], pp. 189–90.

[5] Bernays [1939], p. 83.

[6] Mostowski [1955], p. 42.

[7] Kalmár [1967], pp. 192–3.

in dramatic terms. Russell, in his autobiography, says: 'The splendid certainty which I had always hoped to find in mathematics was lost in a bewildering maze.'[1] Von Neumann writes: 'I know myself how humiliatingly easily my own views regarding the absolute mathematical truth changed...and how they changed three times in succession!'[2] Weyl, recognizing before Gödel that classical mathematics was *unrescuably* fallible, refers to [this state of affairs as] 'hard fact'.[3]

We could go on quoting; but surely this is enough to show that mathematical empiricism and inductivism (not only as regards the *origin* or *method*, but also as regards the *justification*, of mathematics) is more alive and widespread than many seem to think. But what is the background and what is the *rationale* of this new empiricist–inductivist mood? Can one give it a sharp, *criticizable* formulation?

2 QUASI-EMPIRICAL VERSUS EUCLIDEAN THEORIES

Classical epistemology has for two thousand years modelled its ideal of a theory, whether scientific or mathematical, on its conception of Euclidean geometry. The ideal theory is a deductive system with an indubitable truth-injection at the top (a finite conjunction of axioms) – so that truth, flowing down from the top through the safe truth-preserving channels of valid inferences, inundates the whole system.

It was a major shock for over-optimistic rationalism that science – in spite of immense efforts – could not be organized in such Euclidean theories. Scientific theories turned out to be organized in deductive systems where the *crucial* truth value injection was *at the bottom* – at a special set of theorems. But *truth* does not flow upwards. The important logical flow in such *quasi-empirical theories* is not the transmission of truth but rather the retransmission of *falsity* – from special theorems at the bottom ('basic statements') up towards the set of axioms.[4]

Perhaps the best way to characterize quasi-empirical, as opposed to Euclidean theories, is this. Let us call those sentences of a deductive system in which some truth values are initially injected, 'basic statements', and the subset of basic statements which receive the particular value true, 'true basic statements'. Then a system is Euclidean if it is the [*deductive*] *closure* of those of its basic statements which are assumed to be true. Otherwise it is quasi-empirical.

An important feature of both Euclidean and quasi-empirical systems is the set of particular (usually unwritten) conventions regulating truth value injections in the basic statements.

A Euclidean theory may be claimed to be true; a quasi-empirical theory – at best – to be well-corroborated, but always conjectural. Also,

[1] Russell [1959], p. 212. For further details about Russell's turn, cf. my [1962].
[2] Neumann [1947], p. 190.
[3] Weyl [1928], p. 87.
[4] For an exposition of the story see this volume, chapter 1. The concept and term 'basic statement' is due to Karl Popper; see his [1934], chapter v.

in a Euclidean theory the true basic statements at the 'top' of the deductive system (usually called 'axioms') *prove*, as it were, the rest of the system; in a quasi-empirical theory the (true) basic statements are *explained* by the rest of the system.

Whether a deductive system is Euclidean or quasi-empirical is decided by the pattern of truth value flow in the system. The system is Euclidean if the characteristic flow is the transmission of truth from the set of axioms 'downwards' to the rest of the system – logic here is an *organon of proof*; it is quasi-empirical if the characteristic flow is retransmission of falsity from the false basic statements 'upwards' towards the 'hypothesis' – logic here is an *organon of criticism*.[1] But this demarcation between patterns of truth value flow is independent of the particular conventions that regulate the original truth value injection into the basic statements. For instance *a theory which is quasi-empirical in my sense may be either empirical or non-empirical in the usual sense*: it is empirical only if its basic theorems are spatio-temporally singular basic statements whose truth values are decided by the time-honoured but unwritten code of the experimental scientist.[2] (We may speak, even more generally, of Euclidean versus quasi-empirical theories independently of *what* flows in the logical channels: certain or fallible truth and falsehood, probability and improbability, moral desirability or undesirability, etc. It is the *how* of the flow that is decisive.)

The methodology of a science is heavily dependent on whether it aims at a Euclidean or at a quasi-empirical ideal. The basic rule in a science which adopts the former aim is to search for self-evident axioms – Euclidean methodology is puritanical, antispeculative. The basic rule of the latter is to search for bold, imaginative hypotheses with high explanatory and 'heuristic' power,[3] indeed, it advocates a proliferation of alternative hypotheses to be weeded out by severe criticism – quasi-empirical methodology is uninhibitedly speculative.[4]

The development of Euclidean theory consists of three stages: first the naive prescientific stage of trial and error which constitutes the prehistory of the subject; this is followed by the foundational period which reorganizes the discipline, trims the obscure borders, establishes the deductive structure of the safe kernel; all that is then left is the solution of problems inside the system, mainly constructing proofs or disproofs of interesting conjectures. ([The discovery of] a decision method for theoremhood may abolish this stage altogether and put an end to the development.)

The development of a quasi-empirical theory is very different. It starts with problems followed by daring solutions, then by severe tests,

[1] Cf. Popper [1963*a*], p. 64.
[2] For a discussion cf. volume 1, chapter 3.
[3] For the latter concept cf. volume 1, chapter 1.
[4] The elaboration of empirical methodology – which of course is the paradigm of quasi-empirical methodology – is due to Karl Popper.

refutations. The vehicle of progress is bold speculations, criticism, controversy between rival theories, problemshifts. Attention is always focussed on the obscure borders. The slogans are growth and permanent revolution, not foundations and accumulation of eternal truths.

The main pattern of Euclidean criticism is suspicion: Do the proofs really prove? Are the methods used too strong and therefore fallible? The main pattern of quasi-empirical criticism is proliferation of theories and refutation.

3 MATHEMATICS IS QUASI-EMPIRICAL

By the turn of this century mathematics, 'the paradigm of certainty and truth', seemed to be the last real stronghold of orthodox Euclideans. But there were certainly some flaws in the Euclidean organization even of mathematics, and these flaws caused considerable unrest. Thus the central problem of all foundational schools was: 'to establish once and for all the certitude of mathematical methods'.[1] However, foundational studies unexpectedly led to the conclusion that a Euclidean reorganization of mathematics as a whole may be impossible; that at least the richest mathematical theories were, like scientific theories, quasi-empirical. Euclideanism suffered a defeat in its very stronghold.

The two major attempts at a perfect Euclidean reorganization of classical mathematics – logicism and formalism[2] – are well known, but a brief account of them from this point of view may be helpful.

(*a*) *The Frege–Russell approach* aimed to deduce all mathematical truths – with the help of ingenious definitions – from indubitably true logical axioms. It turned out that some of the logical (or rather set-theoretical) axioms were not only not indubitably true but not even consistent. It turned out that the sophisticated second (and further) generations of logical (or set-theoretical) axioms – devised to avoid the known paradoxes – even if true, were not indubitably true (and not even indubitably consistent), and that the crucial evidence for them was that classical mathematics might be *explained* – but certainly not *proved* – by them.

Most mathematicians working on comprehensive '*grandes logiques*' are well aware of this. We have already referred to Russell, Fraenkel, Quine and Rosser. Their 'empiricist' turn is in fact a quasi-empiricist

[1] Hilbert [1925], p. 35.

[2] Intuitionism is omitted: it never aimed at a reorganization but at a truncation of classical mathematics. *Not all the theorems of intuitionist mathematics are theorems of classical mathematics. In this sense, Lakatos is wrong to describe intuitionism as simply a 'truncation' of classical mathematics. Nevertheless, an important point remains. While Russell's logicism and Hilbert's formalism each regarded its task as the justification of the whole of classical mathematics, Brouwer's intuitionism was willing to jettison large parts of classical mathematics which do not meet its standards of justification. (*Eds.*)

one: they realized (independently even of Gödel's results) that the *Principia Mathematica* and the strong set-theories, like Quine's *New Foundations* and *Mathematical Logic*, are all quasi-empirical.

Workers in this field are conscious of the method they follow: daring conjectures, proliferation of hypotheses, severe tests, refutations. Church's account of an interesting theory based on a restricted form of the law of excluded middle (later shown to be inconsistent by Kleene and Rosser[1]) outlines the quasi-empirical method:

> Whether the system of logic which results from our postulates is adequate for the development of mathematics, and whether it is wholly free from contradiction, are questions which we cannot now answer except by conjecture. Our proposal is to seek at least an empirical answer to these questions by carrying out in some detail a derivation of the consequences of our postulates, and it is hoped either that the system will turn out to satisfy the conditions of adequacy and freedom from contradiction or that it can be made to do so by modifications or additions.[2]

Quine characterized the crucial part of his *Mathematical Logic* as a 'daring structure...added at the constructor's peril'.[3] Soon it was shown by Rosser to be inconsistent and Quine then himself described his earlier characterization as one that had 'a prophetic ring'.[4]

One can never refute Euclideanism: even if forced to postulate highly sophisticated axioms, one can always stick to one's hopes of deriving them from some deeper layer of self-evident foundations.[5] There have been considerable and partly successful efforts to simplify Russell's *Principia* and similar logicistic systems. But while the results were mathematically interesting and important they could not retrieve the lost philosophical position. The *grandes logiques* cannot be proved true – nor even consistent; they can only be proved false – or even inconsistent.

(*b*) While the Frege–Russell approach aimed to turn mathematics into a unified classical Euclidean theory the *Hilbert approach* offered a radically new modification of the Euclidean programme, exciting both from the mathematical and the philosophical points of view.

Hilbertians claimed that classical analysis contains an absolutely true Euclidean kernel. [But alongside this there are 'ideal elements' and 'ideal statements' which, though indispensable for the deductive–

[1] Kleene and Rosser [1935].

[2] Church [1932], p. 348.

[3] Quine [1941*a*], p. 122. Some critics of Quine may say that it is only he who has made a 'daring' structure out of the natural simplicity of mathematics. But surely the Cantorian paradise is a 'bold theoretical construction, and as such the very opposite of analytical self-evidence' (Weyl [1947], p. 64). Also cf. the Weyl quotation in section 2.

[4] Quine [1941*b*], p. 163. By the way, the most interesting feature of Rosser's paper is the search for ways of testing the consistency of *ML*. Rosser shows that 'if one can prove *201 from the remaining axioms, then the remaining axioms are inconsistent' (Rosser [1941], p. 97).

[5] Also, one can choose to cut down a quasi-empirical theory to its Euclidean kernel (that is the essential aspect of the intuitionist programme).

heuristic machinery, are not absolutely true (in fact they are neither true nor false).] But if the whole theory, containing both the concrete–*inhaltlich* and the ideal statements can be proved consistent in a Euclidean meta-mathematics,[1] the entire classical analysis would be saved. That is, analysis *is* a quasi-empirical theory[2] but the Euclidean consistency proof will see to it that it should have no falsifiers. The sophistication of Cantorian speculation is to be safeguarded not by deeper-seated Euclidean axioms *in the theory itself* – Russell has already failed in this venture – but by an austere Euclidean *meta-theory*.*

Eventually, Hilbertians defined the set of statements whose truth values could be regarded as directly given (the set of finitistically true statements) so clearly that their programme could be refuted.[3] The refutation was provided by Gödel's theorem which implied the impossibility of a finitary consistency proof for formalized arithmetic. [The reaction of formalists is well summed up by Curry]:

> This circumstance has led to a difference of opinion among modern formalists, or rather, it strengthened a difference of opinion which already existed. Some think that the consistency of mathematics cannot be established on *a priori* grounds alone and that mathematics must be justified in some other way. Others maintain that there are forms of reasoning which are *a priori* and constructive in a wider sense and that in terms of these the Hilbert program can be carried out.[4]

That is, either meta-mathematics was to be recognized as a quasi-empirical theory or the concept of finitary or *a priori* had to be stretched. Hilbert chose the latter option. According to him the class of *a priori* methods was now to include, for example, transfinite induction up to ϵ_0, used in Gentzen's proof of the consistency of arithmetic.

But not everybody was happy about this extension. Kalmár, who applied Gentzen's proof to the Hilbert–Bernays system, never believed that his proof was Euclidean. According to Kleene: 'To what extent

[1] Originally the meta-theory was not to be axiomatized but was to consist of simple, protofinitary thought-experiments. In Bologna (1928) von Neumann even criticized Tarski for axiomatizing it. (The generalization of the concept of 'Euclidean theory' to informal, unaxiomatized theories does not constitute any difficulty.)

[2] To quote Weyl again: 'whatever the ultimate value of Hilbert's program, his bold enterprise can claim one merit: it has disclosed to us the highly complicated and ticklish logical structure of mathematics, its maze of back-connections, which result in circles of which it cannot be gathered at a first glance whether they might not lead to blatant contradictions' (*op. cit.*, p. 61).

* Hilbert's philosophy, at least as here presented, cannot be subsumed so easily under Euclideanism. Meta-mathematics is an informal unaxiomatized theory and such theories do not have the required deductive structure to be candidates for Euclidean status. Informal theories can obviously be axiomatized, but one of Hilbert's central claims was that there was no need for this in the case of meta-mathematics (cf. n. 1 *above*). Each principle assumed in a meta-mathematical proof was to be so obviously true as not to be in need of justification (or, rather, to be immediately justified by the so-called 'global intuition'). (*Eds.*)

[3] Herbrand [1930], p. 248. It took three decades to arrive at this definition.

[4] Curry [1963], p. 11.

the Gentzen proof can be accepted as securing classical number theory...is...a matter for individual judgment, depending on how ready one is to accept induction up to ϵ_0 as a finitary method.'[1] Or, to quote Tarski:

there seems to be a tendency among mathematical logicians to overemphasize the importance of consistency problems, and the philosophical value of the results so far in this direction seems somewhat dubious. Gentzen's proof of the consistency of arithmetic is undoubtedly a very interesting metamathematical result which may prove very stimulating and fruitful. I cannot say, however, that the consistency of arithmetic is now much more evident to me (at any rate, perhaps to use the terminology of the differential calculus, more evident than by epsilon) than it was before the proof was given. To clarify a little my reactions: let G be a formalism just adequate for formalizing Gentzen's proof, and let A be the formalism of arithmetic. It is interesting that the consistency of A can be proved in G; it would perhaps be equally interesting if it should turn out that the consistency of G can be proved in A.[2]

However, even those who find transfinite induction up to ϵ_0 infallible would not be happy to go on stretching the concept of infallibility so as to accommodate consistency proofs of stronger theories. In this sense 'the real test of proof-theory will be the proof of the consistency of *analysis*',[3] and this has still to be seen.

Gödel's and Tarski's incompleteness results however reduce the chances of the final success of Hilbert's programme still further. For if *extant* arithmetic cannot be proved by the original Hilbertian standards, the gradual, consistent (and indeed, ω-consistent) [augmentation] of theories containing arithmetic by further axioms can only be reached by still more fallible methods. That is, the future development of arithmetic will increase its fallibility. Gödel himself has pointed this out in his paper on Russell's mathematical logic:

[Russell] compares the axioms of logic and mathematics with the laws of nature and logical evidence with sense perception, so that the axioms need not necessarily be evident in themselves, but rather their justification lies (exactly as in physics) in the fact that they make it possible for these 'sense perceptions' to be deduced; which of course would not exclude that they also have a kind of intrinsic plausibility similar to that in physics. I think that (provided 'evidence' is understood in a sufficiently strict sense) this view has been largely justified by subsequent developments, and it is to be expected that it will be still more so in the future. It has turned out that (under the assumption that modern mathematics is consistent) the solution of certain arithmetical problems requires the use of assumptions essentially transcending arithmetic, *i.e.*, the domain of the kind of elementary indisputable evidence that may be most fittingly compared with sense perception. Furthermore it seems likely that for deciding certain questions of abstract set theory and even for certain related questions of the theory of real numbers new axioms based on some hitherto unknown idea will be necessary. Perhaps also the apparently unsur-

[1] Kleene [1952], p. 479.

[2] Tarski [1954], p. 19.

[3] Bernays and Hilbert [1939], p. vii.

mountable difficulties which some other mathematical problems have been presenting for many years are due to the fact that the necessary axioms have not yet been found. Of course, under these circumstances mathematics may lose a good deal of its 'absolute certainty'; but, under the influence of the modern criticism of the foundations, this has already happened to a large extent. There is some resemblance between this conception of Russell and Hilbert's 'supplementing the data of mathematical intuition' by such axioms as, e.g., the law of excluded middle which are not given by intuition according to Hilbert's view; the borderline however between data and assumptions would seem to lie in different places according to whether we follow Hilbert or Russell.[1]

Quine says that in the field of *grande logique* construction 'at the latest, the truism idea received its deathblow from Gödel's incompleteness theorem. Gödel's incompleteness theorem can be made to show that we can never approach *completeness* of elementhood axioms without approaching contradiction.'[2]

There are many possible ways of [augmenting systems including] arithmetic. One is through adding strong, arithmetically testable, axioms of infinity to *grandes logiques*.[3] Another is through constructing strong ordinal logics.[4] A third one is to allow non-constructive rules of inference.[5] A fourth one is the model-theoretic approach.[6] But all of them are fallible, not less fallible – and not less quasi-empirical – than the ordinary classical mathematics which was so much in want of foundations. This recognition – that not only the *grandes logiques*, but also mathematics is quasi-empirical – is reflected in the 'empiricist' statements by Gödel, von Neumann, Kalmár, Weyl and others.

It should however be pointed out that some people believe that some of the principles used in these different methods are *a priori* and they were arrived at by 'reflection'. For instance, Gödel's empiricism is qualified by the hope that set-theoretical principles may be found which are *a priori* true. He claims that Mahlo's 'axioms show clearly, not only that the axiomatic system of set theory as used today is incomplete, but also that it can be supplemented without arbitrariness by new axioms which only unfold the content of the concept of set explained above'.[7] (Gödel, however, does not seem to be very sure of the *a priori* characterizability of the concept of set, as is evident from

[1] Gödel [1944], p. 213.

[2] Quine [1941*a*], p. 127.

[3] Such strong axioms were formulated by Mahlo, Tarski and Levy. As to the arithmetical testability of these axioms: 'It can be proved that these axioms also have consequences far outside the domain of very great transfinite numbers, which is their immediate subject matter: each of them, under the assumption of its consistency, can be shown to increase the number of decidable propositions even in the field of Diophantine equations' (Gödel [1947], p. 520).

[4] This line of research was initiated by Turing ([1939]) and developed by Feferman ([1968]).

[5] Cf. e.g. Rosser [1937]; Tarski [1939]; Kleene [1943].

[6] Cf. Kemeny [1958], p. 164.

[7] Gödel [1964], p. 264 (cf. Gödel [1947], p. 520).

his already quoted quasi-empiricist remarks and also from his hesitation in his [1938], where he says that the axiom of constructibility 'seems to give a natural completion of the axioms of set theory, in so far as it determines the vague notion of an arbitrary infinite set in a definite way'.[1]) Weyl actually made fun of Gödel's over-optimistic stretching of the possibilities of *a priori* knowledge:

Gödel, with his basic trust in transcendental logic, likes to think that our logical optics is only slightly out of focus and hopes that after some minor correction of it we shall see *sharp*, and then everybody will agree that we see *right*. But he who does not share this trust will be disturbed by the high degree of arbitrariness involved in a system like *Z*, or even in Hilbert's system. How much more convincing and closer to facts are the heuristic arguments and the subsequent systematic constructions in Einstein's general relativity theory, or the Heisenberg–Schrödinger quantum mechanics. A truly realistic mathematics should be conceived, in line with physics, as a branch of the theoretical construction of the one real world, and should adopt the same sober and cautious attitude towards hypothetic extensions of its foundations as is exhibited by physics.[2]

Kreisel, however, extols this sort of aprioristic reflection by which, he claims, one gains set-theoretical axioms, and 'right' definitions, and calls anti-apriorism an 'antiphilosophic attitude' and the idea of progress by trial and error empirically false.[3] What is more, in his reply to Bar-Hillel, he wants to extend this method to science, thereby rediscovering Aristotelian essentialism. He adds: 'If I were really convinced that reflection is extraordinary or illusory I should certainly not choose philosophy as a profession; or, having chosen it, I'd get out fast.'[4] In his comment on Mostowski's paper he tries to play down Gödel's hesitation as out of date.[5] But just as Gödel immediately refers to inductive evidence, Kreisel refers (in the Reply) to the 'limitations' of the heuristic of reflection. (So, after all, 'reflection', 'explication' *are* fallible.)

4 'POTENTIAL FALSIFIERS' IN MATHEMATICS

If mathematics and science are both quasi-empirical, the crucial difference between them, if any, must be in the nature of their 'basic statements', or 'potential falsifiers'. The 'nature' of a quasi-empirical theory is decided by the nature of the truth value injections into its potential falsifiers.[6] Now nobody will claim that mathematics is empirical in the sense that its potential falsifiers are singular spatio-temporal statements. But then what is the nature of mathematics? Or, what is the nature of the potential falsifiers of mathematical theories?[7]

[1] Gödel [1938], p. 557.
[2] Weyl, *op. cit.*, p. 235.
[3] Kreisel [1967*a*], p. 140.
[4] Kreisel [1967*b*], p. 178.
[5] Kreisel [1967*c*], pp. 97–8.
[6] See *above*, p. 29.
[7] It is hoped that this Popperian formulation of the age-old question will shed new light on some questions in the philosophy of mathematics.

The very question would have been an insult in the years of intellectual honeymoon of Russell or Hilbert. After all, the *Principia* or the *Grundlagen der Mathematik* were meant to put an end – once and for all – to counterexamples and refutations in mathematics. Even now the question still raises some eyebrows.

[But comprehensive axiomatic set theories and systems of metamathematics, can be, and indeed have been, refuted.] Let us first take comprehensive axiomatic set theories. Of course, they have *potential logical falsifiers*: statements of the form $p \& \neg p$. But are there other falsifiers? The potential falsifiers of science, roughly speaking, express the 'hard facts'. But is there anything analogous to 'hard facts' in mathematics? If we accept the view that a formal axiomatic theory implicitly defines its subject-matter, then there would be no mathematical falsifiers except the logical ones. But if we insist that a formal theory should be the formalization of some informal theory, then a formal theory may be said to be 'refuted' if one of its theorems is negated by the corresponding theorem of the informal theory. One could call such an informal theorem a *heuristic falsifier* of the formal theory.[1]

Not all formal mathematical theories are in equal danger of heuristic refutation in a given period. For instance, *elementary group theory* is scarcely in any danger: in this case the original informal theories have been so radically replaced by the axiomatic theory that heuristic refutations seem to be inconceivable.

Set theory is a subtler question. Some argue that after the total destruction of naive set theory by *logical* falsifiers one cannot speak any more of set-theoretical facts: one cannot speak of an *intended* interpretation of set theory any more. But even some of those who dismiss set-theoretical intuition may still agree that axiomatic set theories perform the task of being the dominant, unifying theory of mathematics in which all available mathematical facts (i.e. some specified subset of informal theorems) have to be explained. But then one can criticize a set theory in two ways: its axioms may be tested for consistency and its definitions may be tested for the 'correctness' of their translation of branches of mathematics like arithmetic. For instance, we may some day face a situation where some machine churns out a formal proof in a formal set theory of a formula whose intended meaning is that there exists a non-Goldbachian even number. At the same time a number theorist might prove informally that all

[1] It would be interesting to investigate how far the demarcation between logical and heuristic falsifiers corresponds to Curry's demarcation between mathematical truth and 'quasi-truth' (or 'acceptability'). Cf. his [1951], especially chapter XI. Curry calls his philosophy 'formalist' as opposed to '*inhaltlich*' or 'contensive' philosophies, like Platonism or intuitionism (Curry [1965], p. 80). However, besides his philosophy of formal structure, he has a philosophy of acceptability – but surely one cannot explain the growth of formal mathematics without acceptability considerations, so Curry offers an '*inhaltlich*' philosophy after all.

even numbers are Goldbachian. If his proof can be formalized within our system of set theory, then our theory will be inconsistent. But if it cannot be thus formalized, the formal set theory will not [have been shown to] be inconsistent, but only to be a *false* theory of arithmetic (while still being possibly a true theory of some mathematical structure that is not isomorphic to arithmetic). Then we may call the informally proved Goldbach theorem a *heuristic falsifier*, or more specifically, an *arithmetical falsifier* of our formal set theory.[1] The formal theory is false in respect of the informal ***explanandum*** that it had set out to explain; we have to replace it by a better one. First we may try piecemeal improvements. It may have been only the definition of 'natural number' that went wrong and then the definition could be 'adjusted' to each heuristic falsifier. The axiomatic system itself (with its formation and transformation rules) would become useless as an explanation of arithmetic only if it was altogether 'numerically insegregative',[2] i.e. if it turned out that no finite sequence of adjustments of the definition eliminates *all* heuristic falsifiers.

Now the problem arises: *what class of informal theorems should be accepted as arithmetical falsifiers of a formal theory containing arithmetic?*

Hilbert would have accepted only finite numerical equations (without quantifiers) as falsifiers of formal arithmetic. But he could easily show that *all* true finite numerical equations are provable in his system. From this it followed that his system was complete with regard to true basic statements, therefore, if a theorem in it could be proved false by an arithmetical falsifer, the system was also inconsistent, for the formal version of the falsifier was already a theorem of the system. Hilbert's reduction of falsifiers to logical falsifiers (and thereby the reduction of truth to consistency) was achieved by a very narrow ('finitary') definition of arithmetical basic statements.

Gödel's informal proof of the truth of the Gödelian undecidable sentence posed the following problem: is the *Principia* or Hilbert's formalized arithmetic – on the assumption that each is consistent – true or false if we adjoin to it the negation of the Gödel sentence? According to Hilbert the question should have been meaningless, for Hilbert was an instrumentalist with regard to arithmetic outside the finitary kernel and would not have seen any difference between systems of arithmetic with the Gödel sentence or with its negation as long as they both equally implied the true basic statements (to which, by the way, his implicit meaning-and-truth-definition was restricted). Gödel proposed[3] to extend the range of (meaningful and true) basic

[1] The expression 'ω-consistency', is as Quine pointed out (Quine [1953*a*], p. 117), misleading. A demonstration of the 'ω-inconsistency' of a system of arithmetic would in fact be a *heuristic* falsification of it. Ironically, the historical origin of the misnomer was that the phenomenon was used by Gödel and Tarski precisely to divorce truth ('ω-consistency') from consistency.

[2] See Quine, *loc. cit.*, p. 118.

[3] See his intervention in 1930 in Königsberg; recorded in Gödel [1931].

statements from finitary numerical equations also to statements with quantifiers and the range of proofs to establish the truth of basic statements from 'finitary' proofs to a wider class of intuitionistic methods. It was this methodological proposal that divorced truth from consistency and introduced a new pattern of conjectures and refutations based on arithmetical falsifiability: it allowed for daring speculative theories with very strong, rich axioms while criticizing them from the outside by informal theories with weak, parsimonious axioms. *Intuitionism is here used not for providing foundations but for providing falsifiers, not for discouraging but for encouraging and criticizing speculation!*

It is surprising how far constructive and even finite falsifiers can go in testing comprehensive set theories. Strong axioms of infinity for instance are testable in the field of Diophantine equations.*

But comprehensive axiomatic set theories do not have only arithmetical falsifiers. They may be refuted by theorems – or axioms – of naive set theory. For instance Specker 'refuted' Quine's *New Foundations* by proving in it that the ordinals are not well-ordered by '⩽' and that the axiom of choice must be given up.[1] Now *is* this 'refutation of the *New Foundations*, even a heuristic refutation? Should the well-ordering theorem of shattered naive set theory overrule Quine's system? Even if, with Gödel and Kreisel, we consider naive set theory as re-established by Zermelo's correction,[2] we could admit the well-ordering theorem and the axiom of choice as heuristic falsifiers only if we again extend the class of (intuitionistic) heuristic falsifiers to (almost?) *any* theorem in corrected naive set theory. (We may call the former the class of *strong heuristic falsifiers* and the latter the class of *weak heuristic falsifiers*). But this would surely be irrational: at best we have to consider them as two rival theories (*strictly* speaking *no* heuristic falsifier can be more than a rival hypothesis). After all nothing prevents us from forgetting about naive sets and focussing our attention on the new unintended model of *New Foundations*![3]

Indeed, we can go even further. For instance, if it turned out that all strong set-theoretical systems are arithmetically false, we may modify our arithmetic – the new, non-standard arithmetic may possibly serve the empirical sciences just as well. Rosser and Wang, who – three years before Specker's result – showed that in no model of *New Foundations* does '⩽' well-order both finite cardinals and infinite ordinals as long as we stick to the intended interpretation of '⩽', discuss this possibility:

> One may well question whether a formal logic which is known to have no standard model is a suitable framework for mathematical reasoning. The

* See *above*, p. 34, n. 3. (*Eds.*)

[1] Specker [1953]: also cf. Quine [1963], p. 294ff.

[2] Cf. Gödel [1947], p. 518 and Kreisel [1967].

[3] For philosophers of science after Popper it should anyway be a commonplace that *explanans* and *explanandum* may be rival hypotheses.

proof of the pudding is in the eating. For topics in the usual range of classical mathematical analysis, the reasoning procedures of Quine's *New Foundations* are as close to the accepted classical reasoning procedures as for any system known to us. However, in certain regions, notably when dealing with extremely large ordinals, the reasoning procedures of Quine's *New Foundations* reflect the absence of a standard model, and appear strange to the classically minded mathematician. However, since the theory of ordinals is suspect when applied to very large ordinals, it is hardly a serious defect in a logic if it makes this fact apparent.

We suspect that the idea that a logic must have a standard model if it is to be acceptable as a framework for mathematical reasoning is merely a vestige of the old idea that there is such a thing as absolute mathematical truth. Certainly the requirements on a standard model are that it reflect certain classically conceived notions of the structure of equality, integers, ordinals, sets, *etc.* Perhaps these classically conceived notions are incompatible with the procedures of a strong mathematical system, in which case a formal logic for the strong mathematical system could not have a standard model.[1]

This of course [amounts to the claim] that the only real falsifiers are logical ones. [But other mathematicians,] Gödel for example, would surely reject the *New Foundations* on Specker's refutation: for him the axiom of choice and the well-ordering of ordinals are self-evident truths.[2]

No doubt the problem of basic statements in mathematics will attract increasing attention with the further development of comprehensive set theories. Recent work indicates that some very abstract axioms may soon be found testable in most unexpected branches of classical mathematics: e.g. Tarski's axiom of inaccessible ordinals in algebraic topology.[3] The continuum hypothesis also will provide a testing ground: the accumulation of further intuitive evidence against the continuum hypothesis may lead to the rejection of strong set theories which imply it. Gödel [1964] enumerates quite a few implausible consequences of the continuum hypothesis: a crucial task of his new Euclidean programme is to provide a self-evident set theory from which its negation is derivable.[4]

1 Rosser and Wang [1950], p. 115.

2 In his original paper [1947], Gödel says that the axiom of choice is exactly as evident as the other axioms 'in the present state of our knowledge' (p. 516). In the 1964 reprint (Gödel [1964]) this has been replaced by 'from almost every possible point of view' (p. 259, n. 2). He proposed, after some hesitation, a further extension of the range of set-theoretical basic statements that in fact amounted to a new Euclidean programme – but immediately proposed a quasi-empirical alternative in the case of failure. (See especially the supplement to his [1964].)

3 Cf. Myhill [1960], p. 464.

4 Kreisel criticizes Gödel (Kreisel [1967*a*]) for not discussing his turn from proposing the constructibility axiom as a completion of set theory in 1938 to surreptitiously withdrawing it in 1947. One would think the reason for the turn is obvious: in the meantime he must have studied the work done on the consequences of the continuum hypothesis (mainly by Lusin and Sierpinski) and must have come to the conclusion that a set theory in which the hypothesis is deducible (like the one he suggested in 1938) is false. It may be interesting to note that according to Lusin a simple

If one regards comprehensive set theories – and mathematical theories in general – as quasi-empirical theories, a host of new and interesting problems arise. Until now the main demarcation has been between the proved and the unproved (and the provable and unprovable); radical justificationists ('Positivists') equated this demarcation with the demarcation between meaningful and meaningless. [But now there will be a new demarcation problem]: *the problem of demarcation between testable and untestable (metaphysical) mathematical theories with regard to a given set of basic statements.* Certainly one of the surprises of set theory was the fact that theories about sets of very high cardinality are testable in respect to a relatively modest kernel of basic statements (and thus have arithmetical content).[1] Such a criterion will be interesting and informative – but it would be unfortunate if some people should want to use it again as a meaning criterion as happened in the philosophy of science.

[Another problem is that] testability in mathematics rests on the slippery concept of a heuristic falsifier. A heuristic falsifier after all is a falsifier only in a Pickwickian sense: it does not falsify the hypothesis, it only suggests a falsification – and suggestions can be ignored. It is only a rival hypothesis. But this does not separate mathematics as sharply from physics as one may think. Popperian basic statements too are only hypotheses after all. *The crucial role of heuristic refutations is to shift problems to more important ones,* to stimulate the development of theoretical frameworks with more content. One can show of most classical refutations in the history of science and mathematics that they are heuristic falsifications. The battle between rival mathematical theories is most frequently decided also by their relative explanatory power.[2]

Let us finally turn to the question: *what is the 'nature' of mathematics,* that is, on what basis are truth values injected into its potential falsifiers? This question can be in part reduced to the question: What is the nature of *informal* theories, that is, what is the nature of the potential falsifiers of *informal* theories? Are we going to arrive, tracing back problemshifts through informal mathematical theories to empirical theories, so that mathematics will turn out in the end to be *indirectly empirical,* thus justifying Weyl's, von Neumann's and – in a certain sense – Mostowski's and Kalmár's position? Or is *construction* the only source of truth to be injected into a mathematical basic statement? Or *platonistic intuition*? Or *convention*? The answer will scarcely be a monolithic one. Careful historico-critical case-studies will probably lead to a sophisticated and composite solution. But whatever the

proposition in the theory of analytic sets which Sierpinski showed to be incompatible with the continuum hypothesis is 'indubitably true' – indeed he puts forward an impressive argument (Lusin [1935] and Sierpinski [1935]).

[1] The term 'content' is here used in a Popperian sense: the 'arithmetical content' is the set of arithmetical potential falsifiers.

[2] Cf. *below,* chapter 3.

solution may be, the naive school concepts of static rationality like *apriori–aposteriori*, *analytic–synthetic* will only hinder its emergence. These notions were devised by classical epistemology to classify Euclidean certain knowledge – for the problemshifts in the growth of quasi-empirical knowledge they offer no guidance.*

5 PERIODS OF STAGNATION IN THE GROWTH OF QUASI-EMPIRICAL THEORIES

The history of quasi-empirical theories is a history of daring speculations and dramatic refutations. But new theories and spectacular refutations (whether logical or heuristic) do not happen every day in the life of quasi-empirical theories, whether scientific or mathematical. There are occasional long *stagnating periods* when a single theory dominates the scene without having rivals or acknowledged refutations. Such periods make many forget about the criticizability of the basic assumptions. Theories, which looked counterintuitive or even perverted when first proposed, assume authority. Strange methodological delusions spread: some imagine that the axioms themselves start glittering in the light of Euclidean certainty, others imagine that the deductive channels of elementary logic have the power to retransmit truth (or probability) 'inductively' from the basic statements to the extant axioms.

The classical example of an abnormal period in the life of a quasi-empirical theory is the long domination of Newton's mechanics and theory of gravitation. The theory's paradoxical and implausible character put Newton himself into despair: but after a century of corroboration Kant thought it was self-evident. Whewell made the more sophisticated claim that it had been solidified by 'progressive intuition',[1] while Mill thought it was inductively proved.

Thus we may name these two delusions 'the Kant–Whewell delusion', and the 'inductivist delusion'. The first reverts to a form of Euclideanism; the second establishes a new – inductivist – ideal of deductive theory where the channels of deduction can also carry truth

* Since this paper was written a good deal of further work has been done on testing proposed set-theoretical axioms, like the continuum hypothesis and strong axioms of infinity. (A good survey is to be found in Fraenkel, Bar Hillel and Levy [1973]. See also Shoenfield [1971] for the axiom of measurable cardinals.) Levy and Solovay's work ([1967]) indicates that large cardinal axioms will not decide the continuum problem. As another line of attack, alternatives to the continuum hypothesis have been formulated and tested. An example is 'Martin's axiom', which is a consequence of the continuum hypothesis, but consistent with its negation (see Martin and Solovay [1970] and Solovay and Tennenbaum [1971]). Of the six consequences of the Continuum Hypothesis which Gödel regarded as highly implausible, three follow also from Martin's Axiom. But Martin and Solovay take a different attitude to that taken by Gödel. They have, they say, 'virtually no intuitions' about the truth or falsity of these three consequences. (*Eds.*)

[1] E.g. Whewell [1860], especially chapter XXIX.

(or some quasi-truth like probability) upwards, from the basic statements to the axioms.

The main danger of both delusions lies in their methodological effect: both trade the challenge and adventure of working in the atmosphere of permanent criticism of quasi-empirical theories for the torpor and sloth of a Euclidean or inductivist theory, where axioms are more or less established, where criticism and rival theories are discouraged.[1]

The gravest danger then in modern philosophy of mathematics is that those who recognize the fallibility and therefore the science-likeness of mathematics, turn for analogies to a wrong image of science. The twin delusions of 'progressive intuition' and of induction can be discovered anew in the works of contemporary philosophers of mathematics.[2] These philosophers pay careful attention to the degrees of fallibility, to methods which are *a priori* to some degree, and even to degrees of rational belief. But scarcely anybody has studied the possibilities of refutations [in mathematics].[3] In particular, nobody has studied the problem of how much of the Popperian conceptual framework of the logic of discovery in the empirical sciences is applicable to the logic of discovery in the quasi-empirical sciences in general and in mathematics in particular. *How can one take fallibilism seriously without taking the possibility of refutations seriously?* One should not pay lip-service to fallibilism: 'To a philosopher there can be nothing which is absolutely self-evident' and then go on to state: 'But in practice there are, of course, many things which can be called self-evident...each method of research presupposes certain results as self-evident.'[4] Such *soft fallibilism* divorces fallibilism from criticism and shows how deeply ingrained the Euclidean tradition is in mathematical philosophy. It will take more than the paradoxes and Gödel's results to prompt philosophers to take the empirical aspects of mathematics seriously, and to elaborate a philosophy of critical fallibilism, which takes inspiration not from the so-called foundations but from the *growth* of mathematical knowledge.

[1] Cf. Kuhn, especially his [1963].

[2] The main proponents of Whewellian progressive intuition in mathematics are Bernays, Gödel, and Kreisel (see *above*, pp. 34–5). Gödel also provides an inductivist criterion of truth, should progressive (or as Carnap would call it 'guided') intuition fail: an axiomatic set theory is true if it is richly verified in informal mathematics or physics. 'The simplest case of an application of the criterion under discussion arises when some set-theoretical axiom has number-theoretical consequences verifiable by computation up to any given integer' (Supplement to Gödel [1964], p. 272).

[3] Kalmár – with his criticism of Church's thesis – is a notable exception (see Kalmár [1959]).

[4] Bernays [1965], p. 127.

3

Cauchy and the continuum: the significance of non-standard analysis for the history and philosophy of mathematics[1]

Non-standard analysis is a fascinating topic for the historian and philosopher of mathematics. First, it revolutionizes the historian's picture of the history of the calculus. Also, it is one of the most interesting signs that meta-mathematics is turning away from its originally philosophical beginnings and is growing into an important branch of mathematics.

1 NON-STANDARD ANALYSIS SUGGESTS A RADICAL REASSESSMENT OF THE HISTORY OF THE INFINITESIMAL CALCULUS

The history of mathematics has been distorted by false philosophies even more than has the history of science.[2] It is still regarded by many as an accumulation of eternal truths;[3] false theories or theorems are banished to the dark limbo of prehistory or recorded as regrettable mistakes, of interest only to curiosity collectors. According to some

[1] The author is indebted to Professor Abraham Robinson for instructive discussions.
* This paper was delivered at the International Logic Colloquium, Hanover, 1966 – European meeting of the Association for Symbolic Logic. It was accepted for publication by the *British Journal for the Philosophy of Science* in 1966 but Lakatos withheld it. A number of marginal notes on the original typewritten manuscript indicate that he was not satisfied with some of the statements made there. There are, however, no indications that Lakatos contemplated alterations to the main points of his arguments. – (J. P. C.)

[2] The dismal effect of false philosophies upon the historiography of mathematics is discussed in my [1963–4], especially on pp. 2–6 and then *passim*. (For the impact of false philosophies of science upon the historiography of science see Agassi [1963], and especially volume 1, chapter 2.)

[3] A remark by Duhem, the most important and influential historian of science at the beginning of our century is very characteristic. In his [1906] (p. 269 of the English translation), he talks about 'an additional mark of the great difference between physics and geometry.

'In geometry, where the clarity of deductive method is fused directly with the self-evidence of common sense, instruction can be offered in a completely logical manner. It is enough for a postulate to be stated for a student to grasp immediately the data of common-sense knowledge that such a judgment condenses; he does not need to know the road by which this postulate has penetrated into science. The history of mathematics is, of course, a legitimate object of curiosity, but it is not essential to the understanding of mathematics.

'It is not the same with physics...'

historians of mathematics the 'proper' history of mathematics starts with those works which conform to the standards they regard as being ultimate. Others descend to the prehistoric ages only to pick out shining fragments of eternal truth from among the rubbish. Both miss some of the most exciting patterns of conjectures and refutations in the history of mathematical thought. Still worse, interesting inconsistent theories are distorted into 'correct' but uninteresting precursors of up-to-date theories. Efforts to save the authority of the giants of the past by giving them a polished modern look have gone further than one would imagine.

All this applies especially to the historiography of the infinitesimal calculus. Some of the most interesting features of the pre-Weierstrass era have gone unnoticed or remained un-understood (if not misunderstood) because of 'rational reconstructions'. Robinson's work revolutionizes our picture of this most interesting and important period. It offers a rational reconstruction of the discredited infinitesimal theory which satisfies modern requirements of rigour and which is no weaker than Weierstrass's theory. This reconstruction makes infinitesimal theory an almost respectable ancestor of a fully fledged, powerful modern theory, lifts it from the status of pre-scientific gibberish and renews interest in its partly forgotten, partly falsified history.

In the last chapter of his [1966], Robinson himself outlines some of the most important changes which his non-standard analysis suggests in the historiography of the calculus ('Concerning the History of the Calculus'). I shall discuss only one example in detail: Cauchy and the problem of uniform convergence. First I shall show that some very interesting historical problems related to the emergence of uniform convergence have never been satisfactorily solved (section 2: 'Cauchy and the problem of uniform convergence'). This will be followed by a section outlining how a rational reconstruction in Robinson's spirit can illuminate them (section 3: 'A new solution'). Then I shall discuss the merits and limitations of rational reconstructions for the understanding of real history (section 4: 'Rational reconstruction versus history'). After this I discuss further related problems (section 5: 'What caused the downfall of Leibniz's theory?'). Finally there will be two sections discussing some problems in the philosophy of the history of mathematics (section 6: 'Metaphysical versus technical'; and section 7: 'Appraisal of informal mathematical theories').[1] I shall argue that the Robinsonian approach to the history of the calculus cannot be fully exploited within the framework of the formalist philosophy of mathe-

[1] I should perhaps mention here that I studied these problems first in 1957–8 and discussed them at some length in my doctoral thesis *Essays in the Logic of Mathematical Discovery*, 1961. The reason why I had not published my results was that I had the uneasy feeling that something was wrong with my discussion. Having read Robinson I realized my mistake: I misread Cauchy as a direct precursor of Weierstrass. *This material has now been published as Appendix 1 of Lakatos [1976*c*]. (*Eds.*)

matics which is today the main impediment to the study and understanding of the history of mathematics.[1]

2 CAUCHY AND THE PROBLEM OF UNIFORM CONVERGENCE

Cauchy has been commonly regarded by historians of mathematics as the person who gave the calculus its 'final foundation'[2] and put it 'on solid ground'.[3] One historian's eulogy is worth quoting in full:

> Modern mathematics is indebted to Cauchy for two of its major interests, each of which marks a sharp break with the mathematics of the eighteenth century. The first was the introduction of rigor into mathematical analysis. It is difficult to find an adequate simile for the magnitude of this advance; perhaps the following will do. Suppose that for centuries an entire people has been worshipping false gods and suddenly their error is revealed to them. Before the introduction of rigor mathematical analysis was a whole pantheon of false gods.[4]

The worst 'false gods' were surely the infinitesimals. But Cauchy uses the term 'infinitesimal' all the time. Historians interpreted his repeated blasphemy as a manner of speech: he used 'infinitesimal', they say, to denote 'nothing more than a variable converging toward zero'.[5] The progress from Cauchy to Weierstrass was cumulative: Weierstrass *added* the arithmetization of analysis, i.e. the theory of real numbers, to Cauchy's theory without refuting anything in Cauchy's work.[6]

But what about Cauchy's well-known 'mistakes'? How could he prove in his celebrated *Cours d'Analyse* [1821] – fourteen years after the discovery of the Fourier series – that any convergent series of continuous functions always has a continuous limit function?* How could he prove the existence of the Cauchy integral for any continuous function?[7] Was all this just carelessness, oversight, a series of 'unfortunate' technical mistakes?[8]

[1] The term 'formalism' is used here *not* to designate the school in meta-mathematics associated with Hilbert but that philosophy of mathematics which identifies mathematics with its formalized meta-mathematical abstraction (and the philosophy of mathematics with meta-mathematics). See Kreisel and Krivine [1967], appendix II.

[2] Klein [1908], volume I, part III, I.2 (p. 154).

[3] Bourbaki [1960], p. 218.

[4] Bell [1937], p. 271.

[5] Cf. Boyer [1939], p. 273. As is clear from the context, Boyer means here a Weierstrassian real variable.

[6] Cajori even goes further: 'With Cauchy begins the process of "arithmetisation"' (Cajori [1919], p. 369).

* Cauchy [1821], p. 131: 'lorsque les différents termes de la série

$$u_0, u_1, u_2, \ldots, u_n, u_{n+1}, \ldots,$$

sont des fonctions d'une même variable x, continues par rapport à cette variable, dans le voisinage d'une valeur particulière pour laquelle la série est convergente, la sommes de la série est aussi, dans le voisinage de cette valeur particulière, fonction continue de x.' (J.P.C.)

[7] Cauchy [1823], pp. 81–4.

[8] According to Bourbaki (*op. cit.*, p. 219): 'Unfortunately for him, Cauchy claimed to prove [more than he did prove]...' But talking about Cauchy's 'unfortunate mistake'

But if Cauchy's 'mistakes' were sheer oversights, why was one of them put right only in 1847 (by Seidel) and the other only as late as 1870 (by Heine)?

There are some other queer facts. For example, Fourier's counterexamples* were well known when Cauchy wrote his book: it seems that Cauchy proved a theorem which many people, including himself, knew to be false or at least problematic. Abel's footnote to the effect that Cauchy's theorem 'suffers exceptions'[1] only put into print part of the 'folklore' of the experts: as he himself says, after giving an example from Fourier's published work, 'it is *well known* that there are many series with similar properties'.[2]

If a counterexample was well known, why wasn't the proof immediately checked, the hidden lemma discovered and made explicit, the validity of the proof restored, and, by incorporating the lemma into the original theorem, a more correct one formulated? In particular, why did Abel make no effort to find out what was amiss in the proof? Why was he content to copy Cauchy's original proof without alteration, restating it for the safe domain of power-series? Abel, a typical rigourist, was ready to forego difficult terrain rather than risk his standards of rigour; he boldly proposed to restrict the domain of validity of all theorems in analysis to power-series. In withdrawing behind the safe boundary of power-series, he banished Fourier's series, as an uncontrollable jungle of exceptions, from the field of rational investigation.[3]

But it was not only Abel who displayed such a curiously confused attitude when facing the clash between Cauchy's theorem and Fourier's counterexamples. Dirichlet certainly must have seen the problem; but he clearly decided that he would not mention it in his celebrated paper on the convergence of Fourier series, in which he showed some subtle details of how convergent series of continuous functions converge – defying Cauchy's proof – to discontinuous functions. Seidel, who, twenty-six years after Cauchy's proof, finally solved the problem with the discovery of uniform convergence,[4] was Dirichlet's pupil and probably inherited the problem from him.

Why this delay of twenty-six years? Today, if one gave Cauchy's false proof to a bright undergraduate, it would not take him long to put it right; and indeed, Seidel himself did not find the problem at all

explains nothing, although it is better than to hush up, as some historians do, the 'errors' of the great mathematician, especially of one who boasted (in the Introduction to his [1812]) that he would 'dispel all uncertainty'.

*I.e. the trigonometric series, see n. 3, p. 48 (J.P.C.).

1 Abel [1826*a*], p. 316.

2 *Ibid.*

3 I gave a detailed description of this puzzling methodological attitude of 'exeption-barring' – which so frequently replaces the search for hidden lemmas – in my [1963–4], pp. 124 and 234–5. * Cf. Lakatos [1976*c*], pp. 24–30, and 133–6. (*Eds.*).

4 We know now from Weierstrass's manuscripts that he had known about uniform convergence – and lectured about it with textbook clarity – since 1841.

difficult![1] What inhibited a whole generation of the best minds from solving an easy problem?

One may of course point out that many problems look simple only after they have been solved. But then why could Cauchy, even after Seidel's paper, not understand uniform convergence, which, according to Seidel, was an obvious hidden lemma in Cauchy's own proof? In a paper read to the Academy in March 1853,[2] Cauchy stubbornly restated his theorem, and claimed that the recalcitrant sequences do *not* converge everywhere, and in particular not in the infinitely small neighbourhood of the points of discontinuity.

Cauchy's story is a fairly mysterious one, everywhere dense with problems. Robinson's theory, however, gives us the crucial clue to their solution.

3 A NEW SOLUTION

The gist of the solution suggested by Robinson is that in the history of the calculus from Leibniz to Weierstrass there were two rival theories of the continuum. On the one hand there was the now accepted Weierstrassian theory and on the other there was the Leibnizian theory of the continuum: the Archimedean continuum extended to a non-Archimedean one by adding infinitesimals and infinitely large numbers. Leibniz's theory was the dominant one until the Weierstrassian revolution, and Cauchy himself was completely in the Leibnizian tradition. What was revolutionary about Weierstrass's theory was that the known calculus could be fully explained, and even further developed, with Weierstrassian real numbers only – the set of which was a mere skeleton of what Leibnizians regarded as the set of

[1] Seidel wrote in his [1847], p. 383: 'When, following the discovery that the theorem cannot be generally valid – that is, there must be somewhere in the proof some hidden lemma – one scrutinizes the proof thoroughly, it is not difficult at all to spot the hidden lemma; and after having spotted it one can conclude that the so found lemma cannot possibly be satisfied for series representing discontinuous functions, since this is the only way to reconcile the otherwise correct proof with the established results.'

[2] Cauchy [1853], p. 454. *Cauchy says 'Au reste, il est facile de voir comment on doit modifier l'énoncé du thèoréme, pour qu'il n'y ait plus lieu à aucune exception. C'est ce que je vais expliquer en peu de mots.' At first (Weierstrassian!) sight it seems that Cauchy adds the condition of uniform convergence. Robinson [1967], p. 273, states that this is an added condition, but it is clear that Cauchy regards this as a trivial consequence of his notion of convergence. His argument is entirely consistent with his 1821 conceptions. His 'modified' theorem is clearly as suspect on Weierstrassian grounds as his original theorem: 'Si les différents termes de la série

$$u_0, u_1, u_2, \ldots, u_n, u_{n+1}, \ldots, \quad (1)$$

sont des fonctions de la variable réelle x, continues, par rapport à cette variable, entre des limites données, si, d'ailleurs, la somme

$$u_n + u_{n+1} + \ldots + u_{n'-1} \quad (3)$$

devient toujours infiniment petite pour des valeurs infiniment grandes des nombres entiers n et $n' > n$, la série (1) sera, entre les limites données, fonction continue de la variable x.'

See n. *, p. 52. (J.P.C.)

real numbers. Cauchy's real 'variables' ran through Weierstrassian real numbers *and* infinitesimals *and* those numbers which differed from Weierstrassian real numbers by infinitely large numbers and/or infinitesimals: the later Weierstrassian points were *finite* Leibniz–Cauchy points deprived of their infinitesimal neighbourhoods (or *monads*, to use Robinson's picturesque expression).*

In this light one can now understand the story of Cauchy's 'mistakes', and also other aspects of the story of uniform convergence and uniform continuity. It will be worth while to recall a few details.

The quasi-empirical thesis that the limit of any convergent sequence of continuous functions is continuous was taken for granted, and thought to be in no need of proof, throughout the eighteenth century. It was regarded as a special case of Leibniz's '*principe de continuité*'[1] and, in particular, of the principle that 'if a variable quantity at all stages enjoys a certain property, its limit will enjoy this same property'.[2] Cauchy was the first to try to prove the thesis; perhaps this was because he construed irrational numbers as limits of convergent sequences of rational numbers, and this was already one refutation of Leibniz's general principle, or perhaps it was because Fourier, in 1807, seemed to have produced counterexamples to the thesis, and Cauchy may have thought that his proof would show that Fourier's series cannot converge properly.[3]

* Strictly, Cauchy's variables are sequences of Weierstrassian reals. 'A variable is a quantity which is thought to receive successively different values.' His infinite numbers are unbounded sequences of reals. The infinitely small quantities are sequences which converge (in the Weierstrass sense) to zero: 'when the successive numerical values of a variable decrease indefinitely so as to become smaller than any given number, this variable becomes what is called an *infinitesimal*, or infinitely small quantity' (Cauchy [1821], pp. 4, 5). Although Cauchy did not explicitly use the notion of sequence for his variables, this idea is implicit in his actual usage.

It is interesting that as late as 1878 Cauchy's notions of *variable* and *infinitely small quantity* appeared in texts on the calculus. For instance, Houël [1878], p. 106, said: 'un quantité *infinitement petite* étant essentiellement *variable*, n'a pas de valeur fixe, et consequement sa grandeur n'est liée en rien a nos appréciation physique. L'essence d'un infinitement petit n'est pas d'être imperceptable, mais de *pouvoir* décroître autant que l'on voudra.' (J. P. C.)

1 Leibniz [1687], p. 744.

2 Lhuilier [1787], p. 167. It is interesting that Whewell as late as 1858 accepts this as being 'involved in the very conception of a limit' (Whewell [1858], p. 152).

3 The usual statement 'Abel was *the first to note* that Cauchy's announced theorem is not in general valid' (Smith [1929], p. 287) is palpably false and only obscures the intriguing fact that Cauchy proved the theorem *knowing* about the counterexamples.

On the other hand, one wonders whether it ever occurred to Fourier that some of his series contradict the principle. In his [1822], he states that the function

$$\cos x - \tfrac{1}{3}\cos 3x + \tfrac{1}{5}\cos 5x - \ldots$$

'is composed of separate straight lines, each of which is parallel to the axis, and equal to the circumference. These parallels are situated alternately above and below the axis, at the distance $\pi/4$ and joined by perpendiculars which themselves make part of the line' (section 178). That is, Fourier may have regarded this function as continuous, the perpendicular lines forming part of it.

However, my friend Dr J. R. Ravetz kindly drew my attention to an unpublished

In fact *Cauchy's theorem was true and his proof as correct as an informal proof can be.* Following Robinson[1] we can show that Cauchy's argument, if not interpreted as a proto-Weierstrassian argument but as a genuine Leibniz–Cauchy one, runs as follows:

Let $\lim s_n(x) = s(x)$, where $s_n(x)$ are continuous. Then in order to prove that $s(x)$ is continuous at some x_1, we have to show that $s(x_1+\alpha)-s(x_1)$ is infinitesimal for all infinitesimal α. (This uses Cauchy's concept of continuity, which would be equivalent to Weierstrass's concept only if any proposition that is true for all infinitesimal quantities is also true for sufficiently small finite quantities, and vice versa).[2]

Now

$$\begin{aligned} |s(x_1+\alpha)-s(x_1)| &= |s_n(x_1+\alpha)-s_n(x_1)+r_n(x_1+\alpha)-r_n(x_1)| \\ &\leqslant |s_n(x_1+\alpha)-s_n(x_1)|+|r_n(x_1+\alpha)|+|r_n(x_1)|, \end{aligned}$$

where r_n are the remainders. Cauchy thought the left-hand side was infinitesimal for all infinitesimal α since $|s_n(x_1+\alpha)-s_n(x_1)|$ is infinitesimal for all n because of Cauchy's definition of *continuity*; and $|r_n(x_1+\alpha)|$ and $|r_n(x_1)|$ are again infinitesimal for all infinitely large n because of Cauchy's definition of the *limit*: $a_n \to 0$ if a_n is infinitesimal for infinitely large n.

Of course this argument implies that $s_n(x)$ should be defined and continuous and converge not only at standard Weierstrassian points but at *every* point of the 'denser' Cauchy continuum, and that the sequence $s_n(x)$ should be defined for infinitely large indices n and represent continuous functions at such indices.* Cauchy's 'limit' and 'continuity' are only defined for 'transfinite' sequences of functions defined on his overdense continuum. For such sequences of functions Cauchy's theorem is indeed true; and Fourier–Abel counterexamples

manuscript of Fourier (dated 1809) which uses the term 'discontinuous' in the modern sense. Did he then draw the perpendiculars *after* the publication of Cauchy's [1821] only to comply (rather naively) with Cauchy's theorem? Or did he use the term 'discontinuous' in one sense when he thought of temperature and in another sense when he thought of vibrating strings? After all, the modern definition of continuity is strongly counter-intuitive, e.g. it is not invariant to rotation! Fourier's perpendiculars survived – in spite of Dirichlet's 1829 paper – in the notion that the value of the function is 'indefinite' at the points of discontinuity: Dirichlet was still criticized in 1870 by Schläfli (in *Crelle's Journal*, p. 284) and, indirectly, in 1874 by Du Bois Reymond (in *Math. Annalen*, p. 244). *See also Grattan-Guinness, I. and Ravetz, J. R. [1972]. (*Eds.*)

[1] Robinson [1967], p. 272. My reconstruction will differ slightly from Robinson's.

[2] Pringsheim's authoritative account of the history of the calculus in *Encyclopädie der mathematischen Wissenschaften* credits Cauchy with the Weierstrassian concept of continuity (volume 2, II.1, p. 17). Bell follows suit: 'The definitions of limit and continuity current today in thoughtfully written texts are substantially those expounded and applied by Cauchy' (Bell [1940], p. 292).

* Cauchy's notion of convergence can be interpreted in Robinson's non-standard analysis in the following way. Let $*R$ be an elementary extension of the real number system R and $*N$ the corresponding extension of the natural numbers N. Cauchy's proof of his 'continuity' theorem requires the convergence of the 'transfinite' (Lakatos) sequence

$$\{s(n)\colon n \in *M\} \quad \text{where} \quad s(n) \in *R:$$

are not continuous either at infinitely large indices or at non-Weierstrassian points of functions with finite indices. Alternatively, these recalcitrant sequences may not *S*-converge in the entire monad of points of discontinuity, and it is very likely that Cauchy may have suspected something like this in 1821.[1] Indeed, in his 1853 paper where he reiterates his theorem without any alteration, he pointedly stresses that the sequence of functions *must* converge at *every* point!

This interpretation sheds completely new light on Cauchy's famous 'mistake': *Cauchy made absolutely no mistake, he only proved a completely different theorem, about transfinite sequences of functions which Cauchy-converge on the Leibniz continuum.*

Thus the appraisals of Felix Klein, Pringsheim, Cajori, Boyer, Bourbaki, Bell[2] and others which give Cauchy credit for initiating the Weierstrassian revolution are utterly false and are nothing but a 1984-wise rewriting of history according to the latest party line. And to reproach Cauchy for his 'deplorable error' again completely misses the point: the 'error' appeared only in Seidel's 'translation' of Cauchy's proof into Weierstrassian theory.

But even if Cauchy had devised his proof in Weierstrass's theoretical framework, to say that he made the 'erroneous supposition that pointwise convergence *eo ipso* means uniform pointwise convergence'[3]

that is, by proceeding sufficiently (finitely) far along the sequence, the values of $s(n)$ become arbitrarily close to the limit. Thus the sequence $\{s(n): n \in {}^*N\}$, where $s(n) \in {}^*R$, Cauchy-converges to the limit t ($\in {}^*R$) if there exists a function $M(n)$ *in* R such that for all m *in* N and n *in* ${}^*N$

$$n > M(m) \rightarrow |s(n) - t| < m^{-1}.$$

With this definition Cauchy's theorem is then correct – but about ${}^*R$, not R. The crucial thing to observe is that Cauchy assumes convergence (in this sense) of the series in the *infinitesimal* neighbourhood of x_1 – for Cauchy, 'neighbourhood' in this theorem means 'infinitesimal neighbourhood'. (J. P. C.)

1 Fourier himself was doubtful about the convergence of his series in these critical cases. He noticed that 'The convergence is not sufficiently rapid to produce an easy approximation, but it suffices for the truth of the equation' (Fourier [1822], section 177). (This remark, of course, was a far cry from Stokes's discovery that the convergence in these places was infinitely slow, which was made only after 40 years experience in calculating Fourier-series. And this discovery could not possibly be made before Dirichlet's decisive improvement (in 1829) on Fourier's conjecture, showing that only those functions can be represented by Fourier series whose value at the discontinuities was $\frac{1}{2}\{f(x+0)+f(x-0)\}$.)

2 We find a typical false appraisal by E. T. Bell [1940], p. 292: 'indicative of the subtleties inherent in consistent thinking about the infinite and the continuum, even so cautious a mind as Cauchy's went astray when it surrendered itself to intuition'. This comment, apart from its completely mistaken appraisal, is also a good example of the dangers inherent in the concept of 'intuition'.

3 A. Pringsheim [1916], p. 34.* If the interpretation in terms of non-standard analysis, as given in n. *, p. 49, is correct it can be shown that Cauchy's notion of convergence implies uniform convergence in the following sense.

Let $\{f(n, x): n = 0, 1, 2, \ldots\}$ be a sequence of functions in R. Let $\{{}^*f(n, x): n \in {}^*N\}$ be its extension in ${}^*R$ (so that ${}^*f(n, x) = F(n, x)$ for $n \in N$, $x \in R$). If $\{{}^*f(n, x): n \in {}^*N\}$ Cauchy-converges in the neighbourhood of x_0 to a function ${}^*F(x)$, where $F(x)$ is a function in R, then x_0 is a point of uniform convergence of $\{f(n, x)\}$.

Proof. As $\{{}^*f(n, x)\}$ Cauchy-converges to ${}^*F(x)$, by definition of Cauchy-convergence

would still be a characteristic piece of justificationist reconstruction. It comes from the widespread view according to which an informal proof is a formal proof with gaps, with the 'hidden lemmas' carelessly omitted. This view does not allow for any genuine evolution of conceptual frameworks, and it has caused as much damage to the historiography of mathematics as the analogous view that a child is a small adult has caused in the theory of education.

We can also understand now why Abel did not discover the hidden lemma of uniform convergence: for he never freed himself from the Leibniz–Cauchy conceptual framework. If we look at the proof of his restricted theorem we find that he, like Cauchy, operates with infinitesimals, with 'quantities smaller than any *given* quantity'.* He uses for 'given' the term '*angebar*'[1] (specifiable): this of course suggests Weierstrassian numbers which indeed are the only 'specifiable' (or as Bolzano would say 'measurable') quantities of the Leibnizian continuum. Abel uses here the letter ω for infinitesimals. Sylow, the editor of the second edition (1881) of Abel's Collected Works was very unhappy about Abel's proof, since he took ω to mean a Weierstrassian ϵ.[2] Pringsheim, with characteristic self-assurance, states that Abel, in a special case, 'gave a straightforward proof of the existence of the property now called uniform convergence';[3] Hardy follows suit: 'The idea is present implicitly in Abel's proof of his celebrated theorem'.[4] Bourbaki gives a similarly false account:

Cauchy first did not notice the difference between simple and uniform convergence...but the error was almost immediately discovered by Abel, who

(n. *, p. 49) there exists an $r>0$ in $^*R$ such that for each x satisfying $|x-x_0|<r$ there exists a function $M_x(n)$ in R for which

$$n > M_x(m) \rightarrow |{}^*f(n,x) - {}^*F(x)| < m^{-1}$$

for all $m \in N$ and all $n > m$ in $^*R$. In particular, for each infinite integer ∞, and each positive ϵ in R

$$|x-x_0| < r \rightarrow |{}^*f(\infty, x) - {}^*F(x)| < \epsilon.$$

Fix an infinite integer ∞_0. Then for all $n \epsilon {}^*N$ and all $x \in {}^*R$

$$n > \infty_0 \,\&\, |x-x_0| < r \rightarrow |{}^*f(n,x) - {}^*F(x)| < \epsilon.$$

Thus

$$(Em \in {}^*N)(Er \in {}^*R)(n \in {}^*N)(x \in {}^*R)(n > m \,\&\, r > 0 \,\&\, |x-x_0| < r \rightarrow |{}^*f(n,x) - {}^*F(x)| < \epsilon)$$

holds in $^*R$ for each positive ϵ in R. As $^*R$ is an elementary extension of R, for each positive ϵ in R

$$(Em \in N)(Er \in R)(n \in R)(x \in R) \quad (n > m \,\&\, r > 0 \,\&\, |x-x_0| < \rightarrow |f(n,x) - F(x)| < \epsilon)$$

holds in R. Thus x_0 is a point of uniform convergence of the sequence $\{f(n,x): n \in N\}$. (J.P.C.)

* This is particularly clear in Houel's explanation of '*infinitement petite*' (n. *, p. 48). It motivates the definition of Cauchy-convergence in n. *, p. 49: $s(n)-t$ *can* be made smaller than m^{-1} for any *given* number m (i.e. $m \in N$) by taking n beyond the specifiable (i.e., in N) stage $M(n)$ – thus M is a function in R. (J. P. C.)

[1] It should be said that the German expression is Crelle's who translated it from the French.

[2] Cf. his analysis in volume II, p. 303. [3] *Loc. cit.*, p. 35. [4] Hardy [1918], p. 148.

proved that all power series are continuous in the open interval of their convergence... For this particular case he, essentially, makes use of the idea of uniform convergence. All that was left to be done was to apply this idea in general; this was done, independently, by Stokes and Seidel in 1847–8 and by Cauchy himself in 1853.[1]

Alas, a historical blunder in each sentence. Abel could not possibly 'reveal' Cauchy's 'mistake'. His proof does not 'exploit the concept of uniform convergence' which is alien to his infinitesimal theory. Abel's and Seidel's results are not related as 'special' and 'general' – they are on quite different levels, parts of totally different theories. By the way, Bourbaki does not even notice that Abel restricts the domain of eligible functions, and not the way they converge (as Seidel does!). Finally, to say that Cauchy's 1853 paper contains the independent rediscovery of uniform convergence, is not a remark that can be made without serious qualifications.*

We also see now why Seidel found it so easy to discover the hidden lemma in what he thought was Cauchy's proof: because he scrutinized his own Weierstrassian reconstruction of Cauchy's theorem and proof. In this reconstruction the theorem is false and the guilty lemma can in fact easily be found.

[1] Bourbaki [1949*b*], p. 65. See also Bourbaki [1960], p. 228.

* Cauchy's proof of his 1853 theorem (see n. 2, p. 47) is given in the following words (the brackets [ ], { } have been added by the editor): 'Soient alors

s la somme de la série
s_n la somme de ses n premiers termes;
et $r_n = s - s_n = u_n + u_{n+1} + \ldots .$ le reste de la série

indéfiniment prolongée à partir du terme général u_n.
Si l'on nomme n' un nombre entier supérieur à n, le reste r_n ne sera autre chose que la limite vers laquelle convergera, pour des valeurs croissantes de n', la différence

$$s_{n'} - s_n = u_n + u_{n+1} + \ldots + u_{n'-1}. \qquad (3)$$

[Concevons, maintenant, qu'en attribuant à n une valeur suffisamment grande, on puisse rendre, pour toute les valeurs de x comprises entre les limites données, le module de l'expression (3) (quel que soit n'), et, par suite, le module de r_n, inférieur à un nombre ϵ aussi petit que l'on voudra.] Comme un accroissement attribué à x pourra encore être supposé assez rapproaché de zéro pour que l'accroissement correspondant de s_n offre un module inférieur à un nombre aussi petit que l'on voudra, {il est clair qu'il suffira d'attribuer au nombre n une valeur infiniment grande, et à l'accroissement de x une valeur infiniment petite, pour démontrer, entre les limites données, la continuité de la fonction

$$s = s_n + r_n.\}$$

Mais cette démonstration suppose évidement que l'expression (3) remplit la condition ci-dessus énoncée, c'est-à-dire que cette expression devient infiniment petite pour une valeur infiniment grande attribuée au nombre entier n. D'ailleurs, si cette condition est remplie, la série (1) sera évidement convergente.'

The sentence [ ] shows that Cauchy recognized that uniform convergence was sufficient to ensure the continuity of s. But the passage { } shows that Cauchy regarded this condition as a trivial consequence of his notion of *convergence in a neighbourhood* by taking n infinite. (This step is similar to the first step in the proof in n. 2, p. 47.) Thus uniform convergence is implicit in Cauchy's 1821 ideas and not an extra condition added in 1853. (J. P. C.)

Finally, we understand why as late as 1853 Cauchy did not understand uniform convergence, even if he was informed (as he probably was) of Seidel's result: because he did not understand Weierstrass's theory, just as Seidel, having no idea of the Leibniz–Cauchy infinitesimal theory, misunderstood Cauchy's proof.

So slowly everything seems to fall into place, and an exciting story of two rival theories of the calculus is revealed at, however, a surprisingly low degree of articulation. It is a most interesting historical fact that Bolzano, the best logical mind of the generation, made a real effort to clarify matters. He was possibly the only one to see the problems related to the difference between the two continuums: the rich Leibnizian continuum and, as he called it, its 'measurable' subset – the set of Weierstrassian real numbers. Bolzano makes it very clear that the field of 'measurable numbers' constitutes only an Archimedean subset of a continuum enriched by non-measurable – infinitely small or infinitely large – quantities.[1] The editor makes a misguided attempt to reconstruct Bolzano's theory as a mere precursor of Cantor's theory of real numbers (cf. his dictionary of the two theories on p. 98); one wonders whether he has omitted some crucial passages from those parts of the manuscript which try to set up a consistent theory of the Leibniz–Cauchy continuum. No doubt, since Robinson has shed new light on the latter, historians will approach the Bolzano manuscripts with new eyes and the relation between Bolzano's measurable and non-measurable quantities and Robinson's standard and non-standard numbers will be clarified.

4 WHAT CAUSED THE DOWNFALL OF LEIBNIZ'S THEORY?

There are, however, some problems that have not yet been explained: first of all, why Abel's 'exception-barring'? If the informal Leibniz theory of the continuum existed, did it not contain enough strength to suggest suitable hidden lemmas which would account for the counterexamples? Why did Abel not say that the counterexamples show that the freakish function sequences cannot possibly converge in the monads of their points of discontinuity? Was it because this would have raised the question of defining the functions at non-standard points: this extension may be done in each case in many different ways! But even so, Abel should have at least concluded that in the exceptional cases there could be no possibility of continuously extending the functions defined at standard points in such a way that they should converge also at the non-standard points. Then he could have reasserted Cauchy's theorem confidently, simply stressing that the function sequences must converge *everywhere* (i.e. on the entire Leibniz con-

[1] Bolzano worked on this analysis ('Theory of quantities') in the years 1830–5 but never finished it. Parts of the manuscript were recently published under the misleading title 'Theory of real numbers' (K. Rychlik [1962]).

tinuum). *Why did he not do this?* The solution may be that there is certainly not much point in making a hidden lemma explicit if it is *untestable.* Untestable, that is, in the mathematical sense, namely that possible counterexamples should be – in some sense – *constructible* or *definable* and thus open up a new field of research, as happens in Weierstrass's theory with easily specifiable non-uniformly convergent sequences. The decisive difference between Leibniz's and Robinson's theory of infinitesimals is exactly this: Robinson devises a *particular* non-standard analysis which is an elementary extension (in Tarski's sense) of real analysis, and where there are important bridges between the two analyses which make non-standard analysis testable. But this progress could not have been made before and without Weierstrass and Tarksi.*

The downfall of Leibnizian theory was then not due to the fact that it was inconsistent,[1] but that it was capable only of limited growth. It was the heuristic potential of growth – and explanatory power – of Weierstrass's theory that brought about the downfall of infinitesimals. The crucial hidden lemmas which emerged under the pressure of the criticism of its proofs were not independently testable: this discouraged the advocates of infinitesimals, led some of them, like Cauchy, to the belief that infinitesimals are admissible in proofs but not in the formulation of theorems, and made them finally disappear for a century from the history of mathematics.[2]

* A construction of non-standard analysis is given in Chwistek [1948] which is derived from a paper published in 1926. It is basically the reduced power R^N/F where F is the Frechet filter on the natural numbers (the collection of cofinite sets of natural numbers) (see Frayne, Morel and Scott [1962–3]). It is not difficult to prove the theorem of n. *, p. 47 for R^N/F. This particular construction is not an elementary extension of R but there are sufficiently powerful transfer properties to enable some non-standard analysis to be performed. It may be observed that the elements of R^N/F are equivalence classes of sequences of reals, two sequences $s_1, s_2, \ldots$ and $t_1, t_2, \ldots$ being counted equal if for some n, $s_m = t_m$ for all $m \geqslant n$. The relation of these classes to Cauchy's variables is obvious. (J. P. C.)

1 Competent mathematicians even after Weierstrass (like Dubois-Reymond and Stolz) thought a consistent infinitesimal theory of the calculus was perfectly possible. In Felix Klein's words: 'the question naturally arises whether...it would be possible to modify the traditional foundations of infinitesimal calculus, so as to include infinitely small quantities in a way that would satisfy modern demands as to rigour; in other words, to construct a non-Archimedean analysis. I will not say that progress in this direction is impossible, but it is true that none of the investigators who have busied themselves with actually infinitely small quantities have achieved anything positive' (Klein [1908], p. 219). I think that Robinson's explanation of the defeat of infinitesimal theory by its inconsistency is untenable. (He says in the introduction to his [1966] that 'Neither Leibniz nor his disciples and successors were able to give a rational development' leading up to a consistent non-Archimedean system. 'As a result, the theory of infinitesimals gradually fell into disrepute and was replaced eventually by the clasical theory of limits.')

2 This section indicates how a theory of the growth of informal mathematical theories may be inspired by suitable use of Popperian ideas.

5 WAS CAUCHY A 'FORERUNNER' OF ROBINSON?

There can be no doubt that Robinson's contribution is going to be epoch-making in the historiography of the calculus. But two cautionary points have to be made.

The first point is this. We have already seen the danger in interpreting Cauchy as a Weierstrassian. The lesson of this should be that past theories should be regarded as respectable even if they are defeated rivals of modern theories; the criterion for a place in history should not be continuity with the modern theories of the day. It would be a mistake to pay new attention to the infinitesimal theory only because Robinson gave a reconstruction of it which is respectable by present-day standards, and instead of treating Cauchy as an inarticulate Weierstrass, now to treat him as an inarticulate Robinson. This would alter the *pattern* of justificationist historiography but not its basic tenet – to reconstruct history as a mixture of meaningless gibberish and continuous growth towards the up-to-date theories. This would still make the historian blind to the real dialectical (i.e. critical) pattern of historical progress, of conjectures, proofs and refutations and of the struggle of competing theories. Unfortunately Robinson himself seems occasionally to be tempted in the wrong direction. He says that modern non-standard analysis provides 'precise explications' of Cauchy's notions; in the Introduction to his book he claims to have shown that 'Leibniz's ideas can be fully vindicated'. This overemphasis on the continuity between the Leibniz–Cauchy theory and his has, however, led him to a false reconstruction of Cauchy's 1821 proof of which we read the following account in his book:

Interpreted in terms of non-standard analysis, the argument runs as follows. Let x_1 be a standard number, $a < x_1 < b$. In order to prove that $s(x)$ is continuous at x_1 we attempt to show that $s(x_1+\alpha)-s(x_1)$ is infinitesimal for all infinitesimal α. Now

(10.5.3) $$s(x_1+\alpha)-s(x_1) = (s_n(x_1+\alpha)-s_n(x_1))+(r_n(x_1+\alpha)-r_n(x_1))$$

Following Cauchy's argument we might be inclined to claim that the left hand side is infinitesimal since $s_n(x_1+\alpha)-s_n(x_1)$ is infinitesimal for all n, and $r_n(x_1+\alpha)$ and $r_n(x_1)$ are infinitesimal for all infinite n. However this is erroneous, for although $r_n(x_1)$ is infinitesimal for all infinite n, $r_n(x_1+\alpha)$ has to be infinitesimal only for sufficiently *high* infinite n; while $s_n(x_1+\alpha)-s_n(x_1)$ is infinitesimal only for all *finite* n, and hence, by one of our basic lemmas for sufficiently *small* infinite n. In order to prove that the left hand side of 10.5.3 is infinitesimal we have to ensure that there exists an n for which $r_n(x_1+\alpha)$ and $s_n(x_1+\alpha)-s_n(x_1)$ are infinitesimal simultaneously. Two natural alternatives offer themselves, (i) to assume that $u_0(x)+u_1(x)+\ldots$ is uniformly convergent in the interval $a<x<b$, so that $r_n(x_1+\alpha)$ is infinitesimal for *all* infinite n, or (ii) to assume that the family $\{s_n(x)\}$ is equicontinuous in the interval, so that $s_n(x_1+\alpha)-s_n(x_1)$ is infinitesimal for all infinite n (Robinson [1966], p. 272).

According to this account Cauchy's original theorem refers only to convergence at standard points; therefore his theorem is false and he *did* make a mistake in his proof.[1] But the mistake is only in the framework of the Robinsonian reconstruction which assumes a *particular* non-standard analysis of which Cauchy could not possibly have dreamed. For instance, there is no reason why Cauchy could not think that $s_n(x_1+\alpha)-s_n(x_1)$ or $r_n(x_1+\alpha)$ are infinitesimal for all infinitely large indices. Robinson's analysis of Cauchy's 'mistake' is certainly reminiscent of H. Liebmann's analysis (in 1900)[2] of the same mistake where he carefully reconstructs – following Seidel – Cauchy's 'mistake' in Weierstrass's framework.

However, the continuity between Cauchy and Robinson is *much less* than the continuity between Cauchy and Weierstrass.

Hitherto we assumed that the infinitesimal theory of the calculus consisted of one single school according to which the continuum was a non-Archimedean extension of the field of real numbers by infinitesimals and infinitely large numbers. We assumed that this continuum was *static* in the sense that the quantities were fixed, and *variables* were only a – not very fortunate – manner of speech to describe *functions* in the modern sense. In particular we assumed that Cauchy's theory was static in this sense and we analysed his work according to this interpretation.

But a more careful analysis of Cauchy's term 'variable' soon shatters our conception. It will turn out that it is impossible to maintain that 'variable' is only a manner of speech for Cauchy.[3] Robinson[4] is right in pointing out that it indicates a systematic effort to get away from the actual infinitesimals and infinities whose logical weaknesses were so forcefully shown up by Berkeley. However this widens dramatically the gulf between Robinson's and Cauchy's theory. To understand this fully, we have to analyse Cauchy's 1853 paper. According to the history textbooks, Cauchy discovered in this paper – 32 years after his *Cours d'Analyse* – the concept of uniform convergence. According to Robinson, in this paper he correctly asserts that his theorem is faultless provided only that the requirement that the sequence should converge *everywhere* (that is, also at the non-standard points) is fulfilled.[5] Who is right? The school historians can hardly be right because uniform convergence is a theoretical concept of the Weierstrass theory. Without the Weierstrass theory, no uniform convergence. But is Robinson right? If he is right, Cauchy must have changed his mind in 1853 about banning theoretical (non-standard) terms from theorems, for

[1] This would certainly be the case if Cauchy allowed the use of infinitesimals only in proofs but not in theorems.

[2] In the editorial comments to Dirichlet's [1829] and Seidel's [1847] paper in the *Ostwald's Klassiker* edition, p. 51.

[3] According to Felix Klein, the expression that 'ϵ becomes infinitely small' is 'since Cauchy only a convenient expression implying that the quantity decreases without bound towards zero' (*loc. cit.*, p. 219). This, of course, is another instance of projecting Weierstrass back into Cauchy.

[4] Robinson [1967], p. 35.

[5] Cf. Robinson [1966], p. 273.

everywhere is a term which according to his philosophy should only be used in proofs, not in theorems. But Robinson is wrong. To understand this, let us quote Cauchy's argument in defence of his theorem. He takes a Fourier counterexample and shows that it does *not* converge everywhere. His example is the series:

$$\sin x + \frac{\sin 2x}{2} + \frac{\sin 3x}{3} + \ldots$$

He shows that in the neighbourhood of zero, where the limit function is discontinuous, 'the value of the remainder for xs very near to zero, for instance for $x = (1/n)$ where n is a very large number, can differ considerably from zero', that is, the function is not continuous at points very near to zero. (Cauchy points out that the remainder at $x = (1/n)$ tends to

$$\int_1^\infty \frac{\sin x}{x} dx.)$$

Now this is a curious argument. It shows that our Robinsonian interpretation of the Cauchy continuum was not quite correct. Cauchy's continuum (perhaps unlike Leibniz's) is not a set of *actual* points but a set of *moving* points. His 'variables' are not Weierstrassian 'variables'; the latter can be eliminated without loss, since the Weierstrass theory of motion explains motion, change, variables in terms of an infinitistic algebra of *actual* quantities: this is one of its most important achievements. Not so Cauchy's theory, where 'variable quantity' is not simply a manner of speech but a vital part of the theory. The 'point' at which he shows that

$$\sin x + \frac{\sin 2x}{2} + \ldots$$

does not converge is a *moving* point $x = (1/n)$ where $n \to \infty$. The fact that the sequence does not converge at this *moving* point is in fact what later came to be known as the Gibbs-phenomenon and the corresponding condition – namely that $\Sigma f_n(x)$ is uniformly convergent in I if for all $\{x_n\}$ in I the corresponding remainders $r_n(x_n)$ tend to zero – can be shown to be equivalent to Weierstrassian uniform convergence. But then 'everywhere' in Cauchy's theorem does *not* mean '*at all points, whether standard or non-standard*' as Robinson would have it, but '*at all standard and at all Cauchy-wise moving points*'. So Cauchy's continuum is a rather 'dynamic' one. (It would be interesting to investigate whether Bolzano's continuum was more similar to Robinson's, and what Abel's and Dirichlet's conceptions were like.)[1]

To conclude: Robinson's non-standard analysis provides a powerful stimulus to the historian to have a fresh look at history. But it would

[1] After all this one cannot help being intrigued by Pringsheim's account according to which 'Cauchy himself corrected later independently his false theorem and on this occasion he characterized the nature of uniform convergence with perfect

be a mistake to expect that the continuous history of the calculus will now turn out to converge to Robinson, not to Weierstrass as hitherto assumed; our historian should rather give up his theory of continuous, monolithic streams of history altogether.

6 METAPHYSICAL VERSUS TECHNICAL

Justificationist historiography – as we already said – would present the history of mathematics as an accumulation of eternal truths. This leads one either to date the history of mathematics from the date of the last 'revolution in rigour', or to falsify the history of mathematics and reconstruct it in the up-to-date pattern. A very widespread device to try to maximize continuity is to demarcate in mathematical theories a *hard formal core* which is uncontested and indubitable, eternal; and a '*metaphysical' interpretation* of the formalism which is controversial, 'soft', changing. It is an intriguing challenge to historians of thought to make a detailed study of the origin of this theory. It probably comes from the seventeenth and eighteenth centuries when outstanding mathematicians could handle formulas in analysis to get correct results – they could differentiate and integrate with remarkable success – but when it came to interpreting their formulae, they got into inconsistencies. So they claimed that the successful manipulation of the formulas constitutes infallible mathematics, while interpretation, 'foundations' belong to unprovable, fallible, controversial, philosophical belief. Thus Baumann, in 1869, saved the respectability of the Leibnizian calculus by this demarcation: 'Thus we discard the logical and metaphysical justification, which Leibniz gave to calculus, but we decline to touch the calculus itself.'[1] This is the source of Cauchy's and others' view that infinitesimal theory could be used as an *instrument* in proofs, but that its terms could not appear in theorems. One can easily show that in growing, informal mathematics it is hardly possible to carry out this separation by considering a simple characteristic pattern of growth in mathematics. This – triadic – pattern consists of a '*naive conjecture*' as first step (one may call it *thesis*), of a '*proof' and counterexamples* as second step (one may call them *the two poles of antithesis*) and finally the '*theorem*' (one may call it the *synthesis* of the triad). For instance, take as naive conjecture Cauchy's original thesis;

sharpness' (*loc. cit.*, p. 35). Also, one is intrigued as to why Pringsheim was so sure that Cauchy's discovery was independent. In fact this is a very unlikely assumption.

[1] Baumann [1869], volume II, p. 55. It is interesting how widely accepted this demarcation is. For instance Russell too takes it for granted: 'The interpretation of the infinitesimal calculus was for nearly two hundred years a matter as to which mathematicians and philosophers debated; Leibniz held that it involved actual infinitesimals, and it was not till Weierstrass that this view was definitely disproved. To take an even more fundamental example: there has never been any dispute as to elementary arithmetic, and yet the definition of the natural numbers is still a matter of controversy' (Russell [1948], p. 362). So Russell takes the calculus to be as theory-unimpregnated as elementary arithmetic!

take, say, Weierstrass's proof and Fourier's counterexamples as antithesis. The synthesis, the improved conjecture, will be the Weierstrass–Seidel theorem at which one arrives by finding the 'guilty lemma' in the informal proof and incorporating it into the thesis. Now this triad shows that the standard procedure of lemma-incorporation *carries the theoretical concepts of the proof into the theorem*: improvement by lemma-incorporation means theory-impregnation. Also, one can easily see that proving the original thesis in different conceptual frameworks leads to different theorems.[1] Thus one cannot separate in informal mathematics 'metaphysics' from 'technicality'. (Nor can one separate them in 'formal' mathematics – but that is not now our concern.)

All these considerations do not constitute any criticism of Robinson's approach – they only suggest a slight change in emphasis.

7 APPRAISAL OF MATHEMATICAL THEORIES

The reappraisal of the infinitesimal theory of the calculus raises the problem of how to appraise informal theories and of how to appraise inconsistent theories like Leibniz's calculus, Frege's logic, and Dirac's delta function. Are invalid theories beneath contempt? Must all inconsistent theories be ruthlessly eradicated as utterly useless for rational argument? Can they be appraised only if a posthumous reconstruction has saved them and proved them to be, if not respectable, at least excusable ancestors of respectable theories which look consistent and rigorous by the standards of the day? There surely must be rational standards for the appraisal of informal and inconsistent* mathematical theories; but for such standards one needs a philosophy whose inspiration comes from the study of the growth of informal mathematics rather than from the study of foundations and formal systems, the current trends in the philosophy of mathematics. Also, according to logical positivism, informal mathematics, being neither analytical nor empirical, must be meaningless gibberish; so logical positivism cannot be a philosophy that can provide guidance for the historian.

Historians and philosophers of mathematics will surely note non-standard analysis not only because it is a powerful challenge to the reappraisal of the history of the calculus and a powerful stimulus for a philosophical study of mathematical growth but because *non-standard analysis, together with non-standard arithmetic, represents a radical switch in the purpose and function of meta-mathematics.*

Until recently meta-mathematics has been regarded by most as synonymous with the study of the foundations of mathematics.

[1] This triad is discussed in detail in my [1963–4], especially on pp. 130–9 and 318–23. How different proofs yield different theorems is discussed especially on pp. 236–45. *See now Lakatos [1976*c*], pp. 144–5, 149 and 65–6 (*eds.*).

* Lakatos's manuscript has at this point the following handwritten remark: 'if inconsistent systems are not good for rational discussions what about natural language?' (J. P. C.)

Ordinary working mathematicians with no philosophical inclinations were not interested in it. Its original objective was limited: to prove the consistency of classical mathematics. Its original method was restricted to austere, simple means. Now it has become a live, growing mathematical discipline with unlimited objectives and tools.* It is acquiring an ever more important and possibly crucial influence in the growth of classical mathematical disciplines. Non-standard analysis signifies that meta-mathematics, having failed to achieve its original *intended* purpose, to erect ultimate and infallible foundations for the whole of mathematics, now may make up for its fascinating failure by fascinating *unintended* contributions to the growth of fallible mathematics. Infallible foundations to be gained by restricted methods have given way to unharnessed fallible growth with rich content.

This is not the first time that foundational studies have petered out without achieving their purpose, ultimate rigour, but have stimulated and suggested further growth. The 'cunning of reason' turns each increase in *rigour* into an increase in *content.* This also happened to the Weierstrassian theory of real numbers: first it was treated by the vast majority of ordinary working mathematicians as uninteresting pedantry, until it turned (not without a sharp struggle) into a theory with immense heuristic power, a theory indispensable for the creative mathematician.[1]

One should mention what may be regarded from this point of view as a 'shortcoming' of non-standard analysis: according to Luxemburg's theorem everything that can be proved by non-standard analysis can also be proved by classical analysis. The scope of the two theories are then the same: non-standard analysis opened up a new channel of growth but only within the old country: the Berstein–Robinson theorem too will be proved one day by classical methods. Non-standard arithmetic in this sense may seem more promising: (it aims immediately for results beyond the scope of classical arithmetic).† However, this advantage of non-standard arithmetic over non-standard analysis may lead to some spectacular growth beyond its present scope and, on the other hand, classical arithmetic may still produce informal proofs leading beyond the present Dedekind–Peano framework. There is no way of predicting how the relative strength of theories may change in the course of their growth.

* A modern view of the significance of meta-mathematics has been expressed by G. Sacks in his [1972]: 'The subject of mathematical logic splits fourfold into: recursive functions, the heart of the subject, proof theory which includes the best theorem in the subject, sets and classes whose romantic appeal far outweight their mathematical substance, and model theory, whose value is its applicability to, and roots in, algebra.' An earlier, but more penetrating estimate of the importance of meta-mathematical methods in mathematics is contained in a review-article by G. Kreisel ([1956–7]). (J. P. C.)

[1] Cf. *above,* chapter 1.

† Lakatos later underlined the bracketed passage in pencil and added an emphatic 'No' in the margin of the manuscript. (J. P. C.)

4
What does a mathematical proof prove?*

On the face of it there should be no disagreement about mathematical proof. Everybody looks enviously at the alleged unanimity of mathematicians; but in fact there is a considerable amount of controversy in mathematics. Pure mathematicians disown the proofs of applied mathematicians, while logicians in turn disavow those of pure mathematicians. Logicists disdain the proofs of formalists and some intuitionists dismiss with contempt the proofs of logicists and formalists.

I shall begin with a rough classification of mathematical proofs; I classify all proofs accepted as such by working mathematicians or logicians under three heads:

(1) pre-formal proofs
(2) formal proofs
(3) post-formal proofs.

Of these (1) and (3) are kinds of informal proofs.

I am afraid that some ardent Popperite may already be rejecting all that I am about to say on account of my classification. He will say that these misnomers clearly prove that I really think that mathematics has some necessary, or at least standard, pattern of historical development – pre-formal, formal and post-formal stages, and that I am already showing my hand – that I want to inject a disastrous historicism into sound mathematical philosophy.

It will turn out in the course of my paper that this, in fact, is just what I should like to do; I am quite convinced that even the poverty of historicism is better than the complete absence of it – always providing of course that it is handled with the care necessary in dealing with any explosives.

As a consequence of the unhistorical conception of 'formal theory' there has been a lot of discussion as to what constitutes a respectable formal system out of the immense multitude of capriciously proposed consistent formal systems which are mostly uninteresting games. Formalists had to disentangle themselves from these difficulties. They could of course have done this by dropping their basic outlook, but they have

* This paper seems to have been written some time between 1959 and 1961 for Dr T. J. Smiley's seminar at Cambridge. Lakatos's own copy contains several handwritten corrections; some by himself and some by Dr Smiley. We have incorporated them into the text. There is no indication that Lakatos ever returned to this paper after 1961. He subsequently changed his mind on some of the points made in the paper and had no plans to publish it himself. (*Eds.*)

tended to prefer complicated *ad hoc* corrections. They look for criteria distinguishing those formal systems which are '*interesting*' or '*acceptable*' and so on, thus betraying their bad consciences in accepting the pure formalist conception according to which mathematics is the set of *all consistent* formal systems. For instance, Kneale says that a mathematical system should be 'interesting'. His definition runs as follows: 'A possible – [possible means complying with some usual concept of modern rigour – i.e. consistent] system is interesting mathematically if it is rich in theorems and has many connections with other parts of mathematics, and in particular with the arithmetic of natural numbers.'[1] Curry, who is a most extreme representative of formalism, introduces the notion of 'acceptability'. He says: 'The primary criterion of acceptability is empirical; and the most important considerations are adequacy and simplicity.'[2] I fear there is a point on which I slightly disagree with their approach: they select from a previously given set of formal systems those which are interesting or acceptable. I should like to reverse the order: we should speak of formal systems only if they are formalizations of established informal mathematical theories. No further criteria are needed. There is indeed no respectable formal theory which does not have in some way or another a respectable informal ancestor.

Now I come back to our original subject: proofs. Most of the students of the modern philosophy of mathematics will instinctively define proof according to their narrow formalist conception of mathematics. That is, they will say that a proof is a finite sequence of formulae of some given system, where each formula of the sequence is either an axiom of the system or a formula derived by a rule of the system from some of the preceding formulae. 'Pure' formalism admits any formal system, so we must always specify in which system *S* we operate; then we speak only about an *S*-proof. Logicism admits essentially only one large distinguished system, and so essentially admits a single concept of proof.

One of the most outstanding features of such a formal proof is that we can mechanically decide of any given alleged proof if it really was a proof or not.

But what about an *in*formal proof? Recently there have been some attempts by logicians to analyse features of proofs in informal theories. Thus a well known modern text-book of logic says that an 'informal proof' is a formal proof which suppresses mention of the logical rules of inference and logical axioms, and indicates only every use of the specific postulates.[3]

Now this so-called 'informal proof' is nothing other than a proof in an axiomatized mathematical theory which has already taken the shape of a hypothetico-deductive system, but which leaves its under-

[1] Kneale [1955], p. 106.
[2] Curry [1958], p. 62.
[3] Suppes [1957], p. 128.

lying logic unspecified. At the present stage of development in mathematical logic a competent logician can grasp in a very short time what the necessary underlying logic of a theory is, and can formalize any such proof without too much brain-racking.

But to call this sort of proof an informal proof is a misnomer and a misleading one. It may perhaps be called a quasi-formal proof or a 'formal proof with gaps' but to suggest that an informal proof is just an incomplete formal proof seems to me to be to make the same mistake as early educationalists did, when, assuming that a child was merely miniature grown-up, they neglected the direct study of child-behaviour in favour of theorizing based on simple analogy with adult behaviour.

But now I should like to exhibit some truly informal, or, to be more precise, pre-formal proofs.

My first example will be a proof of Euler's well-known theorem on simple polyhedra.[1] The theorem is this: Let V denote the number of vertices, E the number of edges and F the number of faces of a simple polyhedron; then invariably

$$V-E+F=2$$

By a polyhedron is meant a solid whose surface consists of a number of polygonal faces, and a simple polyhedron is one without 'holes', so that its surface can be deformed continuously into the surface of a sphere. The proof of this theorem runs as follows:

Let us imagine a simple polyhedron to be hollow, with a surface made of thin rubber (see Figure 1 (*a*)). Then if we cut out one of the faces of the hollow polyhedron, we can deform the remaining surface until it stretches out flat on a plane (see Figure 1 (*b*)). Of course, the areas of the faces and the angles between the edges of the polyhedron will have changed in this process. But the network of vertices and edges in the plane will contain the same number of vertices and edges as did the original polyhedron, while the number of polygons will be one less than in the original polyhedron, since one face was removed. We shall now show that for the plane network, $V-E+F=1$, so that, if the removed face is counted, the result is $V-E+F=2$ for the original polyhedron.

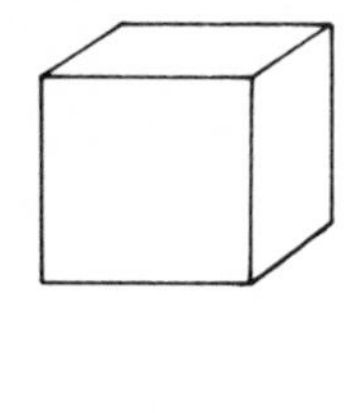

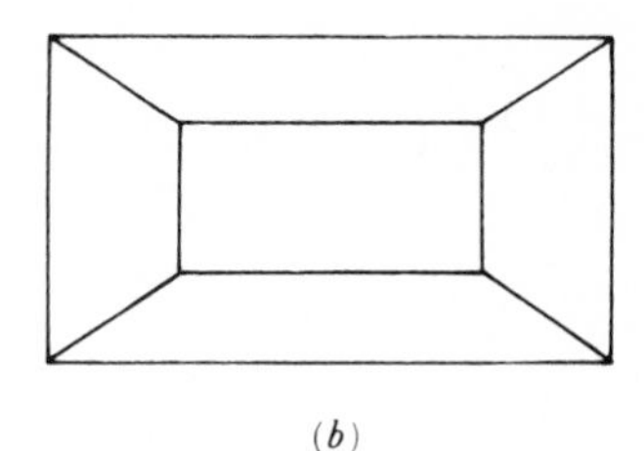

(*a*) (*b*)

Figure 1

[1] For a full discussion of the history of this theorem, see Lakatos [1976*c*].

We 'triangulate' the plane network in the following way: in some polygon of the network which is not already a triangle we draw a diagonal. The effect of this is to increase both E and F by 1 thus preserving the value of $V-E+F$. We continue drawing diagonals joining pairs of points until the figure consists entirely of triangles, as it must eventually (see Figure 2(*a*)). In the triangulated network, $V-E+F$ has the value that it had before the division into triangles, since the drawing of diagonals has not changed it. Some of the triangles have edges on the boundary of the plane network. Of these some, such as ABC, have only one edge on the boundary, while other triangles may have two edges on the boundary. We take any boundary triangle and remove that part of it which does not also belong to some other triangle. Thus, from ABC we remove the edge AC and the face, leaving the vertices A, B, C, and the two edges AB and BC [see Figure 2(*a*)]; while from DEF we remove the face, the two edges DF and FE, and the vertex F [see Figure 2(*b*)]. The removal of a triangle of type ABC decreases E and F by 1, while V is unaffected, so that $V-E+F$ remains the same. The removal of a triangle of type DEF decreases V by 1, E by 2 and F by 1, so that $V-E+F$ again remains the same. By a properly chosen sequence of these operations we can remove triangles with edges on the boundary (which changes with each removal) until finally only one triangle remains, with its three edges, three vertices and one face. For this simple network $V-E+F=3-3+1=1$. But we have seen that by constantly erasing triangles $V-E+F$ was not altered. Therefore in the original plane network $V-E+F$ must equal 1 also, and thus equals 1 for the polyhedron with one face missing. We conclude that $V-E+F=2$ for the complete polyhedron.

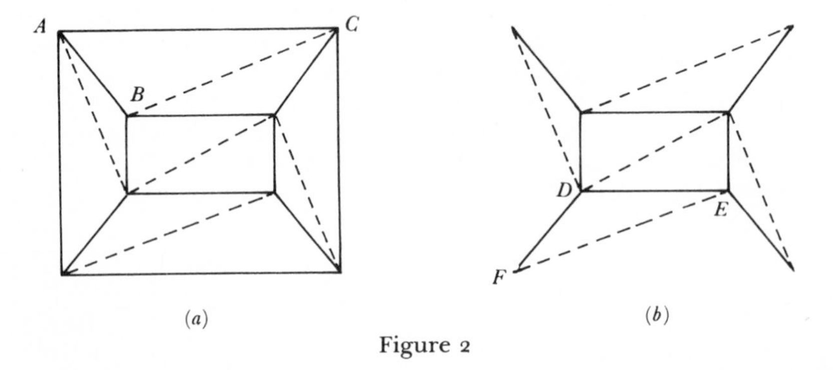

Figure 2

I think that mathematicians would accept this as a proof, and some of them will even say that it is a beautiful one. It is certainly sweepingly convincing. But we did not *prove* anything in any however liberally interpreted logical sense. There are no postulates, no well-defined underlying logic, there does not seem to be any feasible way to formalize this reasoning. What we were doing was *intuitively showing* that the theorem was true. This is a very common way of establishing mathematical *facts*, as mathematicians now say. The Greeks called this process *deikmyne* and I shall call it *thought experiment*.

Now is this a proof? Can we give a definition of proof which would allow us to decide at least *practically*, in most cases, if our proof is really a proof or not? I am afraid the answer is 'no'. In a genuine low-level pre-formal theory proof cannot be defined; theorem cannot be defined. There is no method of verification. As a strict logician like Dr Nidditch would surely say, it is – I quote – '*mere* persuasive argumentation, rhetorical appeal, reliance on intuitive insight or worse'.[1]

But if there is no method of verification, there is certainly a method of falsification. We can point out some hitherto unthought of possibilities. For instance assume that we had omitted to stipulate that the polyhedron be simple. We may not have thought of the possibility of the polyhedron having a hole in it (in which case the theorem would be subject to many counterexamples).* Actually Cauchy made this 'mistake'.[2] This is the frequently occurring phenomenon of mathematical theorems being 'stated in a false generality'.

For the sake of a better and simpler illustration let me quote another famous thought experiment with a celebrated falsification. The problem is to find the two points P and Q that are as far apart as possible on the surface or boundary of any triangle. The answer is easy to guess; P and Q are the ends of the longest side. This can easily be proved by the sort of thought experiment which we just used: no axioms, no rules, but convincing force. Let us see:

If one of the points, say P, lies on the *inside* of the triangle, then PQ obviously does not have its maximum length. For on the extension of the line PQ there is obviously a point P' that is further from Q than P is, and that is still inside the triangle. If both P and Q lie on the *boundary* of the triangle, but one of them, say P, is not a vertex, then we can obviously find a nearby point P' on the boundary that is further from Q than the distance PQ. Therefore PQ can be a maximum only if both P and Q are vertices; otherwise it certainly is not. Thus PQ is a side of the triangle and must obviously be the longest side.

It is obvious that the same thought experiment can be accomplished for polygons to 'prove' the following theorem: in order that two points on the surface of a polygon be farthest apart, they must be two of the vertices that are farthest apart.

I think this should be quite convincing. Nevertheless there is an unthought-of possibility which may spoil our pleasure. Apply the same thought-experimental procedure to this figure:

[1] Nidditch [1957], p. 5.

* One such counterexample is the 'picture frame' (Lakatos [1976*c*], p. 19) (*eds.*).

[2] Cauchy [1813].

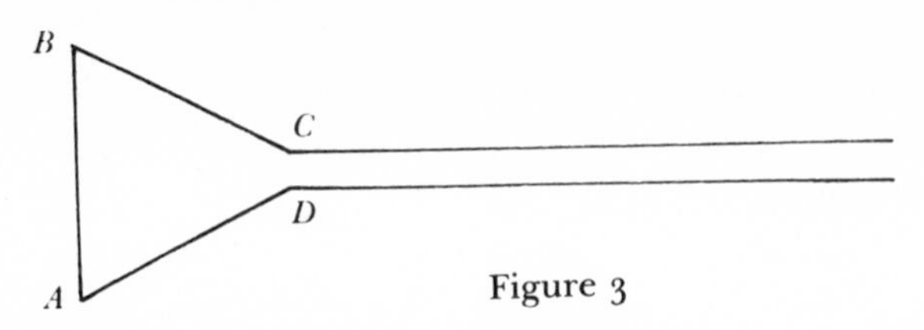

Figure 3

Suppose P and Q lie anywhere inside the figure or on the boundary, even including the possibility that they may be at any of the four vertices A, B, C, D. [Unless PQ is exactly the side AB, a nearby point P' can be found within the figure such that the distance $P'Q$ is greater than the distance PQ.] Just as in the earlier cases, for each pair of points P, Q we can find a nearby pair that are further apart in every case except when the pair is A, B. No pair other than A, B can give a maximum. If we now follow the previous argument strictly, we must conclude that AB is the maximum.

The falsification of our argument ran along the same lines as in the case of Euler's theorem for *all* polyhedra. We thought we showed more than we actually did. In our second case, we showed only that the maximum must be such and such *if the maximum exists at all.* In the case of Euler's theorem we only showed the truth of the theorem for the case where our rubber sheet could really be stretched out to the plane without any holes in it.

I should like to emphasize that the correction of such mistakes can be accomplished on the level of the pre-formal theory, by a new pre-formal theory.

The thought experiments I have just presented constitute only one type of pre-formal proof. There are others, basically different; ones for instance with the rather exciting property that in a certain sense we may say that contrary to the thought experiments we have just considered, they may be verified but not falsified. They give quite an insight into the nature of rules in a pre-formal theory and in pre-formal rigour.*

But now let us turn to axiomatized theories. Up to now no informal mathematical theory could escape being axiomatized. We mentioned that when a theory has been axiomatized, then any competent logician can formalize it. But this means that proofs in axiomatized theories can be submitted to a peremptory verification procedure, and this can be done in a foolproof, mechanical way. Does this mean that for instance if we prove Euler's theorem in Steenrod's and Eilenberg's fully formalized postulate system[1] it is impossible to have any counterexample? Well, it is certain that we won't have any counterexample formalizable in the system [assuming the system is consistent]; but we

* We have been unable to find out what Lakatos had in mind here (*eds.*).

[1] Eilenberg and Steenrod [1952].

have no guarantee at all that our formal system contains the full empirical or quasi-empirical stuff in which we are really interested and with which we dealt in the informal theory. There is no formal criterion as to the correctness of formalization.

Well-known examples of 'falsified' formalizations are (1) the formalization of the theory of manifolds by Riemann, where there is no account of Möbius-strips; (2) the Kolmogorov-axiomatization of probability theory, in which you cannot formalize such intuitive statements as 'every number turns up in the set of natural numbers with the same probability'.* As a final but most interesting example I should mention (3) Gödel's opinion that the Zermelo–Fraenkel and kindred systems of formalized set theory are not correct formalizations of pre-formal set theory as one cannot disprove in them Cantor's continuum-hypothesis.†

I will show with a trivial example how little formalization may add to the demonstrative or convincing force of informal thought experiments. You remember the proof of Euler's theorem? A formalist will certainly reject it. But it won't be easy for him to reject the following 'proof': set up a formal system, with one axiom: A; no rules [except that all axioms are theorems!]. The interpretation of A is Euler's theorem. This system I think complies with the strictest demands of formalism.

Does all this mean that proof in a formalized theory does not add anything to the certainty of the theorem involved? Not at all. [In the informal proof it may turn out that we failed to make some assumption explicit which results in there being a counterexample to the theorem. But, on the other hand, if we manage to *formalize* a proof of our theorem within a formal system, we know that there will never be a counterexample to it which could itself be formalized within the system, as long as that system is consistent.] For instance, if we had a formal proof of Fermat's last theorem, then if our formalized number theory is consistent it would be impossible for there to be a counterexample to the theorem formalizable within the system.

Now we see that if formalization (we shall use this term from now on as essentially having the same meaning as axiomatization) conforms with some informal requirements, such as enough intuitive counterexamples being formalized in it and so on, we gain quite a lot in the value of proofs. But if we try to formalize a pre-formal theory too early, there can be unfortunate results. I wonder what would have happened if probability theory had been axiomatized just in order to supply 'foundations' for probability theory, before the discovery of Lebesgue-measure. Or, to take another example, it is clear that it would have been wasted time and effort to formalize meta-mathematics at the time

* See Renyi [1955] (*eds.*).

† For more detail on this point and references to Gödel's opinions, see this volume, chapter 2 (*eds*).

of finitary illusionism, because later it turned out that the only useful methods must reach not only just beyond finitary tools but even beyond the object-theory in question. In an immaturely axiomatized algebra – axiomatized so as not to allow for complex numbers, say – we could never prove for instance that an equation of *n*th degree cannot have more than *n* real roots. Sometimes a well-formed formula of a theory *T* may be undecidable in the theory, but it may well be decided if suitably interpreted in a different theory, which may not even be an extension of the original theory. It is very difficult to decide in which theory a mathematical statement is really provable: for instance just take some theorems formalizable in the theory of real functions but provable only in the theory of complex functions, or theorems formalizable in measure theory, but provable only in the theory of distributions and so on. Even after a theory has been fruitfully axiomatized, there may arise issues which can bring about a change in axiomatization. This is now going on in probability theory. Axiomatization is a big turning point in the life of a theory, and its importance surpasses its impact on proofs; but its impact on proofs is immense in itself. While in an informal theory there really are unlimited possibilities for introducing more and more terms, more and more hitherto hidden axioms, more and more hitherto hidden rules in the form of new so-called 'obvious' insights, in a formalized theory imagination is tied down to a poor recursive set of axioms and some scanty rules.

Let me finally turn to the third part of my classification: to *post*-formal proofs. Here I shall just make a few programmatic remarks.

Two types of post-formal proofs are well-known. The first type is represented by the Duality Principle in Projective Geometry which says that any properly-worded valid statement concerning incidences of points and lines on a projective plane gives rise to a second valid statement when the words 'point' and 'line' are interchanged. For instance if the statement 'Any two distinct lines in the same plane determine a unique point' is valid, then so is the statement 'Any two distinct points in the same plane determine a unique line'. But then in proving the second statement we use a theorem of the system and another theorem, a meta-theorem, which we cannot specify, and still less prove, without specifying the concepts of provability in the system, theorem in the system and so on. This meta-theorem which we use like a lemma in our proof of an informal mathematical theory is not just about lines or points but about lines, points, provability, theoremhood and so on. Although projective geometry is a fully axiomatized system, we cannot specify the axioms and rules used to prove the Principle of Duality, as the meta-theory involved is informal.

The second class of post-formal proofs I should mention is the class of proofs of undecidability. As students of mathematical logic know, in the last few years it has turned out that formal proofs really prove much more than we want them to prove. Namely, to put it very roughly indeed, axioms in the most important mathematical theories implicitly define not just one, but quite a family of structures. For instance, Peano's axioms may be satisfied not only by our familiar natural numbers, but by some quite queer structures, Skølem's functions, which are far from being isomorphic with the set of natural numbers. Thus it turns out that when we fight hard to prove an arithmetical theorem, we prove at the same time some theorem in this other absolutely unintended structure. Now there are always statements, which are true in one structure but false in the other. Such statements are undecidable in the common formal structure. Are we helpless in such a situation? To see the point better, let us take a concrete, though hypothetical example. If we could prove that Fermat's theory is undecidable, then are we forever helpless to say anything about the truth of Fermat's theorem? Not at all. We may again call informal reasoning to our help, and try to operate informally *only* in the intended model. A concrete example of this is Gödel's proof [that his undecidable sentences are *true* (i.e. true in the standard model)]. But such post-formal proofs are certainly informal and so they are subject to falsification by the later discovery of some not-thought-of possibility.

Now at the present stage of our mathematical knowledge undecidable sentences occur only in rather artificial examples and do not affect the bulk of mathematics. But this situation may turn out similar to the case of transcendental numbers, which occurred first rather as exceptions and later turned out to be the more general case. So post-formal methods may gain in importance as undecidability encroaches more and more on mathematics.

And now a brief summary. We saw that mathematical proofs are essentially of three different types: pre-formal; formal; post-formal. Roughly the first and third prove something about that sometimes clear and empirical, sometimes vague and 'quasi-empirical' stuff, which is the real though rather evasive subject of mathematics. This sort of proof is always liable to some uncertainty on account of hitherto unthought-of possibilities. The second sort of mathematical proof is absolutely reliable; it is a pity that it is not quite certain – although it is approximately certain – what it is reliable about.

5
The method of analysis–synthesis*

1 ANALYSIS–SYNTHESIS: A PATTERN OF EUCLIDEAN HEURISTIC AND ITS CRITICISM

(a) *Prologue on analysis and synthesis*

Psi: Teacher, I should like to come back to your proof of the Descartes–Euler conjecture. It seems to me that you just cheated.

Teacher: Really?

Psi: You claimed that you proved the Descartes–Euler conjecture from subconjectures like 'all polyhedra are simple' and 'all polyhedra have only simply-connected faces'. Though you did not put it in these words, you in fact criticized those who thought they could prove the conjecture, and showed that it cannot be *proved*, only *deduced* from certain subconjectures. The *theorem*, your improved conjecture, was nothing but *a disguised inference: 'From the lemmas the original conjecture follows.'* I admit that you added that this inference may be regarded as invalid if we stretch some of its concepts, but this is a minor issue. You certainly claimed that your 'proof' was a deduction of the original conjecture from certain lemmas – not all of which may have been specified.

Alpha: What are you driving at? Come to the point – if you have one at all.

* This chapter (the title of which we have supplied) consists of two papers written at widely different times. Section 1 is the final chapter of Lakatos's Cambridge PhD thesis written between 1956 and 1961. Section 2 is based on an address given at a conference in Jyväskylä, Finland in 1973, replying to a paper by Professor Hintikka. (See Hintikka and Remes [1974].) Parts of the typescript of Lakatos's Jyväskylä address were in the form of notes. At these points we have made various interpolations which occur between square brackets. There is some overlap between sections 1 and 2.

In the acknowledgments to his PhD thesis Lakatos remarked that 'The three major – apparently quite incompatible – "ideological" sources of the thesis are Pólya's mathematical heuristic, Hegel's dialectic and Popper's critical philosophy.' In addition to this he expressed his gratitude to the following people for their helpful advice and criticism: J. Agassi, W. W. Bartley, R. B. Braithwaite, Lucien Foldes, R. Gandy, J. Giedymin, I. Jarvie, W. C. Kneale, Margaret Masterman, G. Morton, G. Polya, K. R. Popper, H. Post, J. Ravetz, J. E. Reeve, T. J. Smiley, R. C. H. Tanner, and J. W. N. Watkins.

The first part of section 1 is, like the earlier chapters of Lakatos's PhD thesis, in dialogue form. These earlier chapters form the basis of the book *Proofs and Refutations*. There Cauchy's proof of the Descartes–Euler conjecture is discussed. Here this proof is subjected to a new line of attack. (For a brief account of the proof, see section 2, p. 94, *below*.) (*Eds.*)

Psi: Your claim is false. *You in fact deduced from the main conjecture and from the lemmas that, for a triangle,* $V-E+F=1$. But this we knew anyway!

Alpha: What?

Psi: First it was assumed (P) that '$V-E+F=2$ for all polyhedra'. This is the very assertion we set out to prove. From this we inferred (P_1) that '$V-E+F=1$ for all flat polygonal networks': we noted that in this inference we also used the lemma Q_1 that 'all polyhedra are simple' as a premise. Then from this we inferred P_2 that '$V-E+F=1$ for all triangular networks' – in this inference we used also the lemma Q_2 that 'all faces are simply-connected'. From this we finally inferred P_3 that 'for a triangle $V-E+F=1$'. And this trivial finding was accepted with joy. I wonder why. Because we arrived at something which is indubitably true? But from false premises we can validly deduce true conclusions: so we cannot conclude anything about the truth of the premises. Anyway we *know* that in our case all the premises *are* false.

Alpha: To be frank, I am struck by your argument.

Gamma: [But surely there is no real difficulty here.] This chain of inferences – I call it an 'analysis' – can be trivially reversed and by this we can validly deduce P from the indubitably *true* premise P_3 and from the false Q_1 and Q_2; that is, we can *prove* $(Q_1 \,\&\, Q_2) \rightarrow P$.* This inversion I call 'synthesis'. This diagram may help you:

Analysis:
$$\begin{array}{ccccccc} P & \rightarrow & P_1 & \rightarrow & P_2 & \rightarrow & P_3 \\ & \uparrow & & \uparrow & & & \\ & Q_1 & & Q_2 & & & \end{array}$$

Synthesis:
$$\begin{array}{ccccccc} P_3 & \rightarrow & P_2 & \rightarrow & P_1 & \rightarrow & P \\ & \uparrow & & \uparrow & & & \\ & Q_3' & & Q_2' & & & \end{array}$$

Alpha: This inversion is not so trivial. The inferences leading back will differ from our original inferences. For instance from P and Q_1 we inferred P_1. But does the fact that we can infer P_1 from P and Q_1 guarantee that we can infer P from P_1 and Q_1? Not at all. If P is false, but P_1 true and Q_1 true, we cannot possibly infer P from P_1 and Q_1, even though we may be able validly to infer the true P_1 from the false P and the true Q_1. So the inversion is not trivial.†

* Lakatos seems to us to be misportraying here his own method of 'proofs and refutations'. What is assumed proved is not the truth-functional compound '$(Q_1 \,\&\, Q_2) \rightarrow P$' (whose truth is anyway automatically established by the falsity of $(Q_1 \,\&\, Q_2)$), but rather the truth-functionally simple sentence '$\forall x$ (if x is a simple polyhedron and x's faces are simply-connected then $V-E+F=2$ holds for x)'. The assumptions Q_1 and Q_2 are turned into predicates which pick out those polyhedra to which the improved proof applies. (This is essentially what Lakatos says in section 2 of this paper, *below*, p. 95.) (*Eds.*)

† Alpha's assertion that the inversion is not (necessarily) trivial is certainly true. The same lemmas which made the inference from P to P_1 valid, will not always guarantee the validity of the reverse inference from P_1 to P. (Clearly the lemmas Q_i will perform this role when and only when $Q_i \vdash P \leftrightarrow P_1$.) However, Alpha's argument for this correct conclusion seems to us invalid. For the same reasons as indicated in the

Beta: So in proving our theorem we must try to reverse this 'proof' and we may very well fail.

Psi: Indeed we may.

Teacher: When your science teacher 'proves' you his scientific theories by deducing undebated facts from them he follows this same pattern. I wonder why you do not protest to him too.

Psi: We shall.[1]

(*b*) *Analysis–synthesis and heuristic*

Euclidean heuristic separates the processes of finding the truth and of proving it.* But this does not exclude heuristic playing a role in either the process of discovery or of proof.[2]

Proof implies finding lemmas. But where do the lemmas come from? The primitive mind does not like proof which requires a jump to unknown lemmas – even if the lemmas are the listed axioms of a theory; for how can one know *which* trivial truth implies the dubitable one? One would have to guess, to fall back on trial and error. But primitive man shrinks from guessing. He abhors freedom, he feels unsafe if he moves beyond the bounds of ritual. If he guesses, he does it surreptitiously.[3]

Primitive men prefer decision-procedures. With the help of a decision-procedure one can decide mechanically whether a conjecture is true or false. Primitive men worship algorithms. Their concept of rationality, like that of Leibniz, of Wittgenstein and of modern formalists, is essentially algorithmic.

But the Greeks did not find a decision-procedure for their geometry, although they certainly dreamt of one. They did, however, find a compromise solution: a heuristic procedure, which is not quite algorithmic, which does not always yield the desired result, but which is still a heuristic *rule*, a standard pattern of the logic of discovery.

This heuristic method was the method of analysis and synthesis. Let me state it as a rule:

Rule of analysis and synthesis: *Draw conclusions from your conjecture, one after the other, assuming that it is true. If you reach a false conclusion, then*

previous footnote, the problem is *not* to infer (the admittedly false) P from P_1 and Q_1, rather it is thus to infer a qualified version of P which will not in general be known to be false. (*Eds.*)

[1] Psi is an advanced student. Most physicists would not protest – they are verificationists out of sheer ignorance about logic. Such 'proofs' of course constitute a serious flaw in deductivist presentation. It would be interesting to check how many proofs in textbooks and periodicals are like this. It is remarkable that all this was noticed neither by Cauchy, nor by any of his successors, including Courant and Robbins.

* This separation is discussed in Lakatos [1976*c*], pp. 137–8 (*eds.*).

[2] In fact, these processes are represented in the two branches of Pappusian heuristic: the *problematical* and the *theoretical* (see p. 73). However, this separation in practice is not as rigid as it should be in theory.

[3] This primitive heritage is the main obstacle to heuristic style. *Cf. Lakatos [1976*c*], Appendix 2. (*Eds.*)

your conjecture was false. If you reach an indubitably true conclusion, your conjecture may have been true. In this case reverse the process, work backwards, and try to deduce your original conjecture via the inverse route from the indubitable truth to the dubitable conjecture. If you succeed, you have proved your conjecture.

The first part was called the analysis, the second part the synthesis. The heuristic rule shows at once why the Greeks held *reductio ad absurdum* in such exceptionally high esteem: it saved them the labour of the synthesis, analysis alone having proved the case.

The characterization of the method can be found in Euclid's *Elements*, Book XIII. The text is corrupted, but the examples of analysis which follow the definition make the method clear. The best preserved ancient exposition is by Pappus. It has been translated into English by Sir T. L. Heath ([1925], volume 1, pp. 138–9):

The so-called ἀναλυόμενος ('Treasury of Analysis') is, to put it shortly, a special body of doctrine provided for the use of those who, after finishing the ordinary Elements, are desirous of acquiring the power of solving problems which may be set them involving (the construction of) lines, and it is useful for this alone. It is the work of three men, Euclid the author of the Elements, Apollonius of Pergo, and Aristaeus the elder, and proceeds by way of analysis and synthesis.

Analysis then takes that which is sought as if it were admitted and passes from it through its successive consequences to something which is admitted as the result of synthesis: for in analysis we assume that which is sought as if it were (already) done (γεγονός), and we inquire what it is from which this results, and again what is the antecedent cause of the latter, and so on, until by so retracing our steps we come upon something already known or belonging to the class of first principles, and such a method we call analysis as being solution backwards (ἀνάπολιν λύσιν).

But in synthesis, reversing the process, we take as already done that which was last arrived at in the analysis and, by arranging in their natural order as consequences what were before antecedents, and successively connecting them one with another, we arrive finally at the construction of what was sought; and this we call synthesis.

Now analysis is of two kinds, the one directed to searching for the truth and called *theoretical*, the other directed to finding what we are told to find and called *problematical*. (1) In the *theoretical* kind we assume what is sought as if it were existent and true, after which we pass through its successive consequences, as if they too were true and established by virtue of our hypothesis, to something admitted; then (*a*), if that something admitted is true, that which is sought will also be true and the proof will correspond in the reverse order to the analysis, but (*b*), if we come upon something admittedly false, that which is sought will also be false. (2) In the *problematical* kind we assume that which is propounded as if it were known, after which we pass through its successive consequences, taking them as true, up to something admitted: if then (*a*) what is admitted is possible and obtainable, that is, what mathematicians call *given*, what was originally proposed will also be possible, and the proof will again correspond in reverse order to the analysis, but if (*b*) we come upon something admittedly impossible, the problem will also be impossible.

This method has several peculiarities. One is that a false conjecture can be disproved, but not improved by it. Another is that, on the face of it, the only proofs that can be found by using it are those which involve one single axiom or a single proposition already proved. But this is not a serious restriction because the Greeks freely introduced any axiom or proved proposition into the deductive argument both in the analysis and in the synthesis.[1] The main restriction of this method lies in the following consideration: If we deduce from C the basic statement[2] P and from P again C, then P is necessary and sufficient condition for C and vice versa. But this is not always the case. Some axioms e.g. may entail the conjecture under discussion but the conjecture may not entail the axioms. In such cases the method of analysis–synthesis does not work. However we have no clear statement from antiquity about the possible failure of the method in these cases.[3]

This failure to stress the obvious limitations of the method needs explanation. The Greeks must have come across lots of theorems unprovable through analysis–synthesis (though the proportion of theorems with necessary and sufficient conditions is surprisingly high in the *Elements*). Hankel has noticed the following example:

The theorem: 'The vertices of all triangles with common bases, which are such that their angles at the vertex in question are the same, lie on a circle', may not be inverted to give the statement, 'All triangles with a common base, whose vertices lie on a circle, have the same angles'. For on the one hand this is valid only when the circle passes through the endpoints of the base, and on the other hand only for those angles which lie on the same side of the base. If one added these conditions to the former theorem then it will be necessarily reversible. This is similar to the theorem: 'If A, B, D, F lie on a circle, then $EA: EB = ED: EF$ (where E is the intersection of AB and DF)', the condition for the reversal of which is that if A and B lie on the same or opposite sides of E, then ED and EF have the same or opposite direction.[4]

[1] Cf. e.g. Richard Robinson's discussion of a Euclidean analysis in his [1936], pp. 470–1. How the surreptitious introduction of these auxiliary axioms and lemmas (*enthymeme*) was connected with the idea that one single *arche* was enough to deduce the whole of knowledge is a subtle question, nicely formulated but I think unsatisfactorily dealt with by Robinson in his [1953], pp. 168–9. Robinson does not seem to realize that all the theorems of e.g. the *Principia Mathematica* can be deduced from a single axiom which we can construct by connecting all the axioms by '*ands*'.

[2] By basic statement in this context is meant either an indubitably true statement like a Euclidean axiom, or a statement which has already been proved, or a statement which originally has been admitted as a condition. This latter case was very frequent in geometrical constructions.

[3] Robinson says in a footnote in his [1936]: 'I have noticed two passages that may *possibly* be references to this point.' (Italics mine.) I would add a third possible reference. When Geminus argues against Proclus that Geometry *does* investigate causes, he says: 'When geometers reason *per impossibile* they are content to discover the property, but when they argue by direct proof, if such proof be only partial, this does not suffice for showing the cause; if however it is general and applies to all like cases, the why is at once concurrently made evident.' (Cf. Heath [1925], volume 1, p. 150.)

[4] Hankel [1874], p. 139.* We have translated this passage from the original German (*eds.*).

The embarrassed silence of the Greeks about such failures was, I suppose, at least partly due to the central doctrine of Aristotelian essentialism that genuine proofs (or explanations) have to be final and certain (e.g. *Posterior Analytics*, book I, chapter 6) which cohered so well with a heuristic that claimed to devise precisely such proofs.[1] These finality–certainty requirements survive in mathematics until today as the requirement of necessary and sufficient conditions.

Let us raise here still another question: Why did the Greeks not adopt a heuristic style in mathematics? Why did they conceal their analyses and present only syntheses?[2] We do not know. Probably there is at least something to Descartes's guess: 'It was this synthesis alone that the ancient Geometers employed in their writings, not because they were wholly ignorant of the analytic method, but, in my opinion, because they set so high a value on it that they wished to keep it to themselves as an important secret.'[3]

(*c*) *The Cartesian Circuit and its breakdown*

The classical Euclidean programme is anti-empiricist; it is highly critical of the senses. Indubitable propositions can be guaranteed only by the intellect's infallible intuition. Facts have to be proved from indubitable first principles or essential definitions. In such a framework the method of analysis–synthesis may work quite well – just as in Euclidean geometry.

In modern science however two new factors enter. One is a new kind of indubitably true proposition: the reasoned fact. Reasoned facts may run counter to sense-experience. They may – as Galileo put it – 'rape the senses'. Examples of reasoned facts are: 'the earth is round', 'all bodies fall with the same acceleration in a vacuum', and so on. The

[1] For the equivalence of finality and certainty on the one hand and necessary and sufficient conditions on the other hand cf. my [1963–4], section 6*b*.

[2] It should be mentioned here that we find heuristic style in the first part of Book XIII of Euclid's *Elements*. According to Bretschneider and Heiberg this part is not likely to have been written by Euclid (cf. Heath [1925], p. 137). Another example of heuristic style is Archimedes's *On the Sphere and Cylinder*, where he gives both the analyses and the syntheses when he solves the *problems*. (The intriguing point in Archimedes's analyses is that if we translate problems into conjectures then his method looks very much like proof-procedure. The Archimedean analyses frequently end up in a 'διορισμός' which seems to correspond to the lemma converted into a condition. According to Heath this was general in the analyses of problem-solving: 'In cases where a διορισμός is necessary, i.e. where a solution is only possible under certain conditions, the analysis will enable those conditions to be ascertained' (*op. cit.*, volume I, p. 142).)

[3] Haldane and Ross [1912], p. 49. Hankel explains Greek deductivist style as a 'national characteristic of the Greeks' (*op. cit.*, p. 148). My guess is that the main explanation is the infallibilist prejudice against mere guessing. Analysis is just groping, while synthesis is proving; analysis is fallible, synthesis is 'infallible'. To be fallible was thought to be 'beneath one's dignity', so it was omitted. * Lakatos's quotations from Descartes were all in the original French or Latin. Except where Lakatos expressed a disagreement with their translation, we have substituted the English version of Haldane and Ross (*eds.*).

other new factor which modern science introduced is a new kind of dubitable proposition: the occult hypothesis, like 'all bodies attract each other'.

Now these two new factors cause quite a few problems for those who would like to apply the method of analysis–synthesis to facts, occult hypotheses and principles.

The main point about the classical analysis–synthesis is that it links known and unknown together by a chain of deduction, or, by a circuit of truth and/or falsity. We inject truth or falsehood at some point and it will be carried to every part by the circulatory system. Now circulatory troubles arise both near the reasoned facts and near the occult hypotheses.

[Facts and reasoned facts are deductively unconnected. Reasoned facts do not directly entail occult hypotheses, though in this latter case the entailment in the opposite direction, from occult hypothesis to reasoned facts, may be genuine.]

But if we stick to the ideal of scientific infallibilism, the free and safe circulation of truth value from facts to reasoned facts, from reasoned facts to occult hypotheses, from occult hypotheses to first principles and vice versa has to be restored. We have to bring all these different kinds of statements to the same level of certainty.

So infallibilism bridged the gaps by introducing a new sort of truth-transference – 'inductive' inference. [Thus the Pappusian circuit mentioned above might be represented as in the diagram below.]

The Pappusian Circuit

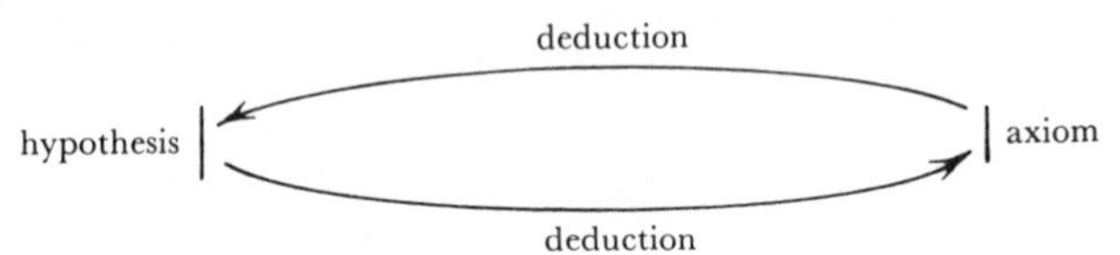

[And the *Cartesian Circuit* may be represented as follows:]

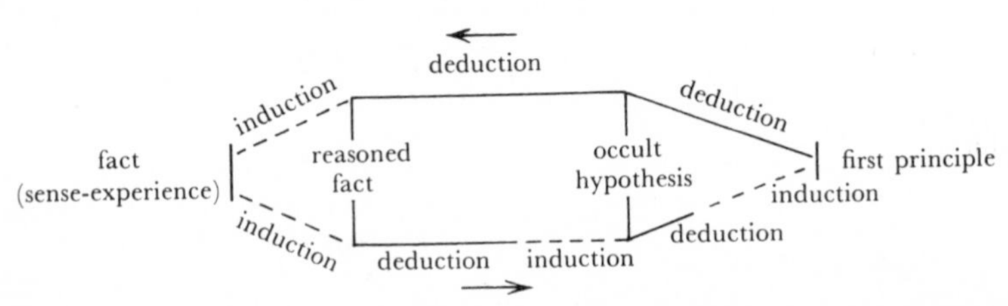

This is an extended version of analysis–synthesis. The extension arose from the Cartesian attempt to adapt the ancient analysis–synthesis to modern science. It is very different from the Pappusian schema: it has quasi-empirical basic statements[1] and it has inductive as well as deductive inferences.

[1] Throughout this section we call 'basic propositions' those propositions, through which some truth value is injected into the Circuit.

The thesis of this section is that a main feature of the story of modern scientific method is the critical elaboration of the ancient Pappusian Circuit into the Cartesian Circuit, followed – in spite of some partial successes and several intriguing rescue-operations – by its breakdown.

Let us first clarify some problems concerning the Cartesian Circuit.

(c1) *The Circuit is neither empiricist nor intellectualist. The source of knowledge is the Circuit as a whole*

In traditional histories of philosophy, empiricism and intellectualism are contrasted. Empiricists are said to inject truth values at the level of factual statements, intellectualists at the first principles: empiricists assent to the authority of the senses, intellectualists to the authority of the intellect.

In fact there are very few, if any, pure-grained empiricists and pure-grained intellectualists. Descartes, Newton and Leibniz certainly all agreed that one can indubitably intuit truth and/or falsehood at both points: on the level of facts and on the level of first principles. Both may serve as basic statements. But everybody also agreed that one cannot talk about true factual statements or true first principles taken in isolation; only fools trust sense-experience, and first principles from out of the blue are just speculations – neither has a place in the perfect, infallible body of Scientific Knowledge. They are only respectable and suitable candidates for truth or falsehood if they are already embedded in the circulatory system of analysis–synthesis. 'Basic statement' is meaningless outside analysis–synthesis.

Both Descartes and Newton were very explicit about the necessity of starting the analysis with facts, from which one proceeded to 'mediate causes' and from there to first principles. They despised those who tried to arrive at first principles with no care for facts, by 'rash anticipation' instead of by laborious analysis.

Some apparently 'puzzling' passages of Descartes and Newton should be interpreted in this light. For instance:

> And I have not named them hypotheses with any other object than that it may be known that while I consider myself able to deduce them from the primary truths which I explained above, yet I particularly desired not to do so, in order that certain persons may not for this reason take occasion to build up some extravagant philosophic system on what they take to be my principles, and thus cause the blame to be put on me.[1]

This passage corresponds very closely to Newton's '*Hypotheses non fingo*'. It means that hypotheses have to be embedded in a Cartesian Circuit and thereby cease to be hypotheses.

[1] Descartes [1637], p. 129. The last words reflect the effect of Galileo's conviction.

The aim of the Cartesian Circuit is to carry truth to all the points in the Circuit, thereby turning hypotheses into facts, and justifying the old Aristotelian claim that the 'conviction of pure science must be unshakable'.[1] The Circuit does not allow unsubstantiated fantasies which would be inconsistent with the dignity of infallible science. That in this set-up effects and causes, and facts and theories, are on the same logical and therefore epistemological (though not heuristical) level (*causa aequat effectu*) is also clear from the following passage from Clarke's *Fifth Reply to Leibniz*:

> The phaenomenon it self, the attraction, gravitation or tendency of bodies towards each other...[is] now sufficiently known by observations and experiments. If this or any other learned author can by the laws of mechanism explain these phaenomena, he will not only not be contradicted, but will moreover have the abundant thanks of the learned world. But, in the mean time, to compare Gravitation, (which is a phaenomenon or actual matter of fact,) with Epicurus's declination of atoms (which, according to his corrupt and Atheistical Perversion of some more ancient and perhaps better philosophy, was an Hypothesis or Fiction only, and an impossible one too, in a World where no intelligence was supposed to be present) seems to be a very extraordinary Method of reasoning.[2]

The heuristic rule that the deductive chain should start with the facts, is an absolute rule for both Descartes and Newton. This has to be stressed repeatedly because it is so counterintuitive in our Popperian age which over-encourages speculation. This rule is the correct interpretation for instance of this statement: 'As in mathematics, so in natural philosophy, the investigation of difficult things by the method of analysis, ought ever to precede the method of composition'.[3] Again it is the correct interpretation of the Fifth Rule in Descartes's *Regulae*:

> he who would approach the investigation of truth must hold to this rule as closely as he who enters the labyrinth must follow the thread which guided Theseus. But many people either do not reflect on the precept at all, or ignore it altogether, or presume not to need it. Consequently they often investigate the most difficult question with so little regard to order, that, to my mind, they act like a man who should attempt to leap with one bound from the base to the summit of a house, either making no account of the ladders provided for his ascent or not noticing them. It is thus that all Astrologers behave, who, though in ignorance of the nature of the heavens, and even without having made proper observations of the movements of the heavenly bodies, expect to be able to indicate their effects. This is also what many do who study Mechanics apart from Physics, and rashly set about devising new instruments for producing motion. Along with them go also those Philosophers who, neglecting experience, imagine that truth will spring from their brain like Pallas from the head of Zeus.[4]

In another passage, in Rule IV, Descartes compares those who look

[1] *Posterior Analytics*, 72[b].
[2] Alexander (*ed.*) [1959], p. 119.
[3] Newton [1717], Query 31.
[4] Descartes [1628], pp. 14–15.

for manifest Truth lying by the wayside without applying the laborious method of the Circuit with 'a man burning with an unintelligent desire to find treasure, who continuously roam[s] the streets, seeking to find something that a passer-by might have chanced to drop'. Descartes admits that this 'method' – pursued by 'most Chemists, many Geometricians, and Philosophers' – may lead to occasional windfalls. But one has to pay dearly for them, since 'unregulated inquiries and confused reflections of this kind only confound the natural light and blind our mental powers'.[1] This was Descartes's real opinion of that brand of intellectualism which was later attributed to him!

But while dismissing the use of First Principles in isolation from the Circuit, both Descartes and Newton thought them to be an essential part of the Circuit which contributed decisively to the safety of this epistemological structure. It is known that Newton was unhappy about the occult character of Gravitation and that he tried to deduce it – by the theory of 'umbrella-effect' – from Cartesian First Principles.

Descartes sharply criticized Galileo for omitting First Principles, 'thus building without a foundation'.[2] And he had a very reasonable point. Facts alone are not reliable enough to guarantee the truth of the Circuit. Mersenne and Rocco simply refused to accept Galileo's 'facts'. (Before the Atwood-machine facts about free fall were very dubitable indeed.) Newton's work was hindered by false astronomical data. The unreliability of empirical evidence was not then concealed by the ritual of statistical decision procedures.

(c2) *Induction and deduction in the Circuit*

Another problem worth clarifying is the relation between induction and deduction in Cartesian logic. Both are inferences based on intuition, which transmit truth [from premises to conclusions] and retransmit falsity [from conclusions to premises]. In the Cartesian set-up they do not differ in any essential respect. To Descartes, and also to Newton, inductive inference is the infallible twin of deductive inference. That induction has nothing to do with fallible conjecturing is made clear in Newton's famous letter to Cotes:

> For anything which is not deduced from phenomena ought to be called a hypothesis, and hypotheses of this kind, whether metaphysical or physical, whether of occult qualities or mechanical, have no place in experimental philosophy. In this philosophy, propositions are deduced from phenomena, and afterward made general by induction.[3]

Or again, in the same letter, equating induction with deduction: 'experimental philosophy proceeds only upon phenomena and deduces general propositions from them only by induction'.[4]

[1] *Ibid.*, p. 9.
[2] Descartes [1638].
[3] Newton [1713], p. 155.
[4] *Op. cit.*

There is certainly a discrepancy between formal Aristotelian logic, with its nineteen valid forms of inference, and induction which has to rely on intuition. But Descartes brushes Aristotelian logic aside with contempt,[1] and replaces the poverty of syllogistic logic by the infinite richness of intuitive deductions, the infallibility of which are guaranteed by God. But if so, why should God not guarantee inductive inference just as He guarantees deductive inference?[2]

Another feature of the Cartesian Circuit which has been frequently misunderstood is the relative length and importance of deductive and inductive passages in the inferential chain which transmits truth from the facts to the occult hypotheses.

Newton was very keen indeed to deduce his theory *fully* from the facts. In his priority quarrel with Hooke he repeatedly stressed that Hooke only *guessed* the inverse square law, but he, Newton, *deduced* it from Kepler's empirical laws. He spurns Hooke's conjecturing: how does he know that the exponent of the radius is 2? Maybe it is a number *near* to 2! But he [Newton] knows it is 2 because he deduced it.[3]

Newton's claim that he deduced his theory from facts has been ridiculed by philosophers from Duhem onwards. The only physicist who came to Newton's defence was Born, the first person in the history of science who reconstructed Newton's deduction.[4] Unfortunately Born missed one important point: Newton's chain of deduction does not and cannot lead up to the law of gravitation, but only to the inverse square law.[5] There is a small, but decisive inductive gap from the inverse square law to the Universal Law of Gravitation. But this gap should not be overestimated. Newton *almost deduced* his theories from facts, and I should not be surprised if the same could be shown to be true of Planck's, Einstein's, or Schrödinger's results too.

Nowadays it has generally been assumed that while deduction leads from theories to facts, it has not the slightest part in the path from facts to theories. Both the Pappusian and the Cartesian Circuits have disappeared into oblivion. The last philosopher who took the Pappusian Circuit seriously enough to criticize it at all was J. M. C. Duhamel. He treated the ancient method with some contempt as something out-of-date and vastly superseded. The *modern* method of analysis is not *deductive*, he claims, but *reductive*, working from the discussed proposition to propositions *from which* it follows, till we get to a proposition which is indubitably true. (Duhamel is still a Cartesian in this sense.) In this pattern, of course, the synthesis is done simultaneously with the analysis; any analysis is mechanically convertible.[6] But nowadays the ancient method is not only not criticized, it has been

[1] Descartes [1637], p. 91.

[2] That deduction and induction are on the same level in Descartes's *Regulae* was realized quite clearly by Joachim. (See his [1906], pp. 71–2.)

[3] Newton [1686], in Brewster [1855], volume 1, p. 441.

[4] Born [1949], Appendix 2.

[5] This was pointed out by Popper, see his [1957], p. 198, n. 8.

[6] Duhamel [1865], pp. 37–57 and 62–8.

almost completely forgotten. Only an occasional student of the history of geometry remembers it.[1] If a scholar stumbles upon it from time to time and learns that the Greeks *deduced* mathematical theories from facts (i.e. axioms from conjectures), he is likely not to believe his eyes. Thus F. M. Cornford not only thought, like Duhamel, that this method is out of date, but he insisted that it is a paradoxical, 'nonsensical' method that the Greeks could not possibly have followed, and argued that Pappusian analysis is in fact identical with what Duhamel called the modern reductive analysis.[2] According to Cornford, everybody who interpreted Pappus in the deductive way, 'misunderstood him lamentably'. He says: 'You cannot follow the same series of steps first one way then the opposite way, and arrive at logical *consequences* in both directions'.[3]

What a change! A few centuries earlier it was taken for granted that all proper proofs (or explanations) are reversible and people were reluctant to notice the recalcitrant cases. Today some people take it for granted that there cannot be reversible proofs at all.[4]

All this only goes to show the modern decline of heuristic. In fact Euclid *deduced* most of his theorems with this method and, as Cauchy's proof of the Euler-formula shows, it is still a major pattern of mathematical heuristic.

But let us come back to Descartes and Newton. Newton certainly argues in some *places* that he *deduced* his theories from facts. This can be interpreted in two equally plausible ways: first, that he thought that the inductive gap is negligible; second, that the separation of the two kinds of intuitive truth-inferences, deduction and induction, was rather blurred in Cartesian philosophy.

There is in fact nothing puzzling at all about the 'inter-deducibility' of facts and theories in the Cartesian Circuit: on the contrary, it is one of its most obvious features.

[1] Cf. Heath, *op. cit.*, I, pp. 137–42.

[2] Cornford [1932], pp. 37ff and 173ff. Robinson's retort to Cornford's argument was published four years later (Robinson [1936]).

[3] Whereas Cornford is utterly mistaken, Kneale is right in stressing that there is no (*completely*) deductive path from facts to transcendental hypotheses. But he does not realize that quite a long way *can* be covered deductively: 'It will be noticed that Newton speaks in a very curious way of *deducing* propositions from phenomena. This expression occurs in other places, and we must assume that Newton used it deliberately; *but it obviously cannot mean what is ordinarily called deduction*, and I can only conclude that Newton meant that the propositions which interested him were derived from observations in a very strict way. Apart, however, from the peculiarity of its phraseology, the passage is fairly clear. Newton seems to be saying in effect that he thinks it should be possible to find an explanation for gravitation but that *this must be discovered by ordinary induction from facts* found in experience, because no other method is admissible in natural science' (Kneale [1949], pp. 98–9, my italics).

[4] For Descartes the main epistemological structure was the *Circuit* – for Braithwaite it is the *Zipfastener* ([1953], p. 352). It should be stressed that in the Circuit not only did the synthesis have epistemological relevance, as is commonly assumed, but the analysis did too. The source and guarantee of the truth is the Circuit as a whole.

As an illustration I quote Descartes's account of his *Dioptrices* and *Meteors* at the end of the *Discours*:

il me semble que les raisons s'y entresuiuent en telle forte que, comme les dernières sont demontrées par les premières qui sont leurs causes, ces premières le sont reciproquement par les dernières, qui sont leurs effets. Et on ne doit pas imaginer qui ie commette en cecy la faute que les Logiciens nomment vn cercle; car l'expérience rendant la plus part de ces effets tres certains, les causes dont ie les déduits ne seruent pas tant à les prouuer qu'à les expliquier; mais, tout au contraire, ce sont elles qui sont prouuées par eux.[1]

Descartes himself frequently uses the Aristotelian formal terms 'deduction' and 'induction' alternatively for 'informal inference'. Thus, when he says in his Rule III of his *Regulae ad Directionem Ingenii*, that there are only two infallible ways of arriving at knowledge – intuition and induction ('*intuitus et inductio*') – he means *intuition*, by which we can infallibly recognize truth, and *inference*, by which we can infallibly transfer truth. In Rule IV he uses the term '*deductio*' again. The interpretation of this alternation by several Descartes commentators (Gouhier, LeRoy and others)[2] as a blunder of the copyist seems to me rather a blunder of the commentators.

Another amusing example of the embarrassment about Descartes's concept of deduction is the translation, again by Haldane and Ross, of the passage in Rule XIII, where Descartes, like Newton, wants theories to be '*deduced*' from facts. The text runs like this: 'Sed insuper vt quaestio sit perfecta, volumus illam omnino determinari, adeo vt nihil amplius quaeratur, quam id quod *deduci* potest ex datis' (my italics).[3] Haldane and Ross who usually translate 'deducere' by 'deduce', this time put 'infer'.[4]

[1] Adam and Tannery [1897–1913], volume 6, p. 76. The English translation of *The Philosophical Works of Descartes* by Haldane and Ross, commits a howler here by translating 'prouuées' by 'explained'. They may have misunderstood the Morin–Descartes controversy in 1638 about this subject and the alteration which followed this controversy in the later Latin version. * Haldane and Ross translate the passage as follows: 'It appears to me that the reasonings are so mutually interwoven, that as the later ones are demonstrated by the earlier, which are their causes, the earlier are reciprocally demonstrated by the later which are their effects. And it must not be imagined that in this I commit the fallacy which logicians name arguing in a circle, for, since experience renders the greater part of these effects very certain, the causes from which I deduce them do not so much serve to prove their existence as to explain them; on the other hand, the causes are explained by the effects' (Descartes [1637], pp. 128–9). (*Eds.*)

[2] Cf. Beck [1951], p. 84.

[3] Adam and Tannery (*eds.*) [1897–1913], volume 10, p. 431.

[4] Haldane and Ross [1911], p. 49. (*They translate the whole sentence as: 'But over and above this, if the question is to be perfectly stated, we require that it should be wholly determinate, so that we shall have nothing more to seek for than what can be inferred from the data' (*eds.*).)

Joachim correctly stresses that the difference between 'deduction' and 'induction' is very slight for Descartes, and their logical character is the same: 'an illative movement from a content or contents intuitively apprehended to another content which follows by direct logical necessity from the first' (Joachim [1906], pp. 71–2). If

(*c3*) *The continuity between Pappus and Descartes*

Two points in my presentation might be questioned by critics:

(1) Did Cartesian analysis–synthesis in fact originate with the Pappusian Circuit? Or is this connection only a rational reconstruction?

(2) Is the Cartesian Circuit a factual description of Cartesian ideas or is it again a rational reconstruction of these ideas?

Our presentation has certainly been a rationally reconstructed one. We stressed the objective connection and development of ideas and did not investigate the fumbling way in which they originally became conscious – or semiconscious – in subjective minds.

Nevertheless, our presentation did not deviate from the course of actual history. The main point of this sub-section will be that Descartes and his contemporaries were very much aware that they were reviving the Pappusian tradition and adapting it to modern science.

Descartes's main interest was to find a method of discovery of infallible knowledge, an infallibilist heuristic. The paragon of infallible knowledge was of course Euclidean Geometry. And the only extant method of discovery in Euclidean Geometry was the Pappusian Circuit. This was Descartes's natural starting point. His programme was to carry the logic of discovery of Euclidean Mathematics into all domains of human knowledge.

That this is a fair account of Descartes's approach can be easily shown. He explains it with unusual force and clarity in Rule IV of his *Regulae*. Even the heading of the rule is characteristic: We need heuristic.[1] The chapter starts with an onslaught on mere conjecturing. We can rely only on *intuition* and *deduction* which will never lead us astray. These two – if not cluttered up with anything else – direct us to infallible, quasi-divine knowledge. Then comes the decisive passage. Descartes looks for predecessors who shared his Method, and, after first disparaging Aristotelian logic,[2] refers to Pappus:

> I am quite ready to believe that the greater minds of former ages had some knowledge of it [the method]...Arithmetic and Geometry, the simplest sciences, give us an instance of this: for we have sufficient evidence that *the ancient Geometricians made use of a certain analysis* which they extended to the resolution of all problems, though they grudged the secret to posterity. At the present day also there flourishes a certain kind of Arithmetic, called Algebra, which designs to effect, when dealing with numbers, what the ancients achieved in the matter of figures. *These two methods* are nothing else than the spontaneous fruit sprung from the inborn principles of the discipline here in question; and I do not wonder that these sciences with their very simple

logical validity is a psychologistic concept, why should there be a sharp difference between the feelings of certainty concerning deductive or inductive validity?

[1] 'There is need for a method for finding out the truth' (Descartes [1628], p. 9).

[2] 'But as for the other mental operations, which Dialectic does its best to direct by making use of these prior ones, they are quite useless here; rather they are to be accounted impediments, because nothing can be added to the pure light of reason which does not in some way obscure it' (*ibid.*, p. 10).

subject matter should have yielded results so much more satisfactory than others in which greater obstructions choke all growth. But even in the latter case, if only we take care to cultivate them assiduously, fruits will certainly be able to come to full maturity.

This is the chief result which I have had in view in writing this treatise.[1]

Nothing can be clearer: the Method is the Pappusian one. The trouble is that inquiry does not usually deal with problems which are as simple and unobstructed and stated with as perfect clarity, as those of Geometry. The reference to Algebra is worth explaining.

Algebra is considered here not as a branch of Mathematics, but as a Method, a twin to the method of analysis. And indeed, Algebra is *par excellence* 'analytic' in the Pappusian sense:

the whole of the device here disclosed will consist in treating the unknowns as though they were known, and thus being able to adopt the easy and direct method of investigation even in problems involving any amount of intricacy. There is nothing to prevent us always achieving this result, since we have assumed from the commencement of this section of our work that we recognise the dependence of the unknown terms in the inquiry on those that are known to be such that the former are determined by the latter. This determination also is such that if, recognising it, we consider the terms which first present themselves and reckon them even though unknown among the known, and thus deduce from them step by step and by a true connection all the other terms, even those which are known, treating them as though they were unknown, we shall fully realise the purpose of this rule. [Rule XVII.][2]

And

employing this method of reasoning we have to find out as many magnitudes as we have unknown terms, treated as though they were known, for the purpose of handling the problem in the direct way; and these must be expressed in the two different ways. For this will give us as many equations as there are unknowns. [Rule XIX.][3]

A glimpse at Pappus's classical text is enough to ascertain that Algebra is indeed 'analytic'. Descartes's version follows rather the pattern of problem-solving and not that of proof.

Descartes explains that he was not interested in mathematics for its own sake: he was interested in the important secrets of the Universe and not in 'the trivialities of Geometry':[4] 'the reader who follows my drift with sufficient attention will easily see that nothing is less in my mind than ordinary mathematics, and that I am expounding quite another science, of which these illustrations are rather the outer husk than the constituents'.[5] He studied mathematics mainly because 'no other science furnishes us with illustrations of such self-evidence and certainty'.[6] But this did not satisfy Descartes. He found certainty in

[1] *Ibid.*, p. 10 (my italics).

[2] *Ibid.*, p. 71.

[3] *Ibid.*, p. 77.

[4] He did not know that Euclid's *Elements* were meant to be a cosmological theory (cf. Popper [1952], pp. 147–8).

[5] *Ibid.*, p. 11.

[6] *Ibid.*

the Geometers, but not how to *attain* certainty. 'They did not seem to make it sufficiently plain to the mind itself why those things are so, and how they discovered them'.[1] This passage contains a most crushing criticism of Euclidean synthetic style[2] which, he says, stifles the mind.[3] But if Mathematics only stifles the mind, why then did Plato refuse to admit to his School those who were unversed in Mathematics? The question confirmed Descartes in his suspicion that in antiquity Geometers 'had knowledge of a species of Mathematics very different from that which passes current in our time'.[4] And it is on this crucial point that he in fact refers to Pappus – and also to Diophantus, the founder of Algebra: 'Indeed I seem to recognise certain traces of this true Mathematics in Pappus and Diophantus'.[5]

After this he repeats his programme of the *Mathesis Universalis*, wondering how it is that although people know the meaning and importance of mathematics, they still neglect it while laboriously pursuing disciplines which in fact depend on it?[6]

There are many other passages in Descartes in which he refers to Pappusian Analysis and, again, to Algebra as the starting point of his Method. In the *Discours* for instance he repeats that he studied three disciplines 'which seemed as though they ought to contribute something to the design I had in view'.[7] He then says that neither of the three are satisfactory: Logic is only for explaining what one knows anyway, and good and bad things are inextricably intertwined in it;

> And as to the Analysis of the ancients and the algebra of the moderns, besides the fact that they embrace only matters the most abstract, such as appear to have no actual use, the former is always so restricted to the consideration of symbols that it cannot exercise the Understanding without greatly fatiguing the Imagination; and in the latter one is so subjected to certain rules and formulas that the result is the construction of an art which is confused and obscure, and which embarrasses the mind, instead of a science which contributes to its cultivation. This made me feel that some other method must be found, which, comprising the advantages of the three, is yet exempt from their faults.[8]

In spite of this criticism, which can be at least partially explained by his eagerness to stress the novelty of his Method, Descartes studied Analysis and Algebra very carefully, for 'two or three months to borrow all that is best in Geometrical Analysis and Algebra, and correct the errors of the one by the other'.[9]

[1] *Ibid.*

[2] 'they preferred to exhibit...certain barren truths, deductively demonstrated, which show enough of ingenuity, as the results of their art, in order to win from us our admiration for these achievements' (*ibid.*, p. 12).

[3] 'those superficial demonstrations, which are discovered more frequently by chance than by skill...that in a sense one ceases to make use of one's reason' (*ibid.*, p. 11).

[4] *Ibid.*, p. 12.

[5] *Ibid.*

[6] 'so many people laboriously pursue the other dependent sciences, and no one cares to master this one' (*ibid.*, p. 13).

[7] Descartes [1637], p. 91.

[8] *Ibid.*, p. 92.

[9] *Ibid.*, p. 93.

The essential identity of Rule IV and this basic passage in the *Discours* is striking: both of them contain Descartes's account of the Three Sources of his Method.

At least two more points should be mentioned here. One is Descartes's interest in Archimedes, the other his interest in Apollonius. Both Archimedes and Apollonius use Pappusian method and terminology.

To conclude: one cannot understand the *Regulae* and the *Discours* – and Descartes's intellectual history – if one ignores the Pappusian Circuit.[1]

(The Pappusian Circuit – translated from Arabic into Latin by Commandinus in 1566 and by Halley in 1706 – was much discussed in the seventeenth century. It figures in Galileo's *Dialogue on the Great World Systems*:

[1] The only Cartesian scholar who seems to have appreciated the impact of the Pappusian heuristic tradition was A. Robert [1937]. Unfortunately even Robert misinterprets this impact. He assumes that Descartes's problem was to get rid of synthesis, to have an analysis which could be demonstrative in itself. Now Robert thought that (*a*) in algebra proofs were reversible, and (*b*) the algebraization of geometry and of the sciences would therefore enable us to have perfect demonstrative analyses. To quote him: (*a*) 'Analysis in algebra is no longer only the invention of a proof but a proof. In fact all algebraic quality, being purely quantitative, is always reciprocal. It is thus useless to verify the results obtained by starting from both simple elements (roots of equations) to reconstruct the complex relations (equations) from which one started. Synthesis became useless. Analysis suffices: it is at once the method of invention and of proof. It is that which Descartes sought' (p. 242). (*b*) 'The introduction of algebra will show that Descartes attributes to analysis a demonstrative value denied it by the Greeks' (p. 230).

Now the first thesis is obviously false, although the myth of the reversibility of algebraic proof is quite widespread. (According to L. Brunschvicg, Greek analysis 'does not suffice in itself: for the ancients chose as their domain not *the field of algebra where the propositions are in general expressed by equations and are reciprocal*, but the field of geometry in which they are usually hierarchically ordered.' ([1912], p. 54, my italics). The same mistake occurs in Robinson's [1936], p. 465 and p. 469.)

The second thesis again unfortunately flatly contradicts several passages in the *Regulae*. According to Rule IV, Algebra 'designs to effect, when dealing with numbers, what the ancients achieved in the matter of figures'. This means that Descartes put Algebra and Geometry on a par. And one may wonder whether his Rule XX does not warn us not to indulge in possibly irreversible algebraic operations the existence of which he well realized: 'Having got our equations, we must proceed to carry out such operations as we have neglected, taking care never to multiply where we can divide.'

Robert's original problem – of which this is an incorrect solution – was why Descartes said in his *2ᵉ Réponses* that both analysis and synthesis were demonstrative, independently of each other. Moreover, he preferred analysis, for according to him only the shallow and unattentive mind needs the synthesis: the deep and attentive one wants analysis, the 'truer' demonstration.

The reason for this preference is that Descartes here starts analysis with reasoned facts and not with occult hypotheses. So he in fact starts – unlike Pappus – with basic statements. Therefore there is nothing peculiar in his claim that analysis is demonstrative. What has to be explained is rather the opposite, why he thought that in this case the synthesis is also demonstrative. The answer to this problem – which was missed by Robert – lies in the nature of the Cartesian Circuit: the occult hypotheses also receive a truth-injection which is independent of the factual basic statements, to wit, the first principles. (*We have translated the above passages by Robert and Brunschvicg from the French (*eds.*).)

Simplicius: Aristotle first laid the basis of his argument *a priori*, showing the necessity of the inalterability of heaven by means of natural, evident, and clear principles. He afterwards supported the same *a posteriori* by the senses and by the traditions of the ancients.
Salviati: What you refer to is the method he uses in writing his doctrine, but I do not believe it to be that with which he investigated it. Rather, I think it certain that he first obtained it by means of the senses, experiments and observations, to assure himself as much as possible of his conclusions. Afterwards he sought to make them demonstrable. That is what is done for the most part in the demonstrative sciences; this comes about because when the conclusion is true, one may by making use of analytic methods hit upon some proposition which is already demonstrated or arrive at some axiomatic principle; but *if the conclusion is false, one can go on forever without ever finding any known truth* – if indeed one does not encounter some impossibility or manifest absurdity.[1]

Or, let us quote Arnauld's version:

We may hence understand that this is the analysis of the geometers; for it proceeds as follows: A question having been proposed to them, in relation to which they are ignorant – if it be a theorem, of its truth or falsehood; if a problem, of its possibility or impossibility – they assume that it is as it is proposed; and examining what follows from this, if they arrive, in that examination, at some clear truth from which what is proposed to them is a necessary consequence, they conclude from this that what is proposed to them is true; and returning then through the way they had come, they demonstrate it by another method which is called *composition*. But if they fall, as a necessary consequence from what is proposed to them, into some absurdity or impossibility, they conclude from this that what is proposed to them is false and impossible.

This is what may be said generally touching analysis, which consists more in judgment and sagacity of mind than in particular rules.[2]

It is not necessary to investigate these more or less corrupted versions now. Our only point was to show that the Pappusian Circuit was highly topical in discussions of heuristic in the seventeenth century: it was indeed part of advanced logic courses. Nor need we now go into the question either of the continuity of the Pappusian tradition in medieval logic or of its place in the Paduan methodology.)

But, granted that Descartes started with the Pappusian Circuit – what did he preserve from it? Did he in fact develop it into what we called the Cartesian Circuit? Or is this a piece of rational reconstruction of history?

In this essay – while leaving these questions open – we shall be content to claim that the Cartesian Circuit in fact *is* the rational reconstruction of the problem in question, and that history can only be rationally understood in the light of such reconstructions.

[1] Galileo [1630], pp. 50–1. * We have substituted Drake's translation, which only became available later, for the one by Santillana which Lakatos gave (*eds.*).

[2] Arnauld and Nicole [1702], p. 315.

(*c4*) *The Cartesian Circuit in mathematics*

Those who – like Descartes – identified mathematics with Euclidean Geometry and Elementary Algebra thought that in mathematics facts are reasoned facts and occult hypotheses do not exist.

In the seventeenth and eighteenth centuries calculus, 'unreasoned' facts, invaded mathematics. How to rationalize them, how to raise them to the level of 'reasoned' facts soon became a central problem. Cauchy and his followers solved the problem by the 'translation-procedure'* which corresponds to the inductive passage from facts to reasoned facts in the Cartesian Circuit. At the same time occult hypotheses also appeared. The explanation of some facts about the *real* line using complex function theory is analogous to the transcendental hypotheses of physics. To deduce these hypotheses from first principles was one of the problems which the arithmetization and then the logicization of mathematics set out to solve.

A detailed discussion of the *Spezialdialektik* of the mathematical Cartesian Circuit may bring to the surface some aspects of the history and philosophy of mathematics which have hitherto been missed.

(*c5*) *The breakdown of the Cartesian Circuit*

(i) *Induction does not transfer truth.* One important stream of criticism was directed against the safety of the intuitive truth-transference along the Circuit. First of all the *inductive passages* came under the attack, and above all the one which led to the occult hypotheses. Taking the passage from reasoned facts to the occult hypotheses in isolation, these critics[1] denied that the truth value injected at the reasoned facts can ever reach the occult hypotheses.

If we accept this criticism we can either (*a*) give up infallibilism and confess the conjectural character of scientific hypotheses or (*b*) replace

* Cf. Lakatos [1976*c*], p. 121 (*eds.*).

[1] The first critics were Leibniz and Huygens (cf. Kneale [1949], pp. 97–8). Leibniz discovered that the Pappusian Circuit breaks down in science. He knew the condition for it to work: 'The propositions must be reciprocal (*réciproques*), in order that the synthetic demonstration be able to retrace the steps of analysis in the opposite direction.' But this condition is not satisfied in science 'in astronomical and physical hypotheses the retracing of the steps (*retour*) does not take place' (Leibniz [1704], IV, XVII, section 6) (* Our translations from the French (*eds.*)).

Huygens in the Preface of his [1690] described the same plight: 'There is to be found here a kind of demonstration which does not produce a certainty as great as that of geometry and is, indeed, very different from that used by geometers, since they prove their propositions by certain and incontestable principles, whereas here principles are tested by the consequences derived from them. *The nature of the subject permits no other treatment*' (my italics).

Newton was equally aware of the problem, but he thought one could fill the gap by infallible intuitive inference: '*In this philosophy propositions are deduced from phenomena and rendered general by induction*' (General Scholium at the end of the Principia).

this particular inductive passage by the infallible *Method of Division*.[1] There is a third, neutral way: (*c*) to introduce a theory of the probability of scientific hypotheses (which however leads inexorably to logically untenable theories of confirmation which try to reintroduce infallibility through the back door).[2]

The infallibility of the Method of Division was crushingly criticized by Catholic logicians from Pope Urban VIII to Duhem[3] and by many others; the probabilistic theories of confirmation were crushingly criticized by Popper.

In the course of these criticisms however the fact that the passage is not entirely inductive, that there is a considerable deductive part (which could be called the 'Newtonian deductive passage') has been completely forgotten. The infallibilist heuristic of deducing theories from facts certainly failed – but to replace it by the Popperian heuristic of speculations and refutations was to pour out the baby along with the bathwater.

(In the seventeenth and especially in the eighteenth centuries induction was just as widespread in mathematics as in science. It was called 'formalism' and it was criticized and banned altogether by Cauchy, Abel and others in the 'critical', 'rigourist' or 'exact' period.)

(ii) *Improved deduction transfers truth perfectly*. The deductive passages too were later sharply criticized, but never abandoned as conveyor belts of truth. They were improved in a piecemeal way by several translation-procedures (arithmetization, set-theoretization) which successively

[1] The Method of Division (e.g. Descartes [1628], *passim*, and [1664], XVIII) was a method of proving occult hypotheses without falling back on induction. We enumerate all the possible conjectures from which the facts to be explained can be derived; we falsify (i.e. deduce from them false factual propositions) all of them but one (by analysis only); and thereby we prove the unfalsified conjecture.

This method of division of course relies on the absolute infallibility of the *intuition* devising this complete enumeration and on the effective constructability of crucial experiments.

[2] Cf. Leibniz [1678], and Huygens, *op. cit.*

[3] Duhem's criticism of the Baconian *experimentum crucis* was in fact the elaboration of Urban VIII's argument put into Simplicius's mouth by Galileo at the end of his [1630]: 'I know that if I asked whether God in his infinite power and wisdom could have conferred upon the watery element its observed reciprocating motion using some other means than moving its containing vessels, both of you would reply that He could have, and that He would have known how to do this in many ways which are unthinkable to our minds. From this I forthwith conclude that, this being so, it would be excessive boldness for any one to limit and restrict the Divine power and wisdom to some particular fancy of his own' (p. 464). Galileo in fact dismissed this recommendation and regarded the Method of Division as perfectly infallible. The Earth either moves or stands still: there is no other possibility. Now he destroyed one by one the arguments for the second possibility, so he was left with the safe first possibility. Descartes was more cautious: 'The earth, properly speaking, does not move, nor do any planets, although they are carried along' ([1644], Part III, XXVIII).

Leibniz too was critical of the Method of Division. But as a heuristic method it was generally accepted; the Port-Royal Logic warns us that 'it is an equal defect to make enough, and to make too many divisions; the one does not sufficiently enlighten the mind, the other dissipates it too much' (Arnauld and Nicole [1724], p. 166).

raised the level of rigour by reducing the number of quasi-logical constants, and at the end arrived at proofs which could be checked by Turing-machines.

But one had to pay for each step which increased rigour in deduction by the introduction of a new and fallible translation. The impact of this fact has not yet been sufficiently appreciated.*

(iii) *There are no first principles and there are no perfect reasoned facts.* The breakdown of inductive logic destroyed the [Cartesian] Circuit. Truth value in the truncated circuit flows only in *one direction.* The Braithwaitian zipfastener takes the place of the Cartesian Circuit. But can the Braithwaitian zipfastener transmit truth? If first principles are admitted: yes. If not, it can at best transmit falsity. If first principles are admitted we can *prove* along the zipfastener. If they are not admitted we can, at best, only *disprove.*

Now the truncated Cartesian Circuit underwent a second attack: not this time against the safety of its truth-value-transmitting channels but rather against the justifiability of its truth-injections. Cartesians injected truth values at two levels: first principles and reasoned facts. The optimistic search for first principles whose truth will grow on us went on for centuries in all branches of human knowledge, e.g. in Mechanics,[1] in Ethics (Spinoza, Kant), in Economics (L. von Mises), in Political Philosophy (Hobbes). It is, however, generally considered today a futile venture; only some neo-Kantian philosophers would today expect or accept truth value injections into the truncated Circuit from intuitively indubitable first principles.

So we cannot prove in science; the most we can do, if we are anti-inductivist empiricists, is to disprove. If, however, we extend our critical attitude to the facts too – and this we have to do, especially following the forceful revival of the ancient Greek criticism of sense-experience by Duhem and Popper – we can allow only tentative recognition to basic statements. The Popperian epistemological zipfastener, unlike the zipfastener of logical positivism, is unsuitable even for infallible disproof. It cannot prove and it can disprove only tentatively. As to the heuristic zipfastener, this may start from a reasoned fact – like Kepler's Law – may move upwards through a deductive passage and then make an inductive jump to the Theory of Gravitation; and then turn back along a purely deductive path, erase the former fact, write down the corrected Newtonian version of Kepler's Laws, and so on. There are no absolutely hard, stubborn, perfectly reasoned facts.

The reader will recognize in this heuristic zipfastener a model of our proof-procedure: Kepler's Laws as the primitive conjectures, and the Newtonian correction as the theorem. The only difference is that

* Cf. Lakatos [1976*c*], chapter 2 (*eds.*)

[1] Stevinus thought he had *proved* his law of inclined pressure, D. Bernoulli the parallelogram of forces, Euler the principles of mechanics, and Lagrange the principle of virtual displacements. This was also Maupertuis's motive in trying to reduce Newtonian mechanics to intuitively obvious principles.

here we may prefer to use the term 'explanation-procedure' – but this difference seems important only to those logical positivists who believe in first principles in mathematics and deny them in science. These positivists thus use the proof versus explanation divide (i.e. first principles versus no first principles divide), as a demarcation criterion between mathematics and science:

there is an essential difference in the way in which we think by means of a mathematical and by means of a scientific deductive system. In the former we start from the beginning and go on to the end, both logically and epistemologically; in the latter we start from the beginning and go on to the end only logically, the epistemological order being from the end to the beginning. To use again the metaphor of the zipfastener, the truth-value of truth (i.e. of formal truth) for mathematical propositions is assigned first at the top and then by working downwards, in a scientific system the truth-value of truth (i.e. of conformity with experience) is assigned at the bottom first and then by working upwards...[I]t took a long time for scientists to realise that the hypothetico-deductive inductive method of science was epistemologically different from the *prima facie* similar deductive method of mathematics; and that, in properly imitating the deductive form of Euclid's System, they were not *ipso facto* taking over his deductive method of proof. The enormous influence of Euclid has been so good in inducing scientists to construct deductive systems as more than to counterbalance his bad influence in causing them to misunderstand what they were doing in constructing such systems; the good genius of mathematics and of unself-conscious science, Euclid has been the evil genius of philosophy of science – and indeed of metaphysics. [Braithwaite, *op. cit.*, pp. 352–3.]

This logicist demarcation between science and mathematics is however unjustified. The arithmetization of mathematics was the basic argument for the neo-Kantians' claim that they could deduce all mathematics from Peano's five synthetic *a priori* axioms. The logicization of mathematics was the basic argument for the logical positivists' claim that they could deduce all mathematics from the analytically true axioms of the *Principia*.

Russell's attempt was genuinely Cartesian.[1] But he failed. His axioms of infinity and choice were anything but analytic,[2] and the analyticity of some of the rest was also problematic. One can of course still fall back on Kantianism, and claim that one's logical axioms are synthetic *a priori*.[3] But this reinterpretation has two fatal weaknesses. One is

[1] 'I hoped sooner or later to arrive at a perfected mathematics which should leave no room for doubts, and bit by bit to extend the sphere of certainty from mathematics to other sciences' (Russell [1959], p. 36).

[2] The Cartesian Ramsey succeeded only in eliminating the Axiom of Reducibility.

[3] I assume it was this that Russell had in mind in the last sentence of the following passage, where he announces the failure of his original brand of Cartesianism: 'Where pure mathematics is organised as a deductive system – i.e. as the set of all those propositions that can be deduced from an assigned set of premises – it becomes obvious that, if we are to believe in the truth of pure mathematics, it cannot be solely because we believe in the truth of the set of premises. Some of the premises are much

that the translation-procedures are again anything but synthetic *a priori*. The other is that the Gödelian argument about incompleteness of reasonably rich theories shows that the infallible air of arithmetic is due to the fact that its presently known part is only a poor fraction of the infinite whole.

One could argue that the Gödelian proofs of undecidability which lead to these new axioms will still be perfectly simple and translucent. But Gödel's results destroyed also these Hilbertian dreams about trivial meta-theories.

The Euclidean–Cartesian dream of the trivialization of knowledge has failed; not only in science but also in logic and mathematics.[1]

less obvious than some of their consequences, and are believed chiefly because of their consequences. This will be found to be always the case when a science is arranged as a deductive system. It is not the logically simplest propositions of the system that are the most obvious, or that provide the chief part of our reasons for believing in the system. With the empirical sciences this is evident. Electro-dynamics, for example, can be concentrated into Maxwell's equations, but these equations are believed because of the observed truth of certain of their logical consequences. Exactly the same thing happens in the realm of logic; the logically first principles of logic – at least some of them – are believed, not on their own account, but on account of their consequences. The epistemological question "Why should I believe this set of propositions?" is quite different from the logical question: "What is the smallest and logically simplest group of propositions from which this set of propositions can be deduced?" Our reasons for believing logic and pure mathematics are, in part, only inductive and probable, in spite of the fact that, in their *logical* order, the propositions of logic and pure mathematics follow from the premises of logic by pure deduction. I think this point important, since errors are liable to arise from assimilating the logical to the epistemological order, and also, conversely, from assimilating the epistemological to the logical order. The only way in which work on mathematical logic throws light on the truth or falsehood of mathematics is by disproving the supposed antinomies. This shows that mathematics *may* be true. But to show that mathematics *is* true would require other methods and other considerations' (Russell [1924], pp. 325–6). The attitude is very similar to Newton's. Newton wanted a Cartesian cosmology, Russell a Cartesian logic. After failing, both believed that they should approach the original – and only worthwhile – problem once again.

[1] We did not discuss here the formalist demarcation criterion between science and mathematics which does not inject any truth value in the mathematical zipfastener at all; it can transmit truth, it can 'derive', but it is essentially neutral, it cannot 'prove'. This also fails because of the Gödelian results. The variety of possible transmissions of truth is much richer than any given logical theory; but even the most trivial-looking logical theory may turn out to be non-trivially inconsistent.

2 ANALYSIS–SYNTHESIS: HOW FAILED ATTEMPTS AT REFUTATIONS MAY BE HEURISTIC STARTING POINTS OF RESEARCH PROGRAMMES

Professor Hintikka's and Mr Remes's essay is an interesting contribution to the vast literature on the problem of Pappusian analysis–synthesis, and I am delighted to have the opportunity to supplement it with the rather different views which I advocated in my 'Proofs and Refutations' and which seem to have escaped their attention.[1]

(*a*) *An analysis–synthesis in topology which does not prove what it sets out to prove*

Like Polya, whose work unfortunately is ignored by Hintikka, I regard analysis as a heuristic pattern which although it may have been *started* by the Greeks has been characteristic of scientific and of mathematical research up to the present day.

I shall begin by reminding you of two classical examples of analysis. The first example is Cauchy's 1811 proof of Euler's celebrated theorem on polyhedra. Euler, in 1751, proved that for all polyhedra $V-E+F=2$, where V is the number of vertices, E is the number of edges and F is the number of faces of the polyhedron.

Cauchy's proof ran as follows. Let us assume that in fact it is the case that $V-E+F=2$ for all polyhedra. Let us then take a particular instance of a polyhedron, for instance, the cube, and perform on it the following experiment.

We first prepare a hollow rubber model of a cube, with the edges brightly painted in red.* If we cut out one of the faces, we can stretch the remaining surface flat on the blackboard, without tearing it. The faces and edges will be deformed, the red-painted edges may become curved, but V, E and F will not alter, so that $V-E+F=2$ for the original polyhedron if and only if $V-E+F=1$ for this flat network – remember that we have removed one face. (Figure 1 shows the flat network for the case of a cube.) *Step 2*: Now we triangulate our map – it does indeed look like a geographical map. We draw (possibly curvilinear) diagonals in those (possibly curvilinear) polygons which are not already (possibly curvilinear) triangles. By drawing each diagonal we increase both E and F by one, so that the total $V-E+F$ will not be altered (Figure 2). *Step 3*: From the triangulated network we

[1] My essay 'Proofs and Refutations' was published in 1963–4 in the *British Journal for the Philosophy of Science*: it appeared in book form in Moscow in 1966 under the title: *Dokazatelstva i Oprovezemia* [*and since then in book form in English as *Proofs and Refutations*, Cambridge University Press, 1976 (*eds.*)]. For further information on my philosophy of mathematics see Lakatos [1961], [1962], [1963–4], and this volume, chapter 2.

* Lakatos's typescript leaves a gap at this point. The description of Cauchy's proof which follows is taken from Lakatos's [1976*c*], pp. 7–8. (*Eds.*)

now remove the triangles one by one. To remove a triangle we either remove an edge – upon which one face and one edge disappear (Figure 3*a*), or we remove two edges and a vertex – upon which one face, two edges and one vertex disappear (Figure 3*b*). Thus if $V-E+F=1$ before a triangle is removed, it remains so after the triangle is removed. At the end of this procedure we get a single triangle. For this $V-E+F=1$ holds true.

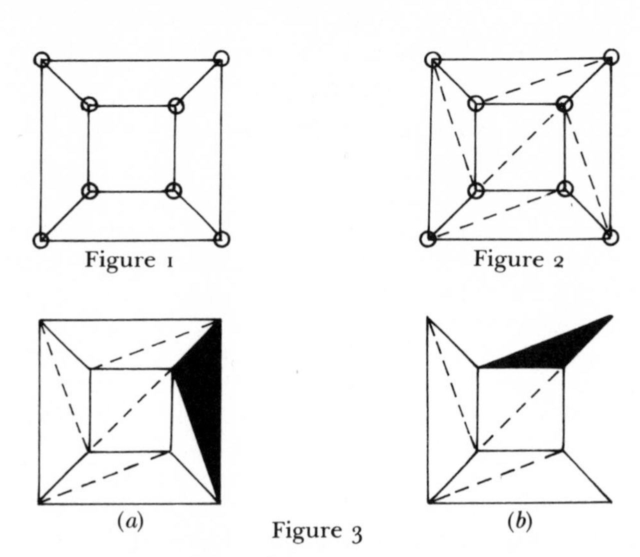
Figure 1

Figure 2

(*a*) Figure 3 (*b*)

It may sound strange that mathematicians accepted this argument as a proof of Euler's conjecture since all that Cauchy did was to prove that if $V-E+F=2$ for a cube, $V-E+F=1$ for a specific triangle. But the latter equation is surely trivially true. However, this curious proof has immense heuristic power. We may of course describe the inference which it involves by

$$E(P_1) \longrightarrow E'(T_{P_1})$$

where P_1 is a special polyhedron (namely the cube) and T_{P_1} is the triangle resulting from the transformation described in the 'proof'. The predicate E stands for Eulerian, the predicate E' stands for quasi-Eulerian [i.e. the property which holds for those objects for which $V-E+F=1$].

However, this trivial derivation strongly *suggests* the more general formula

$$E(P) \longrightarrow E'(T_P)$$

where P is a free variable ranging over all polyhedra. In this case, however, we need very strong auxiliary assumptions to derive from $E(P)$ the conclusion $E'(T_P)$. We need to assume that *all* rubber polyhedra after having a face removed can be stretched, without tear, flat on a blackboard. We need to assume that *all* flat networks at which we arrive in this way can be triangulated without changing $V-E+F$. We need to assume that from *all* such networks *all* triangles can be removed one by one until we have reached the last one again without changing $V-E+F$. Our deductive chain is really more like this:

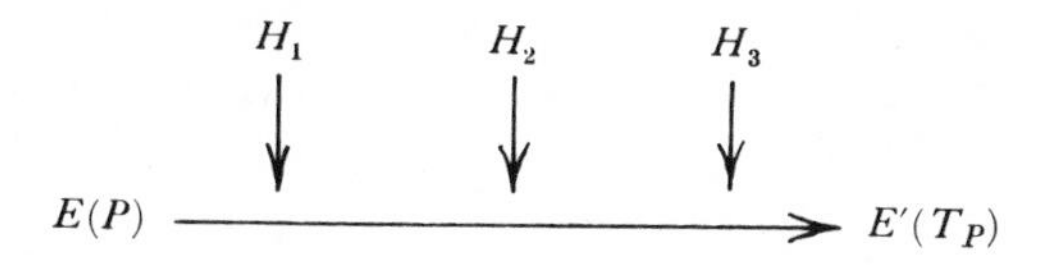

Thus the derivation of a very weak conclusion ($V-E+F=1$ for our triangle) was reached from a strong premise only with the help of some *very strong assumptions*. However, once we make use of these strong assumptions, we can work our way backwards, from the triangle to the polyhedron and derive Euler's theorem from the fact that a triangle has 3 vertices, 3 edges and 1 face. The analysis provides the hidden assumptions needed for the synthesis. The *analysis* contains the creative innovation, the *synthesis* is a routine task for a schoolboy. In this case the creative innovation was the idea that polyhedra are 'really' closed, triangulated, rubber surfaces. The analysis, incidentally, was performed on one specific polyhedron, and therefore the universal lemmas were only suggested but not made explicit.

However, the hidden lemmas are false. Not all polyhedra are homeomorphic with the sphere and not all polyhedral faces are simply connected. Therefore only those polyhedra are Eulerian which satisfy the auxiliary assumptions. Both the analysis and the synthesis are invalid and the theorem which we set out to prove turns out to be a mere 'naive conjecture'. But nevertheless we can extricate from the analysis (or from the synthesis) a 'proof-generated theorem' by incorporating the conditions articulated in the lemmas. Thus we do not prove what we set out to prove. We set out to prove $V-E+F=2$ for all polyhedra and by the critical examination of the analysis and a more explicit synthesis we do not arrive back at the original starting point. We start with the proposition 'All polyhedra are Eulerian' and, after a process of imaginative–critical analysis–synthesis we arrive at the proposition 'All *Cauchy* polyhedra are Eulerian'.

What happens then to non-Cauchy polyhedra? This problem engendered a veritable research programme. It led to a full classification of topologically equivalent closed surfaces, to the classification of *n*-tuply connected polygon sets and to the calculation of $V-E+F$ for a wide range of topological objects. In the course of this investigation a series of further hidden lemmas emerged and finally the research programme evolved a hard core (in the axioms of algebraic topology) and a sophisticated, rich positive heuristic.

But if we start from a proposition P and *draw consequences* from it rather than try to *look for premises* from which it follows, then what we *objectively* do is *test rather than prove*. Thus in analysis we test – Popperianwise – a conjecture; but if we *fail* to refute it, we may succeed in turning it first into a proof and then into a mathematical research programme.

At this point we might raise the problem of how we arrived originally at $V-E+F=2$, at the 'naive conjecture'. It is important to realize that most mathematical conjectures appear *before* they are proved; and they are usually proved before the axiomatic system is articulated in which the proof can be performed in a formalized way. We arrive at mathematical conjectures by tentatively solving mathematical problems by trial and error. Thus we raise the problem as to the relations between the faces, edges and vertices of a polyhedron. We try out different conjectures one after the other. I described in detail, following Polya, this series of conjectures and refutations.* It took in this case nearly 2000 years to reach the naive conjecture by Popperian conjectures and refutations which I called the stage of 'naive trial and error'. This 'naive' period, the first stage of mathematical discovery, lasted in this particular case from Euclid to Descartes. But at some stage the naive conjecture is subjected to a sophisticated attempted refutation; analysis and synthesis starts: this is the *second* stage of discovery which I called 'proof-procedure'. This proof-procedure generates first the brand-new proof-generated theorem and then a rich research programme. The naive conjecture disappears, the proof-generated theorems become ever more complex and the centre of the stage is occupied by the newly invented lemmas, first as hidden (*enthymemes*), and later as increasingly well articulated auxiliary assumptions. It is these hidden lemmas which, finally, become the hard core of the programme. In our case a few hundred years later they (or rather their further 'derivatives') appear as axioms of algebraic topology.

It is important to see that in the analysis I described there is no trace either of an axiomatic system or even of a specified body of knowledge or of a set of lemmas known to be true. We start with a naive conjecture and we have to *invent* the lemmas, and even perhaps the conceptual framework in which the lemmas can be framed. Moreover we find that in a heuristically fruitful analysis most of the hidden lemmas will be found on examination to be *false*, and even known to be false at the time of their conception. All this is very different from Hintikka's (or Pappus's) conception of (theoretical) analysis. In my conception the problem is not to prove a proposition from lemmas or axioms but to discover a particularly severe, imaginative 'test-thought experiment' which creates the tools for a 'proof-thought experiment', which, however, instead of *proving* the conjecture *improves* it. The synthesis is an 'improof', not a 'proof', and may serve as a launching pad for a research programme.

* See Lakatos [1976*c*], chapters 1 and 2 (*eds.*)

There is then a pattern by which one gets from naive Popperian guessing to the method of *proofs* and refutations (*not* conjectures and refutations), and then, one step further, to mathematical research programmes. This pattern refutes the philosophical claim that the heuristic source of research programmes is always some big metaphysical vision. A research programme may be of humbler origin: it may originate in low-level generalizations. My case-study in a sense rehabilitates inductivist heuristic: it is frequently the study of facts and the practice of low-level generalizations which serve as the launching pad for programmes. Mathematics and science are importantly inspired by facts, factual generalizations and then by this imaginative deductive analysis.[1]*

(*b*) *An analysis–synthesis in physics which does not explain what it set out to explain*

I should like to switch briefly to a second example. Before I do so, I should like to draw your attention to the fact that at least up to the stage that I carried it, there was nothing specifically mathematical in my *first* example. Everything that I said can be interpreted as a research programme in networks painted on closed rubber sheets. In this case our analysis leads to an *explanation* rather than to a *proof*, and the emerging hidden premises are *explanatory* propositions. I repeat: in this case *we would have deduced from Euler's formula its own explanation.* This is clearly the case in Newton's celebrated analysis–synthesis which I shall consider in this section, and which is an explanation-procedure rather than a proof-procedure. But before turning to Newton let me emphasize that '$V-E+F=2$ for all polyhedra' is no less and no more of a *fact* than 'all planets move in ellipses'.

Newton started with low-level hypotheses: Kepler's three laws of planetary motion. He took – as one does in analysis – one instance of a planetary system: one in which the Sun is held in a fixed position by an invisible hand and in which there is only one planet orbiting it. He set out to perform an 'analysis' of Kepler's laws for this particular case. First he deduced, in his chosen particular instance, the purely kinematic implication that a plane planetary motion has its acceleration directed toward the Sun, from Kepler's naive conjecture that the *radius vector* covers equal areas in equal times. Unlike the Cauchy case, this end-result about the acceleration is not evidently true, but it certainly has a degree of plausibility in the light of Platonic metaphysics. Then Newton proceeded to the synthesis. Assuming that the acceleration of the plane motion *is* directed towards the Sun, he

[1] This implication was pointed out to me by Spiro Latsis. Also cf. Latsis [1972].

* At this point in Lakatos's typescript there is the new section heading: 'An analysis–synthesis in the calculus which does not prove what it set out to prove'. But there is no accompanying text. (*Eds.*)

deduced *backwards* Kepler's law of equal areas. Having concluded this piece of analysis and synthesis, he took Kepler's law of the ellipse and, by analysis, deduced that the acceleration, which he already proved to be directed towards the Sun, has a magnitude proportional to the inverse square of the planet's distance from the Sun at any given point. The assumption of this analysis was the law that the planet's orbit is an ellipse. But the synthesis does not yield simply an ellipse. Unlike the analysis, the synthesis contains a false lemma, and moving backwards from the inverse square law with improved lemmas Newton arrived at an empirically stronger proposition: the planet moves on some *conic*, and the type of the conic can be predicted with the help of the starting velocity.

Now comes Newton's analysis of Kepler's third law. Before sketching this, we have to remember that what Newton deduced from Kepler's first two laws was that the planet in his particular model moves with an acceleration directed towards the Sun and that this acceleration is inversely proportional to the square of its distance from the Sun. But let us now move the planet further away from the Sun onto a larger ellipse. Will the factor of proportionality be the same? Kepler's two laws do not give us any information on this point. Newton guessed that the factor was the same. He already derived from Kepler's first two laws that the factor of proportionality was

$$\gamma = 4\pi^2 \frac{a^3}{T^2}$$

where a is the semi-major axis of the ellipse and T the period of revolution; and now he derived from Kepler's third law that γ is independent of the distance of the planet from the Sun. Thus Newton derived from Kepler's three laws that, given a fixed Sun and one planet, the acceleration acting on the planet is (γ/r^2), where γ is the same for all such planetary systems with a single planet.

This purely kinematical reconstruction of Newton's analysis and synthesis is due to Toeplitz.[1] It is an open question whether or not this was Newton's actual path of discovery and we may never find out. Let us now assume that it was. The analyses mobilize the mathematical tools used in the syntheses; once the analysis is there, the synthesis is not too difficult. In Newton's case, dramatically *unlike* Cauchy's, the analysis seems to lead from the 'known' to the 'unknown' and not from the 'unknown' to the 'known'. But as a matter of fact Newton knew that planets move only *approximately* in ellipses, just as in Cauchy's time the anomalies to Euler's naive conjecture were well known. Nonetheless, both Cauchy and Newton performed an exact analysis–synthesis procedure without any regard for the anomalies and both were immensely proud of it. *The real achievement in both cases was not the 'proved' end result but the intellectual achievement involved in creating the necessary*

[1] Cf. Toeplitz [1963], pp. 156–61.

mathematical apparatus in the analysis. Newton, of course, also used hidden lemmas, like the one according to which the mass both of the Sun and of the planet were concentrated in geometrical points. It was only later that Newton showed that these analyses can also be carried out if one assumes only that both are perfect and homogeneous spheres. But Newton's analysis and synthesis was still restricted to fixed-Sun, one-planet systems. He then let the Sun move and showed that the planet will still move along a conic with the mobile Sun at a focus – but for this he had to introduce his dynamics.[1] In order to calculate planetary motions with more than two bodies and bulging planets he had to mobilize a complex mathematical apparatus with the full force of his dynamics. At this stage 'analysis' helped him no longer, just as in the creative development of mature algebraic topology we rarely any longer find analyses. Once the lemmas become corroborated and even organized in axiomatic systems, once the mathematical machinery is established, analysis, 'working backwards' may still be applied as a heuristic tool in puzzle-solving, but it becomes clear that its role is only psychological. It helps the imagination to produce valid proofs or explanations in terms of a *given* research programme. Analysis in mature science and mathematics no longer leads to revolutionary progress. *Analysis is only revolutionary when it engineers a breakthrough from a low-level naive conjecture to a research programme.*

I suspect that this was the case in several major developments in physics. Planck's and Einstein's quantum theory, for instance, I think, emerged from deductive probing into low-level radiation laws (although probably not on the lines suggested by Dorling[2]).

(c) *Pappusian analyses–syntheses in Greek geometry*

Let me now turn to ancient Greek Geometry. I propose the following historical thesis. Ancient Geometry first developed empirically through naive trial and error. The Greeks inherited from the Babylonians and Egyptians a body of naive *conjectures* with relatively high truth content (and they invented others). These conjectures were a precondition for later developments. Then came test- and proof-thought-experiments, mainly analyses, without any *known* lemmas, without any safe axiomatic systems. This is what Szabo showed the original concept of Greek proof to be, the *deiknymi. Deiknymi* can proceed in two ways [corresponding to analysis and synthesis]. It was only after hundreds of successful analyses and syntheses, after hundreds of 'proof-procedures' (in the sense of my *Proofs and Refutations*) that certain lemmas kept cropping up, became 'corroborated' (their alternatives remaining sterile) and finally were turned into the

[1] He, of course, had known all the time that there is no invisible hand holding the planet and thus that his third law of motion forbade a fixed Sun.

[2] Cf. Dorling [1971].

hard core of a research programme (an 'axiomatic system') by Euclid. *By this time* if one suggested a geometrical conjecture the question was whether it *followed* from Euclid's postulates and axioms, and not whether it was *true*. Analysis as a means of getting to *novel* lemmas and axioms lost its function; when used at all it was only a heuristic device for mobilizing the – already proven or trivially valid – lemmas necessary for the synthesis. Analysis was not any more a venture into the unknown; it was an exercise in mobilizing and ingeniously connecting the relevant parts of the known. The lemmas which were once daring and often falsified conjectures harden into auxiliary *theorems*.

This is why in Euclid's own analyses we never find 'local counter-examples' which show that hidden assumptions are false and have to be modified and replaced. The period in which some lemmas *were* refuted by concept-stretching had already passed; alternative lemmas, like 'not all entities are greater than their parts' or 'parallels meet' were eliminated because of degeneration: alternatives were proposed but did not lead to interesting results. It would be interesting to see whether Archimedes's and Apollonius's analyses were – to use Kuhn's terminology – part of revolutionary or of normal science.

One final remark should be made. In pre-Euclidean geometry, where Pappusian analyses played a revolutionary role, the parallel axiom kept popping up as a novel and dubious lemma. It took some time before it was *decided* that this lemma should be regarded as an unchallenged axiom of Euclidean geometry: this process was indicated by Professor Szabo in his classic work (Szabo [1969]).

Now my differences with Hintikka's and Remes's rational reconstruction of Greek analysis–synthesis become clear. They base their reconstruction on the assumption that Pappusian analysis was a heuristic pattern in *already axiomatized* Euclidean Geometry and they always assume that the deductive parts of analyses are valid and the adduced lemmas provable. In my view the most exciting analyses of Greek Geometry were pre-Euclidean and their role was to generate Euclid's axiomatic system. Most of Euclidean Geometry existed *before* Euclid's postulates, axioms, definitions and common notions; just as number theory existed before Peano's axioms; as the calculus existed before Dedekind's and others' definitions of real numbers; and just as probability theory existed before Kolmogorov. The question arises why the need for articulating an axiomatic system emerged at all, and what heuristic role such a system played in the further development of the discipline. But this is a *separate* question from the one we are discussing and secondary to it.

(*d*) [*False awareness about analysis–synthesis*]

It is, of course, fascinating to study the different forms of false awareness manifested by great scientists when they, for some reason, try to explain why they think that their contributions are scientific, or, indeed, outstanding. For instance, Newton often wanted to justify his scientific merits as compared with those of Hooke. Newton claimed that while Hooke only *conjectured* that the exponent is exactly two in the inverse square law and not just a number very near two, he *deduced* the exact number two from Kepler's laws. Since Duhem, Newton has been much ridiculed for his claim (for instance by Popper and Feyerabend) and the defences put up for Newton's claim by e.g. Born and Bernard Cohen were based on logical misconceptions.[1] But once we realize that Newton's heuristic path may well have really been an *analysis* leading from Kepler's laws to the inverse square law, and that this analysis, even in its weakest purely kinematic version, was content-increasing and indeed content-improving because of the depth of the auxiliary lemmas used in the analysis, one understands that scientific progress in the modern sense of the word has been made already in the analysis. The progress is not even so much in the actual novel predictions which go beyond the premises – in this case that planets might move also in parabolas – but in the mathematical and physical novelty of problem-solving techniques which later lead to, and form part of, a progressive research programme.

The various forms of confusion concerning analysis–synthesis in Euclid, Pappus, Zabarella, Galileo, Descartes, Newton and others can be traced primarily to some few clearly discernible causes.

First, it was thought that each step the scientist takes has to be epistemologically *justified*. Therefore, if the scientist, through analysis, arrives at B from A, he must *know* more than at the time he started. If the analysis was only of heuristic, as opposed to epistemological, value, then in justificationist terms, analysis is not yet a success worth mentioning. It is no discovery. Therefore analysis was regarded as part of the process of justification and not just of the process of discovery. *These two processes were not and could not be clearly separated before the development of modern logic*. The very word 'analytical method' conflates the two. Heuristics and appraisal are in fact distinct.

Secondly, the difference between deduction and induction was not clear. Within Aristotelian syllogistic it is easy to demarcate the two notions, but in a psychologistic theory of logic (such as Galileo's, Descartes's or Newton's) a valid inference, in general, is 'the illative movement from a content or contents intuitively apprehended to another content which follows by direct logical necessity from the first'.[2] But if so, there need be no difference between a content-increasing and content non-increasing inference: an objectively invalid

[1] Cf. Born [1949], pp. 128–33 and Cohen [1974].

[2] Joachim [1906], p. 71, n. 2.

inductive inference may convey more psychological certainty than some deductively valid inferences. In particular what we today call informal (*inhaltlich*) mathematical inference is inductive, since the adduced hidden premises have logical content not present in the explicit premises. *It is only after Bolzano's theory of logical validity that we can demarcate induction from deduction in general terms.* In Descartes, and also in Newton, the two terms are used as synonyms.

Thirdly, before modern logic it was impossible to sort out the difference between cause and effect.

It is these three sources of confusion – all originating in a justificationist epistemology and a psychological theory of logic – which yielded the vague epistemological ideas of what I used to call the Pappusian–Cartesian Circuit.* Let me illustrate the Pappusian Circuit by a diagram:

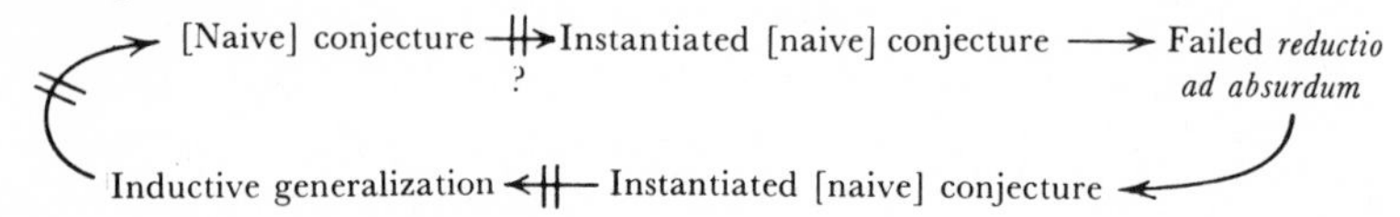

The truth of the conjecture is guaranteed by the *full circuit*, by the cooperation of intellect and experience, again a theme so much emphasized by Bacon, Descartes and Newton. (I do not know who invented the myth of the Bacon–Descartes controversy.)

The trouble with the circuit is, of course, that the arrows do not all represent deductively valid inferences. As a matter of fact, going several times through the circuit, more and more hidden lemmas emerge, and the conjecture is constantly being improved by critical inspection of the circuit. Proof can be proclaimed only by agreeing, by convention, where concept stretching criticism has to stop and by giving, for *mathematical* conjectures, a valid proof in a first- or second-order theory.

When we switch from the Pappusian to the Cartesian Circuit, new aspects of the Circuit, and with them new problems, arise [as indicated *above*, pp. 77–92. The Cartesian Circuit may be represented as follows.]

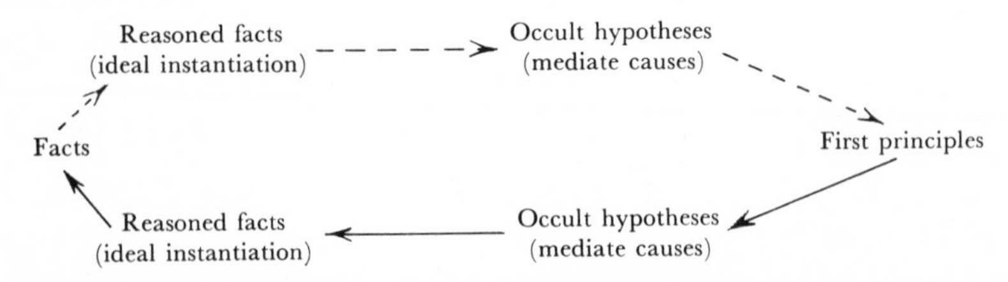

The breakdown of this Cartesian idea of the Circuit has constituted the mainstream of modern philosophy of science. [This breakdown has been effected by] the gradual separation of deduction from induc-

* See *above*, p. 76 (*eds.*).

tion, of psychology of discovery from the logic of justification, by the breakdown of *causa aequat effectu* and, above all, by the gradual separation of mathematical and empirical sciences. This breakdown of the Cartesian circuit took a couple of hundred years. It probably originated with Leibniz, but the real breakthrough came only with Bolzano, Fries, and finally Tarski. [The subsequent emergence of the hypothetico-deductive method of discovery and justification however may blind us to the important role of deductive chains from 'phenomena'. The gap between scientific theories and factual propositions *is* unbridgeable, but occasionally the jump is shorter than Popperian philosophy may lead us to believe.]

Part 2

Critical Papers

6

The problem of appraising scientific theories: three approaches*

One central problem with which philosophy of science has traditionally been concerned is that of the (normative) appraisal of those theories which lay claim to 'scientific' status. Can we specify universally applicable conditions which one theory has to satisfy in order to be a better scientific theory than another? (The demarcation problem, which is now associated with Popper's name, and which is the problem of whether we can specify the conditions a theory must satisfy to be scientific at all, is a sort of 'zero case' of this problem.) The generalized demarcation problem is, it seems to me, the primary problem of philosophy of science. There are three major traditions in the approach to this one problem. [The aim of this paper is to sketch these three traditions and to investigate their strengths and weaknesses.]

1 THREE MAIN SCHOOLS OF THOUGHT CONCERNING THE NORMATIVE PROBLEM OF APPRAISING SCIENTIFIC THEORIES

(a) *Scepticism*

One school of thought on the appraisal problem can be traced back to the Greek tradition of Pyrrhonian scepticism, and is now known as 'cultural relativism'. Scepticism regards scientific theories as just one family of beliefs which rank equal, epistemologically, with the thousands of other families of beliefs. One belief-system is no more 'right' than any other belief-system; although some have more *might* than others. There may be *changes* in belief-systems but no *progress*. This school of thought, temporarily muted by the stunning success of Newtonian science, is today regaining momentum particularly in the anti-scientific circles of the New Left; its most original and colourful version is Feyerabend's '*epistemological anarchism*'. According to Feyerabend, philosophy of science is a perfectly legitimate activity; it may even *influence* science. *Any* belief-system – including those of his opponents – is free to grow and influence any other belief-system; but none has epistemological superiority.[1] Note that this view is different

* This paper originally formed part of Lakatos's review of Toulmin's *Human Understanding*, the history of which is described in the first note to chapter 11, *below*. It was also the basis for some of his lectures at Alpbach in 1973. We have supplied the title. It is published here for the first time. (*Eds.*)

[1] I am here referring to the Feyerabend of the 1970s vintage, as best seen in his [1970], [1972] and [1975].

from Mao's 'let a hundred flowers bloom'. Feyerabend does not wish to impose a 'subjective' distinction between flowers and weeds on anybody. The advice of the sceptic is: *Do your own thing*. This is the sceptic's code of intellectual honesty.

Sceptics make imaginative but unreliable historians. For them history of science can only be a belief about beliefs. One reconstruction differs from another according to the irremediable bias of the historians: and one is no better than another.

The sceptic thus *denies* the possibility of producing any acceptable solution to the problem of appraising scientific theories. Each of the other two schools I shall consider *asserts* the possibility of solving the problem. 'Demarcationists' are preoccupied with trying to produce a *universal* criterion of appraisal which will help us identify scientific progress and will show that, for instance, [Newton's theory is an improvement on Kepler's theory in precisely the same sense as Einstein's is an improvement over Newton's. 'Elitists', as we shall see, agree that Newton's theory is better than Kepler's and that Einstein's theory is better than Newton's, but they deny the possibility of constructing a *universal* criterion of scientific progress which would yield these particular judgments.]

(*b*) *Demarcationism*

The term 'demarcationism' stems from the problem of demarcating science from non-science or from pseudoscience. But I use it in a more general sense. A (generalized) demarcation criterion, a methodology or appraisal criterion, demarcates better from worse knowledge, defines progress and degeneration.

Demarcationist criteria presuppose universal 'third-world' [standards of] logical truth and logically valid inference. The demarcationist research programme aimed at finding such standards began with Aristotelian syllogistic and reached a high-point with Gödel's celebrated completeness theorem. In order to see more clearly the difference between 'demarcationism' and 'élitism' (which I shall outline next) we have to start by reference to Frege's and Popper's distinction between three worlds. The 'first world' is the physical world; the 'second world' is the world of consciousness, of mental states and, in particular, of beliefs; the 'third world' is the Platonic world of objective spirit, the world of ideas.[1] The three worlds interact, but each has considerable autonomy. The *products* of knowledge; propositions, theories, systems of theories, problems, problemshifts, research programmes live and grow in the 'third world'.[2] The *producers* of knowledge live in the first and second worlds.

[1] An exposition of this vital distinction can be found in Popper [1972], pp. 106–90; and especially in Musgrave's important unpublished doctoral thesis [1969].

[2] Most demarcationists agree that propositions are true if they correspond with facts, and thus subscribe to the correspondence theory of truth. (Some conventionalists may

According to 'demarcationists', the products of knowledge can be appraised and compared on the basis of certain *universal* criteria. Theories about these criteria constitute 'methodological' knowledge and also live and grow in the 'third world'.

There are many differences *within* the demarcationist school. These stem from two basic differences. First, different demarcationists may differ in their claims about what the most appropriate *unit* of appraisal is. Leibniz thought that any proposition can be appraised in isolation; Russell thought that only a vast conjunction of propositions can be appraised.[1] In my view it is better to compare problemshifts, and in particular, research programmes for merit.[2] Secondly, demarcationists may agree on the unit of appraisal but still differ over the criterion of appraisal. Some hold that a proposition is acceptable only if true and true only if provable from facts; others hold that a proposition is acceptable only if it has a higher probability than its rivals given the total evidence; yet others hold that a proposition is acceptable only if it has a larger set of potential falsifiers, but a smaller set of actual falsifiers than its rival. Zahar agrees with my choice of research programmes as units of appraisal but he improved on my way of appraising them.[3]

But whatever their differences, all demarcationists agree on some important points. They hold that the question of whether a theory is pseudoscientific or not is a question about the 'third world'. Hence, for demarcationists, a theory may be pseudoscientific even though it is eminently 'plausible' and everybody believes in it and it may be scientifically valuable even if it is unbelievable and nobody believes in it. A theory may even be of supreme scientific value even if no one *understands* it, let alone *believes* it. Thus the *cognitive* value of a theory has nothing to do with its *psychological* influence on people's minds. It matters not whether the theory lures them into intensive belief and vehement commitment, nor whether it induces the euphoric (second-world) mental state called 'understanding'. Belief, commitment, understanding are states of the human mind. They are inhabitants of the 'second world'. But the objective, scientific value of a theory is a 'third world' matter. It is independent of the human mind which creates it or understands it.

Thus demarcationists share *a critical respect for the articulated.* They

prefer the coherence theory.) But most of them carefully distinguish truth and its fallible signs: a proposition may correspond to a fact but there is no infallible way to establish this correspondence. (Cf. Popper [1934], section 84, and Carnap [1950], pp. 37–51.)

1 Cf. Russell [1910], pp. 92–3. He does not say *how*. Incidentally, according to Duhem, an isolated hypothesis may be refuted, but in order to do so, the scientist needs more than deductive logic: he needs *common sense*. (Duhem [1906], chapter VI, section 10.) Popper misread Duhem and his own solution is less clear than Duhem's. For my solution cf. volume 1, chapter 1, pp. 96–101.

2 Cf. Lakatos [1968*c*] and volume 1, chapter 1.

3 Cf. Zahar [1973], especially pp. 99–104.

appraise only what is *articulated* in human knowledge. The demarcationist readily agrees that articulated knowledge is only the tip of an iceberg: but it is exactly this small tip of the human enterprise wherein rationality resides. Demarcationists also share a second important characteristic: *a democratic respect for the layman.*[1] The demarcationist lays down *statute law* for rational appraisal which can direct a lay jury in passing judgment. (One does not, for instance, need to be a scientist to understand the conditions under which one theory is more falsifiable than another.) Of course, no statute law is either unequivocally interpretable or incorrigible. Both a particular ruling and the law itself can be contested. But a statute book – written by the 'demarcationist' philosopher of science is there to guide the outsider's judgment.

In the demarcationist tradition, philosophy of science is a watchdog of scientific standards. Demarcationists reconstruct universal criteria which great scientists have applied sub- or semi-consciously in appraising particular theories or research programmes. But medieval 'science', contemporary elementary particle physics, environmentalist theories of intelligence might turn out not to meet these criteria. In such cases philosophy of science attempts to overrule the apologetic efforts of degenerating programmes.[2]

What advice do demarcationists give to the scientists? Inductivists forbid them to speculate; probabilists to utter a hypothesis without specifying the probability lent to them by the available evidence; for falsificationists *scientific* honesty forbids one *either* to speculate without specifying potentially refuting evidence *or* to neglect the results of severe tests. My methodology of scientific research programmes does not have any such stern code: *it allows people to do their own thing but only as long as they publicly admit what the score is between them and their rivals.* There is freedom ('anarchy' if Feyerabend prefers the word) in creation and over which programme to work on but the products have to be judged. *Appraisal* does not imply *advice.*[3]

Demarcationist historiography recognizes that all histories of science are inevitably methodology-laden and that one cannot avoid 'rational reconstructions'. Each different type of demarcationism leads to a different 'internal reconstruction', with correspondingly different anomalies and different 'external' problems. These 'rational reconstructions', however, can be compared according to well-defined standards and the history of demarcationism – classical inductivism, probabilism, conventionalism, falsificationism, methodology of scien-

[1] Educated laymen, not uneducated sociologists of science.

[2] For such militant demarcationism cf. Newton's well-known fight against hypotheses not deduced from phenomena; Popper on psychoanalysis in his [1963*a*], pp. 37–8; or Urbach on environmentalist theories of intelligence in his [1974]. For an attempt to show up a non-empirical branch of knowledge as degenerating scholasticism cf. my treatment of inductive logic in this volume, chapter 8.

[3] Cf. volume 1, chapter 2, especially p. 117; also cf. Quine [1972].

tific research programmes – itself constitutes a progressive research programme.[1]

(*c*) *Elitism*

Among scientists the most influential tradition in the approach to scientific theories is élitism. Unlike the sceptics – but like the demarcationists – élitists claim that good science *can* be distinguished from bad or pseudoscience, better science from worse science. Elitists acknowledge the vast superiority of Newton's, Maxwell's, Einstein's, Dirac's achievements over astrology, Velikovsky's theories and other kinds of pseudoscience, and they claim to recognize scientific progress. They claim however, that there is, and there can be, no statute law to serve as an explicit, universal criterion (or finite set of norms) for progress or degeneration. In their view, science can only be judged by case law, and the only judges are the scientists themselves. If these authoritarians are right, academic autonomy is sacrosanct and the layman, the outsider, must not dare to judge the scientific élite. If they are right, the demarcationist research programme should be abolished as *hubris*. Recently Polanyi advocated such views (and so did Kuhn).[2] Oakeshott's conservative conception of politics also falls in this third category. According to Oakeshott one can *do* politics, but there is no point in philosophizing about it.[3] According to Polanyi one can *do* science but there is no point in philosophizing about it. Only a privileged élite has the craft of science, just as – according to Oakeshott – only a privileged élite has the craft of politics. This élitist tradition also goes back to antiquity (to some Greek and also some Eastern esotericist philosophies).

While I have characterized élitism as based on the *negative* claim that there is no universal criterion of scientific progress, most élitists present a positive thesis to explain why this is so. The positive thesis is that a large part of scientific knowledge is inarticulable, that it belongs to the 'tacit dimension'. Methodological knowledge also involves a 'tacit dimension'. They give this reason why the layman cannot be a judge in appraising scientific theories: the tacit dimension is shared and understood (*verstanden*) only by the élite.[4] Only they can judge their own work.

[1] This is a difficult issue. For detailed discussions cf. this volume, chapter 8, and volume 1, chapter 2.

[2] Polanyi's original problem was to provide arguments for protecting academic freedom from the communists of the 1930s, 1940s and 1950s; cf. Polanyi [1964], pp. 7–9. Kuhn's problem was very different: cf. Kuhn [1962].

[3] For a critical discussion of Oakeshott's philosophy, cf. Watkins [1952].

[4] Elitism is closely related to the doctrine of *Verstehen*. For this cf. e.g. Jane Martin's [1969]. *Verstehen*, of course, has nothing to do with 'positivistic' criteria for a satisfactory explanation, like the one I offered in volume 1, chapter 1, p. 34, n. 4. ('*Positivismus*' in the German philosophical literature seems to be the swearword for what I call 'demarcationism'.)

Elitism (like scepticism) thrives on the defeats of earlier versions of the demarcationist programme. The downfall of classical inductivism, the apparently incurable poverty of neoclassical inductive logic, the recent degeneration of falsificationism, and, finally, the need for external explanations to resolve some historiographical anomalies in the methodology of scientific research programmes, have all helped the propaganda for the élitist claim that no universal criterion of scientific progress is possible.[1] Elitists generally ascribe the failures and anomalies of demarcationism to the disregard of the tacit dimension. [But élitists should remember that demarcationists may lose a few battles and still win the war.

As I have said, élitism is very powerful and is the dominant tradition among scientists themselves. It is therefore worth analysing at greater length.]

2 ELITISM AND ALLIED PHILOSOPHICAL POSITIONS

Elitists can certainly support their position with some arguments. No doubt, for instance, some demarcationists have tended to overestimate the power of logic,[2] and some did not pay enough attention to questions of actual scientific practice.[3] Moreover, to a very limited extent élitists have a genuine case.[4] Nevertheless, élitism is very closely connected with four abhorrent philosophical doctrines: psychologism, the ideal of an authoritarian closed society (equipped with mental asylums for deviants), historicism and pragmatism.

(a) *Elitists for psychologism and/or sociologism*

The main effect of the élitist solution of the demarcation problem is to shift from the appraisal of third-world products, like propositions, problemshifts, research programmes and their third-world relations, like valid inferences (in Bolzano's and Tarski's sense) to the appraisal of second-world objects like psychological beliefs, mental states, anxieties to solve problems and socio-psychological crises within the scientist's mind or within the scientific community.

According to the demarcationist one theory is better than another if it satisfies certain objective criteria. According to the élitist one theory

[1] Cf. volume 1, chapters 2 and 3, and this volume, chapter 8. The best way to look at demarcationism is as a progressive research programme.

[2] But many more élitists *under*estimate the power of deductive logic.

[3] Cf. my criticism of Carnap, Popper and even of Tarski in this respect; especially my [1963–4], pp. 2–6, this volume, chapter 8, and volume 1, chapter 3, sections 1*c* and 1*d*.

[4] Cf. my plea for a demarcationism which pays serious attention to those scientists' judgments which go against universal demarcationist judgments. But such inconsistencies must in my opinion be resolved *within* the demarcationist programme (volume 1, chapter 3, pp. 151–4). This process is exemplified by Zahar's improvement (in his [1973]) of the methodology of scientific research programmes.

is better than another if the scientific élite prefers it. But then it is vital to know *who* belongs to the scientific élite. While élitists claim that no universal criteria for appraising scientific *achievements* are possible they may admit the possibility of universal criteria for deciding whether *persons or communities* belong to the élite.[1]

Any attempt to appraise persons or communities by their achievements would land the élitist in a vicious circle. So while demarcationists offer rules for assessing the 'third-world' *products* of scientific activity, élitists offer rules to assess the *producers* (primarily their 'second-world' mental states). As a consequence, *while for the demarcationists philosophy of science is the watchdog of scientific standards, for élitists this role is to be performed by the psychology, social psychology or sociology of science.* (Demarcationists deny the autonomy of sociology of science: all accounts of science are rational reconstructions of science.)

For the élitist the attempt to devise a system of quality control over factual and theoretical propositions is hopeless, therefore he must instead devise a system of quality control over élites. If a scientist proposes some theory T then, in order to appraise the epistemological merit of T, the élitist has to decide whether the producer of T, say P, is a genuine scientist: he can only appraise the producer, *not* the product. His approval or acceptance of T follows from his approval of P. If he is faced with two rival theories T_1 and T_2 he investigates the rival producers P_1 and P_2 and concludes from 'P_1 is better than P_2' that 'T_1 is better than T_2'. This is psychologism. If the criteria are to apply to communities rather than individuals, we get sociologism.

Different criteria for scientific minds and scientific communities have been proposed. The first two modern élitists were Bacon and Descartes. Bacon thought that the scientific mind was one purged of 'prejudices'; such a mind became a *tabula rasa* on which Nature would imprint the truth about itself. Descartes thought that the scientific mind was one which had been through the torments of sceptical doubt; such a mind would be rewarded by finding God's hand which would guide him to recognition of the truth.

Other élitists appraise communities rather than individuals. For some pseudo-Marxists (to my knowledge Marx himself never advocated such views) the quality of science depends upon the structure of the society which produced it. Feudal science is better than ancient slave science, bourgeois science is better than feudal science, and proletarian science is true.

Some forms of psychologism and sociologism are more objectionable than others. According to some versions anyone may become a member of the clairvoyant scientific and self-educating community

[1] According to Polanyi one cannot appraise scientific achievements at all without faith in the personal integrity of scientists: 'To speak of science and its continued progress is to profess faith in its fundamental principles and in the integrity of scientists in applying and amending these principles' (Polanyi [1964], p. 16).

provided he has been through certain educational (brainwashing?) therapy. But according to others, the club of genuine scientists is an exclusive one from which one may be permanently barred through reasons of social or racial origin. Again, according to some (e.g. Polanyi, Kuhn and Toulmin) the scientific community is a closed society, according to others (e.g. Popper and Merton) the scientific community is an open one. Merton – who is not an élitist – argued that the ideal scientific community is an open society, whose norms are 'universalism', 'communism', 'disinterestedness' and 'organized scepticism'.[1]

But whatever specific form they take, psychologism and sociologism both seem to me to be open to the following fundamental objection. Everyone, whether élitist or not, is bound to use normative third-world criteria, whether explicit or hidden, in establishing criteria for a scientific community. Merton, for example, no doubt decided what theories to select as scientific before he characterized the institutionalizations of science. He must have already decided that Darwinian biology was scientific, while Catholic theology was not, *before* he specified his four norms. [Similar considerations] apply to Polanyi and Kuhn. But why do Merton, Polanyi, Kuhn and Toulmin all exclude Catholic theology and astrology from science? They certainly do, although Merton does it *before* he constructs his criterion, while Polanyi, Kuhn and Toulmin need *post hoc* adjustments. (Catholic theologians and astrologers clearly do not form an open society.)

But if one must have *some* idea of what constitutes science before one knows which communities ought to count as scientific, then one must first decide what constitutes scientific progress. From the solution of this normative problem one can *then* proceed to the empirical problem of what socio-psychological conditions are necessary (or most favourable) for producing scientific progress. This is precisely how demarcationists approach the sociology of science. They regard the problem of quality control of *products* as primary, the problem of quality control of *producers* as secondary. Since different answers have been proposed to the quality control of products, different problems will face the sociologist according to which answer he presupposes. Moreover, once the normative problem of quality control of products is solved by a definition of scientific progress, the problem of quality control of the producers becomes an empirical one. For a demarcationist, philosophy of science is normative, while sociology of science, although 'norm-impregnated', is empirical.

The derivative nature of the sociology of science can be exemplified by a criticism of Merton's theory of priority disputes. Merton's view of what constitutes scientific progress (seemingly a form of inductivism) led him to regard priority disputes within the scientific community as a 'dysfunction'. But if the view of scientific progress

[1] Merton [1949].

formulated in my methodology of scientific research programmes is accepted, it is vital to know which of two rival programmes *anticipated* facts and which only dealt with them *post hoc*. Hence some priority disputes, far from being anomalous, may be essential and perfectly 'functional'.[1]

Elitists' adoption of psychologism and sociologism has a further unfortunate consequence. The élitist does not have to claim that any change in a community's beliefs constitutes *progress*. Change and progress are only identical within genuinely scientific communities. So for instance, the élitist might explain Lysenko's temporary victory over the Mendelians in the Soviet Union by the destruction of the norms of the scientific community by Stalin.[2] But whether he accepts Merton's or Polanyi's or Kuhn's or Toulmin's criteria for genuine scientific communities, the élitist must claim, once these social norms are fully met, that all change within a scientific community is progress.[3]

But surely degeneration is at least possible even within a 'scientific' community. After all, a victory of Lysenko's research programme might have been brought about in the West too if all Lysenko's opponents had died of natural causes within a couple of months, instead of being sent to concentration camps and killed. Would Lysenko's theory then have been vindicated through having been produced by a Mertonian or Polanyiite scientific community? *Obviously* not. Might is *not* bound to be right even within a perfectly 'rational' scientific community. One may *obviously* even have perfect consensus and degeneration at the same time.[4] But this means that we need (*and use*) criteria to judge scientific achievements rather than communities.

Thus one cannot replace philosophy of science by sociology of science as the supreme watchdog.[5] If both history and sociology of

[1] For a demonstration of how sociologists and historians may be misled by presupposing naive or confused answers to the problem of theory-appraisal, cf. volume 1, chapter 2. For a brilliant case study cf. Worrall's [1976]. Unfortunately even Merton does not seem to recognize that the definition of progress must precede the determination of optimal social norms for bringing it about.

[2] Cf. also Toulmin [1972], p. 259.

[3] This of course is 'historicism'. Cf. below, p. 116.

[4] This is because it is impossible to define scientificness of a community and scientificness of a theory independently; and because, as a matter of fact, scientific research programmes can progress both in a non-Mertonian community (where e.g. substantial energies are devoted to priority disputes or even where 'organized scepticism' temporarily vanishes) and, say, in a non-Kuhnian community (where e.g. there is a balance of power between two 'paradigms').

[5] Sociologists of science willy-nilly use hidden third-world criteria of appraisal. They may naively think that 'Experiment E refuted theory T' or 'Theory T_1 is more probable (or "simpler") than T_2' are empirical statements. But they are norm-laden. And while there *may* be *consensus* about 'T_1 is preferable to T_2', there has never been a consensus, as they themselves point out, concerning a universal criterion of preference. But eliminating expressions like 'proof', 'refutation', 'higher probability' makes bloodless, silly, false history. For instance, historians wanting to avoid 'rational reconstructions', would have to record that 'most experts came to agree by 1830 that Fresnel's theory of light is better than Newton's' instead of saying that 'Fresnel's theory

science are norm-impregnated, rational appraisal of scientific progress must *precede*, not *follow*, full scale empirical history: 'Internal (normative) history is primary and external ("descriptive–empirical") history is secondary'.[1] One cannot write history without *some* rational reconstruction.

In 1970 I put to Kuhn the following point:

> Let us imagine for instance that in spite of the objectively progressing astronomical research programmes, the astronomers are suddenly all gripped by a feeling of Kuhnian 'crisis'; and then they all are converted, by an irresistible *Gestalt*-switch, to astrology. I would regard this catastrophe as a horrifying *problem*, to be accounted for by some empirical externalist explanation. But not a Kuhnian. All he sees is a 'crisis' followed by a mass conversion effect in the scientific community: an ordinary revolution. Nothing is left as problematic and unexplained.[2]

Kuhn replied to the paper,[3] but left this criticism unanswered.

(*b*) *Elitists for authoritarianism and historicism*

According to élitists, only insiders are qualified to judge the products of the scientific community. *But what if the insiders disagree?* We should then get no unequivocal answer to the demarcation problem. But as I pointed out, élitists are convinced that while some theories make good science, others constitute bad or pseudo-science.

Elitists used to obviate this difficulty by simply asserting that such disagreements do not really occur. These élitists claim that scientific communities arrive quickly and easily at consensus concerning scientific knowledge. This, even if accurate, implies that scientists form a *totalitarian society without alternatives*.[4] The most distinguished proponent of this consensus view was Kuhn. His *Paradigm-Monopoly Thesis*[5] implies that the consensus may change during dramatic revolutions, but consensus about the revolutionary change is quickly reached. The wind of change seldom blows but when it does no one can resist it. This is Kuhn's *No-Interregnum* thesis.[6]

If the élitist accepts the opposite empirical view that consensus does not always (or even ever) reign in the scientific community, he is faced with two options. He can claim that there is an authority structure within the scientific élite. Supreme judges are elected (or 'emerge'); they sit *in camera*, and pass judgments according to *case law*. No

of light superseded Newton's by 1830'. Should they fall back on recording beliefs? If they choose to do so they fall back on scepticism. Elitism is untenable: unless one joins the demarcationists, one has to resign oneself to scepticism. There is no consistent élitist historiography of science.

[1] Cf. volume 1, chapter 2.

[2] *Ibid.*, p. 120. Note my redefinitions of the 'internal/external' distinction and of 'rational reconstruction' (*ibid.*, pp. 91–2).

[3] Kuhn [1971].

[4] For an exposition and criticism of these theses cf. Watkins [1970], pp. 34ff.

[5] Watkins, *op. cit.*

[6] *Ibid.*

statute law (or 'demarcation criterion') mitigates their power. But what if the Supreme Court disagrees among itself? Since this problem seems insoluble, most élitists prefer the second solution: any conflict within the élite will be resolved by the survival of the fittest. While the disagreement is unresolved the outsider has to watch, overawed, the struggle of the giants, and accept the victor as the representative of progress. But then, within the élite, might must be right. This is how the Darwinian struggle of ideas and the Hegelian Cunning of Reason are closely linked with élitism. If there are conflicts, the benevolent Cunning of Reason has to appear as *deus ex machina* to provide – even if only in the very long run, or at the End of the Run – the just solution of conflicts in the Holy General Assembly of Scientists. Thus the élitist has to choose between the authority of Infallible Archbishops and that of the Cunning of Reason. Authoritarianism cannot be avoided.[1]

Feyerabend, as an epistemological anarchist, denies that there is 'Right' (i.e. he denies the necessity of appraisal) and so does not need to invoke authority. But the élitist believes in the rationality of science though not in the possibility of a universal appraisal of science: hence he has to invoke either the authority of the Scientific Consensus or, in the case of conflict, that of Great Scientific Archbishops or that of Divine Benevolence. It is only this thin, *ad hoc* authoritarian/historicist doctrine which separates élitists of this kind from the sceptic.

(*c*) *Elitists for pragmatism*

It is possible to hold that scientific theories can only be appraised by the scientific élite; but whether these theories are true or false, or closer to the truth than other theories, are objective matters. Thus one can be an élitist while holding that the products of scientific research exist in Frege's and Popper's 'third world' of ideas. Thus, for instance, one can be an élitist and hold the correspondence theory of truth. In this case the élitist holds that what is true can be universally characterized, but [the signs of truth cannot be].

There is, however, a very influential school of thought which is based on the *denial of the existence of the third world: pragmatism.* Pragmatists do not deny that knowledge exists, but knowledge for them is a state of mind, or even a 'slice of life'.[2] Knowledge thus becomes manifest in (or consists of) behaviour patterns. It is even a way of life which is inexpressible in propositions. How can one judge such 'knowledge'? How is one 'theory' better than another?

Two 'theories' are different if they are 'practically' different. They have different meanings if they have different uses. One 'theory' is

[1] The élitist code of honesty is: *Do your Master's thing.* This is in striking contrast with Feyerabend's code of honesty (and with mine).

[2] Toulmin [1961], p. 99.

better than another for a person or for a community *P* at time *t* if it is more 'pleasing', more 'satisfactory' for *P* at *t*. A 'problem' disturbs the mind. A 'solution' satisfies it. The 'problem' is dissolved rather than solved. Or: 'Pragmatism solves no real problem, it only shows that supposed problems are not real problems' (Peirce). A 'theory' is better than another if it 'works' better, if it has more 'cash value' (James), if it is more efficient in making us successful in life. Meaning and truth of 'propositions' are dependent on their users. What is true for *A* may be false for *B*. What is true today, may be false tomorrow. What is true for Israel, may be false for Egypt. No surprise then that, as F. C. S. Schiller put it, there are as many pragmatisms as pragmatists.

This extreme subjectivism follows simply from the pragmatist denial of the 'third world' together with the empirical fact that different persons and communities have conflicting interests and feelings. Many pragmatists cannot accept this extreme subjectivism and restore the notion of objective truth by invoking, in a completely *ad hoc* way, either Darwinism or historicism. If we *all* agree and *keep on* agreeing, we shall have arrived at the *absolute truth*. Truth is either that which survives in a Darwinian struggle (i.e. the true theory is that into which everyone will [eventually] be terrorized and brainwashed) or that to which everyone is destined to agree. Peirce, adopting the historicist approach, writes 'The opinion which is fated to be ultimately agreed to by all who investigate, is what we mean by the truth.'

Logical empiricists, criticizing pragmatism, pointed out that it conflates truth with its signs, and that it conflates the psychology of discovery, the use and consequences of a discovery, with the appraisal of discovery. Russell devoted considerable time and energy to fighting pragmatism. He abhorred the idea that

> in order to judge whether a belief is true, it is only necessary to discover whether it tends to the satisfaction of desire. The nature of the desire to be satisfied is only relevant in so far as it may conflict with other desires. Thus psychology is paramount not only over logic and the theory of knowledge, but also over ethics. In order to discover what is good, we have only to inquire how people are to get what they want; and true beliefs are those which help in this process.[1]

Russell abhorred the idea that instead of truth being reflected in belief, beliefs should make facts.[2] He abhorred the idea that 'between different claimants for truth, we must provide a struggle for existence, leading to the survival of the strongest.'[3] Russell pointed out that pragmatism is inherently connected with the appeal to force.[4] Unlike Schiller, Russell did not think that truth can be elucidated by opinion polls,[5] or by 'ironclads and maxim guns'.[6] Russell also linked pragmatism with the naive political tenet that democracy can achieve full

[1] Russell [1910], p. 92.
[2] *Ibid.*, p. 102.
[3] *Ibid.*, p. 106.
[4] *Ibid.*, p. 109.
[5] *Ibid.*, p. 107.
[6] *Ibid.*, p. 109.

consensus and that it is all-powerful. Russell points out, that there are non-human limitations to human power. He also points out the 'curious contrast' between the democratic appeal and dictatorial tone of pragmatists. In his beautiful essay *The Ancestry of Fascism*, he claims that pragmatism is the main intellectual source of fascism:

> Hitler accepts or rejects doctrines on political grounds... Poor William James, who invented this point of view, would be horrified at the use which is made of it; but when once the conception of objective truth is abandoned, it is clear that the question 'what shall I believe' is one to be settled...by the appeal of force and the arbitration of big battalions.[1]

Thus, while for élitism, as I defined it, human knowledge cannot be judged independently of its producers, for pragmatism there *is* no product independent of the producers. Truth cannot be predicated of propositions, which anyway do not exist, but only of human beliefs, activities (like 'speech acts'), forms of life, 'paradigms'.[2] The truth relative to P (whether P is an individual or a community) is an activity which relieves P from a (Peircian) anxiety, or from a (Kuhnian) crisis. If scientists (or scientific communities) become ever happier, there is scientific progress. (If science makes mankind ever happier, scientific progress contributes to human progress.) If there are conflicts, they are sorted out by force. All this follows simply from scepticism applied to the third world of ideas. But pragmatism adds to scepticism the idea that in the struggle of beliefs, activities, forms of life the one which *establishes* consensus (or common happiness) by eradicating its rivals is the most progressive. And if it does it with irreversible success it is absolutely true. Pragmatism seems to be separated from scepticism only by this stress on 'absolute truth' and on 'progress' towards it. But this emphasis is nothing but rhetoric.

Although, as I remarked, élitism and pragmatism are logically independent, they are nevertheless natural allies. Indeed if one claims that the tacit dimension is involved in appraising scientific knowledge, it is only a short step to claiming that scientific knowledge is itself not expressible in propositions, and it is 'tacit', and so cannot be regarded as a third-world object. Most élitists do in fact slip down this short step, and, at least on occasions (and perhaps unwittingly) adopt pragmatism.

[1] Russell [1935]. Despite the strength of his arguments and his devotion to 'impersonal reason' (*ibid.*), Russell's only success seems to have been to shake up Dewey. Russell's attack on pragmatism does have certain philosophical weaknesses. These stem from his failure to realize that the pragmatist position hinges on its denial of the existence of the third world and his attendant failure to separate *sufficiently* between (third-world) 'rational belief' and (second-world) actual belief. (It is, however, unfair of Popper to classify Russell as a 'belief philosopher' pure and simple. Cf. Popper [1972], p. 107.)

[2] 'Speech acts', 'forms of life' are Wittgensteinian technical terms. 'Paradigm' is Kuhn's technical term; cf., for instance, definitions 10, 12, 14, 16 in Margaret Masterman's ordering (Masterman [1970], pp. 63-4).

Conversely, by adding to pragmatism two assumptions one arrives at élitism. The two assumptions are the Cunning of Reason and that there is an élite who can sniff out which, possibly tortuous, way leads towards the Ultimate End. Then members of the élite may act as midwives in speeding the delivery.

To conclude: élitism, whether pragmatist or not, has no more problem-solving power than scepticism. For instance, Feyerabend would explain the Velikovsky affair in terms of the superior propaganda of the scientific establishment. The élitist will agree and add his axiomatic approval. So what?

7

Necessity, Kneale and Popper*

The problem [of natural necessity] before us arises on at least two levels: on an ontological and on an epistemological–methodological level. The fact that no proper attention has been given to this distinction is responsible for at least some of the confusion surrounding the problem.

Professor Kneale's paper is a new contribution to a discussion about natural necessity which has been going on in the last ten years between him, Professor Popper, and other philosophers.[1] I shall try to give a critical summary of their discussion.

1 THE ONTOLOGICAL LEVEL

According to Professor Kneale, God may have faced the choice between creating a physical world and not creating a physical world, but once this choice was made, He was no longer free to choose the form, or the structure of the world – just as a poet who has chosen to write a sonnet cannot but write it in sonnet form. So God was not free to determine the laws of nature, just as the sonnet writer is not free to determine the laws of the sonnet. God of course had quite a lot of freedom to fill in the contents of the world within this necessary framework; what He could choose freely – the contents of the sonnet – were later called the initial conditions.[2]

According to Professor Popper, God was completely free to choose any law of nature which occurred to Him at that moment. He dictated – at His pleasure – the Book of Nature, containing the Natural Laws, but left it to his angels to play around with initial conditions, insofar as they were not prohibited by some Natural Law. Now these playful angels may have arranged the set of initial conditions in such a shrewd way that unintended true universal statements emerged. So in Popperian ontology physically necessary statements reflect God's will while accidental universal statements reflect the angels' whim.[3] In the

* This paper was given as a reply to an address by William Kneale at the annual conference of the British Society for the Philosophy of Science in 1960. Professor Kneale's paper was published as 'Universality and Necessity' (Kneale [1961]). Lakatos did not, to our knowledge, intend to publish his paper. (*Eds.*)

[1] Kneale [1949]; Popper [1949]; Kneale [1950]; Popper [1959], Appendix *x.

[2] This view goes back to Descartes.

[3] This view goes back to Leibniz.

Knealian ontology only angels have free will. God acts necessarily. Kneale calls regularities produced by the angels' activity 'historical accidents on a cosmic scale'.[1]

Some of you may be put off by this theological formulation of the controversy. So let me express the Popperian view without involving God and his angels (though the resulting definition is rather poor): 'A statement may be said to be naturally or physically necessary if, and only if, it is deducible from a statement function which is satisfied in all worlds that differ from our world, if at all, only with respect to initial conditions.'[2]

This definition is actually an awkward and only partial expression of the Popperian view as stated above in theological terms. First I show that it is awkward. Imagine a true universal statement like 'All dodos die before sixty'. Now this is a natural law if it is satisfied in all the worlds that differ from our world, if at all, only with respect to initial conditions; if it is satisfied in all the worlds that have the same natural laws as ours. So 'All dodos die before sixty' is a natural law if it is a natural law. This circularity cannot be avoided even if following Professor Popper we change the Knealean dodos for New Zealand moas.[3] But the circularity can be avoided by saying that what God has written into the Book of Nature are Natural Laws, what the angels have scribbled are initial conditions.[4] But even if we don't care about the circularity of the definition we still are left with a puzzle. We may ask ourselves – are there any natural laws or [true] universal statements? A definition of 'law of nature' does not imply any ontological commitment and does not imply that there *are* laws of nature. What if the world was created by the whimsical angels only? Does the Book of Nature consist [only] of angels' scribblings? I hope Professor Popper would agree that the definition should be supplemented by an existential clause stating that in the Creation God uttered at least one sentence. But even in this case it is obvious that in the Popperian ontology the laws of nature, the physically necessary statements, contain an important element of contingency, and I am puzzled how Professor Kneale, who denies God freedom of creation, could accept it.

I happen to agree largely with Professor Popper against Professor Kneale, though with an important modification. God's utterances and the Book of Nature should not be imagined in an anthropomorphic way. I think that the Natural laws uttered by God were of an infinite length. I have good reasons to believe this. Take a statement like 'for all gases $PV = RT$'. Taken strictly, it is false. It could be true only for

[1] Kneale [1950], p. 123.
[2] Popper [1959], p. 433.
[3] Cf. Kneale [1949], p. 75, Popper [1959], p. 427. Popper himself points out the circularity of his definition, p. 435, but this leaves him cold, as in *his* theory of definition this does not give any reason for apprehension.
[4] This shows in an undisguised way the conventional character of the borderline between 'natural laws' and 'initial conditions'.

ideal gases, that is, for gases which consist of completely elastic billiard balls. But near the absolute zero even this breaks down; we can rescue the thesis only by adding more and more qualifications to the formulation. As the universe is infinitely varied, it is very likely that only statements of infinite length can be true.

But my ontological requirement that God's sentences were infinite does not yet guarantee that there are not some accidentally true universal statements of finite length in consequence of the skilful arrangement of initial conditions by our playful angels. But it seems to me that the ontological structure of the universe is such that all universal statements of finite length *are* false. This would certainly be so if all physically possible initial conditions were realized sometime and somewhere, as then only physically necessary statements would be true and these we claimed to be true only if they are of infinite length. But actually we do not need a universe so rich in order to render all universal statements false. What we need is that to any universal statement of a finite length the universe should contain at least one counterexample. (It doesn't need to contain *all* physically possible counterexamples – like the Aristotelean and Popperian universe.[1]) And I believe the universe has this minimal structure by divine order imposed on the angels' activity in arranging the initial conditions.

I think that what I have said so far already shows that there are some basic agreements among the parties to this discussion. *The first common platform* is an interest in metaphysical problems. This discussion is quite meaningless according to the criteria of logical positivism of any vintage and according to Popper's falsifiability criterion [interpreted, *contra* Popper, as a meaning criterion]. So positivists should commit this discussion to the flames as meaningless gibberish. The passion with which both Kneale and Popper have been carrying on this discussion now for about ten years shows their basic agreement about the value of metaphysical speculation.

(Perhaps we should note here that a very respectable English positivist gave the dodo-problem an ironical reformulation which enabled him to regard it as meaningful and to try to solve it. He admitted that there is a problem of linguistic usage: people sometimes distinguish in common language between 'laws' and 'mere generalizations', and offered a witty–sarcastic rationalization of this usage within the Humean ambit. He simply gave those *contingent* universal statements the honorific title of 'law of nature' which occur in a well-embedded position in a respectable, not yet falsified scientific deductive system.[2])

The second common platform among the parties to the discussion is that they are not just metaphysicians but metaphysicians of a special brand, namely, metaphysical realists, or, in Marxist terminology,

[1] Cf. Hintikka [1957]; Popper [1959], p. 436.

[2] Braithwaite [1953], p. 300ff. The idea goes back to Campbell [1920], p. 153.

materialists. They believe that there is a real world, independent of our mind and governed by some sort of natural laws.

There is a *third common platform* as well which will lead us straight into epistemology: all of the parties to the discussion are epistemological optimists; they believe that we can somehow explore the laws of nature and form either an exact or at least an approximate idea of them.

Perhaps I should mention here that metaphysical realism coupled with epistemological optimism amounts to quite a *Weltanschauung*. Its adherents – while aware of their weakness confronted by the vast universe – find the search for truth, for the Laws of Nature a noble challenge and regard the growth of knowledge as the greatest asset of human dignity. For metaphysical idealists the search for truth is a self-scrutiny of their inflated ego. For positivists 'truth', 'natural law' and related concepts are 'sophistry and illusion'.[1] I think that it is only because they are metaphysical realists and epistemological optimists, equally opposed to any sort of metaphysical idealism or to positivism, that Kneale and Popper can discuss the problem of Natural Necessity at all.

But Popper's brand of epistemological optimism is a very far cry from that of Kneale's – and here we shall find the very serious source of their ontological disagreements.

2 THE EPISTEMOLOGICAL–METHODOLOGICAL LEVEL

The heart of the disagreement is indeed epistemological and methodological. Before getting down to its analysis let us repeat some home truths. Metaphysicians may kill each other, but they cannot kill each other's arguments. Epistemological ideas can survive blows which should be mortal, as for instance the Kantian philosophy of geometry survived Bólyai and Lobatschewsky, or as Machism survived Wilson and Millikan. It is only the scientist who according to the cruel scientific tradition must witness the execution of his theories and outlive them. (Though Stalinist Russia may have been an exception.)

At the same time there are strong connections between these three levels. The big metaphysical and the medium-size epistemological cogwheels may turn much slower than the smallish scientific ones, but still they are all organic parts of our huge system of knowledge.

In our problem we can easily show how Popper's metaphysical cogwheels are shrewdly constructed to impede the turning of the Knealean epistemological ones.

As I mentioned, both Kneale and Popper are epistemological optimists, but in very different ways. Popper is a strict fallibilist about scientific knowledge and a rigid infallibilist, and in particular a conventionalist, about mathematical and logical knowledge. Kneale seems

[1] Fortunately none of these types exist puregrained.

to believe that at least a basic part of scientific knowledge is synthetic *a priori* and he thinks we may have certain knowledge in this field; that is, his infallibilism is wider than Popper's; and *no* part of the infallible field does he want to explain by conventionalism. He seems to think that logic and mathematics are certain *and* refer to reality; and he thinks that there are some trivial principles of necessitation, like 'nothing can be both red and green all over', that have the same character. But he is a fallibilist about the actual axioms of *any* physical theory; like Descartes he does *not* claim that 'we can hope to derive laws of nature some day from self-evident truths alone'.[1] Popper's claim, that Kneale wants to 'reduce all the laws of nature to the "true principles of necessitation" – to truisms'[2] is possibly based on some misreading.

Now according to Popper's hypothesis God created the world as it is of His free will. For human beings to have a straightforward intuitive and *infallible* insight into the divine psychology of that great moment is a very unlikely idea, though one can well have guesses and then test these fallible guesses. The only way to make synthetic *a priori* knowledge of the world plausible is to remove any contingency from the Book of Creation. So the previous ontological discussion hides a very serious epistemological discussion where the opposing views already show their hand in methodology as well, as – for instance – no Knealean scientist will waste his time log testing necessary statements about the world which he *knows* are necessary.

Perhaps I should reveal the ulterior motives behind my modification of the Popperian ontology concerning the infinite length of the sentences in the Book of Nature. I didn't like Popper's recent underlining of the possibility that we may unknowingly hit upon the final truth. I was biased against this Xenophanic thesis because it contradicts some of my pet ideas learned from Marxism (and I don't see why I should give these up). Engels says that

> knowledge which has an unconditional claim to truth is realised in a number of relative errors; neither the [absolute truth of knowledge] nor the [sovereignty of thought] can be fully realised except through an endless eternity of human existence... Human thought... is sovereign and unlimited in its disposition, its vocation, its possibilities and its historical goal; it is not sovereign and it is limited in its individual expression and its realisation at *each* particular moment.[3]

So, as Engels explicitly states, final truth can be reached only 'from a practical standpoint, by the endless succession of human generations'. Or to quote Lenin: we may 'draw closer and closer to objective truth (without ever reaching it)'.[4] Now Popper says that here and there,

[1] Kneale [1949], p. 97.
[2] Popper [1959], p. 431.
[3] Engels [1894], pp. 122–3, my italics.
[4] Lenin [1908], p. 137. * The English translation reads 'draw closer and closer to objective truth (without ever exhausting it)' (*eds.*).

though unknowingly, we *may* reach it. I think this is a flaw in his fallibilism, and so I tried to correct it in a true Marxist spirit by my doctrine of infinite sentences in God's Blueprint of the Universe. According to this doctrine there can be no human statements expressing Natural Law. I think it is a bad anthropomorphic feature of both Kneale's and Popper's treatment that they want to find naturally necessary statements among the statements of human language. For me this has a sort of analogy with the Campbell–Braithwaite approach which picks out the necessary statements from among contingent statements.

After having tried to show the dirty epistemological motives behind the lofty ontological disagreement I shall now show that the problem of the difference between physically necessary universal statements and accidental universal statements, or, shortly, the dodo-problem, does not arise at all on epistemological or methodological levels in the Popperian set-up. In this set-up we can falsify the law-statement 'All dodos [necessarily] die before sixty' only by producing a counterexample which will falsify the corresponding weaker universal statement as well. Also it is as impossible to establish that the universal statement 'All dodos die before sixty' is true as it is to establish that it is necessarily true. This is why Popper originally did not feel the need to formulate the difference on the metaphysical level and this was the reason why Kneale labelled – and libelled – him a positivist.[1] Popper's response revealed clearly Kneale's rash anticipation. The Knealean epistemological set-up is different: those sentences which express natural necessitation are not only known to be true but are even known to be necessarily true. The same sentences in the Popperian set-up are not known to be true either necessarily or universally.

In my set-up there cannot be any naturally necessary statements of finite length; moreover all [finitely long] universal statements whether allegedly necessary or accidental are just false.

3 THE CONTINUITY OF LOGICAL AND NATURAL NECESSITY

Popper is an arch-enemy of conventionalism in the field of scientific knowledge, but he is a conventionalist in the field of logical and mathematical knowledge. According to him the source of logical and mathematical necessity is the structure of human language, while natural necessity is God's doing. So they have nothing in common and it is in fact misleading to use the word 'necessary' for both cases.

Kneale is an arch-enemy of conventionalism both in science and in mathematics and logic. He thinks that natural necessity and logical necessity are akin, the first consisting of specific principles of necessitation, the second, general principles of necessitation.

[1] Kneale [1949], p. 76.

So for Kneale it is a crucial problem to make this homogeneity of logical and natural necessity plausible. This is the gist of his present paper which suggests that the two kinds of necessities differ only in the number of logical constants. Unfortunately I do not think he succeeded in substantiating his case. I can see only two possibilities. Either we take the usual logical constants and thus get the division of statements into logically necessary and logically contingent statements, or we regard all terms of the language as logical. The concept of logical truth would then coincide with material truth. Now can we regard *some* terms as logical constants (let us call them quasi-logical) and *some others* as dummies? If this is what Professor Kneale suggests, then *all* statements will be quasi-logically false because we shall easily be able to construct models in which they will be false. So the dichotomy cannot be made to disappear by this method.

Anyway, one of the most essential features of the history of mathematical rigour is the gradual elimination of quasi-logical constants. Weierstrassian rigour still admits natural number as a quasi-logical constant while Russellian rigour eliminates even this last remnant and leaves us with logical constants only. The main point in this progress is that logical constants have to be perfectly-known terms and the quasi-logical terms were discredited one after the other as vague and were substituted by definitions in perfectly known terms. I fully admit that putting the clock back may sometimes be an excellent idea, and bringing back some of the discredited perfectly-known quasi-logical constants may be reasonable, but first let us face the difficulties which led to their elimination.

I am very interested in this problem and actually have been working on it, though it seems to me that the resulting concept of quasi-logical necessity will not coincide with natural necessity but will constitute a sort of non-logical mathematical necessity*

So as you can see, I side with Kneale in assuming some sort of continuity between the different concepts of necessity he advocates and I feel sorry that his present effort has, so far as I can see, failed. But I have very strong feelings against Popper's linguistic conventionalist theory of mathematics and logic. I think with Kneale that logical necessity is a sort of natural necessity; I think that the bulk of logic and mathematics is God's doing and not human convention. We have the huge set of logically possible worlds, we have the subset of mathematically possible worlds, we have the subset of this subset: the physically possible worlds; and then we have the actual world.

But in consequence I am a fallibilist not only in science, but in mathematics and logic as well.

* See Lakatos [1976*c*], chapter 1, section 9 (*eds.*).

8

Changes in the problem of inductive logic*

INTRODUCTION

A successful research programme bustles with activity. There are always dozens of puzzles to be solved and technical questions to be answered; even if *some* of these – inevitably – are the programme's own creation. But this self-propelling force of the programme may carry away the research workers and cause them to forget about the problem background. They tend not to ask any more to what degree they have solved the original problem, to what degree they gave up basic positions in order to cope with the internal technical difficulties. Although they may travel away from the original problem with enormous speed, they do not notice it. Problemshifts of this kind may invest research programmes with a remarkable tenacity in digesting and surviving almost any criticism.[1]

Now problemshifts are regular bedfellows of problem solving and especially of research programmes. One frequently solves very different problems from those which one has set out to solve. One may solve a more interesting problem than the original one. In such cases we may talk about a 'progressive problemshift.'[2] But one may solve some problems less interesting than the original one; indeed, in extreme cases, one may end up with solving (or trying to solve) no other problems but those which one has oneself created while trying to solve

* This paper was originally published in Lakatos (*ed.*) [1968*a*] – part of the Proceedings of the International Colloquium in the Philosophy of Science, London, 1965. Lakatos's paper grew out of a comment on Carnap's address 'Inductive Logic and Inductive Intuition'. Lakatos's acknowledgment reads: 'The author is indebted for criticisms of previous versions to Y. Bar-Hillel, P. Feyerabend, D. Gillies, J. Hintikka, C. Howson, R. Jeffrey, I. Levi, A. Musgrave, A. Shimony and J. W. N. Watkins, but most of all to Carnap and Popper who both spent days on criticizing previous versions and thereby contributed immensely to my understanding of the problem and its history. However I am afraid that Carnap – and possibly also Popper – may disagree with the position at which I have arrived. None of them have seen the latest version.' (*Eds.*)

[1] For a general discussion of research programmes, problem solving versus puzzle solving, problemshifts, cf. volume 1, chapter 1.

[2] A simple example of a 'progressive problemshift' is when we explain more than, or even something inconsistent with, what we set out to explain. This indeed is one of Popper's adequacy requirements for a good solution of an explanatory problem (Popper [1957]).

the original problem. In such cases we may talk about a '*degenerating problemshift*'.[1]

I think that it can do only good if one occasionally stops problem-solving, and tries to recapitulate the problem background and assess the problemshift.

In the case of Carnap's vast research programme one may wonder what led him to tone down his original bold idea of an *a priori*, analytic inductive logic to his present caution about the epistemological nature of his theory;[2] why and how he reduced the original problem of rational degree of belief in hypotheses (principally scientific theories) first to the problem of rational degree of belief in particular sentences,[3] and finally to the problem of the probabilistic consistency ('coherence') of systems of beliefs.

I shall start with a potted version of the problem background of inductive logic.

I THE TWO MAIN PROBLEMS OF CLASSICAL EMPIRICISM: INDUCTIVE JUSTIFICATION AND INDUCTIVE METHOD

Classical epistemology in general can be characterized by its two main problems: (1) the problem of the *foundations* of – epistemic, i.e. perfect, infallible – knowledge (the *logic of justification*); and (2) the problem of the growth of – perfect, well-founded – knowledge or the problem of heuristic, or of method (the *logic of discovery*).

The *empiricist brand of classical epistemology* in particular acknowledged only one *single* source of knowledge about the external world: the natural light of experience.[4] But this light can illuminate at best the meaning and truth-value of propositions expressing 'hard facts': of 'factual propositions'. Theoretical knowledge is left in darkness.

The logics of justification of all kinds of classical epistemology – whether empiricist or rationalist – maintained some strict, black-and-white appraisal of propositions. This amounted to a sharp demarcation between knowledge and non-knowledge. They equated knowledge – *epistēmē* – with the proven; unproven *doxa* was 'sophistry and

[1] The 'degenerating problemshift' can again be illustrated by the example of explanatory problems. An explanation constitutes a degenerating problemshift if it was arrived at by 'conventionalist' (i.e. content-reducing) stratagems. Cf. *below*, p. 172.

[2] Cf. *below*, p. 160, n. 2.

[3] By 'particular sentences' I mean truth-functional compounds of sentences of the form $r(a_1, a_2, \ldots, a_n)$ where r is an n-ary relation and a_i individual constants. Carnap calls such sentences 'molecular'. (Carnap [1950], p. 67.)

[4] *The rationalist brand of classical epistemology*, on the other hand, was less monolithic: Cartesians admitted the evidence of reason, sense experience, and faith, on a par. As for Bacon, he was a confused and inconsistent thinker, and a rationalist. The Bacon-Descartes controversy is a myth invented by the Newtonians. Most empiricists, however – surreptitiously or explicitly – admitted that at least logical knowledge (knowledge about the transmission of truth) was *a priori*.

illusion' or 'meaningless gibberish'. This is how theoretical, non-factual knowledge was bound to become the central problem for classical empiricism: it *had* to be justified – or else scrapped.[1]

In this respect the first generation of empiricists was divided. Newtonians, who soon became the most influential, believed that true theories *can* be proved ('deduced' or 'induced') infallibly from factual propositions but from nothing else. In the seventeenth and eighteenth centuries there was no clear distinction between 'induction' and 'deduction'. (Indeed, for Descartes – *inter alios* – 'induction' and 'deduction' were synonymous terms; he did not think much of the relevance of Aristotelian syllogistic, and preferred inferences which increase logical content. Informal 'Cartesian' valid inferences – both in mathematics and science – increase content and can be characterized only by an infinity of valid patterns.)[2]

This is then the logic of justification of classical empiricism: '*factual propositions' and their informal – deductive/inductive – consequences constitute knowledge: the rest is rubbish.*

Indeed, even meaningless rubbish, according to some empiricists. For, according to an influential trend in empiricism, not only truth but also meaning can be illuminated only by the light of experience. Therefore only 'observational' terms can have primordial meaning; the meaning of theoretical terms can only be derivative, defined (or at least 'partially defined') in terms of observables. But then if theoretical science is not to be branded as altogether meaningless, an inductive ladder not only of propositions but of concepts is needed. In order to establish the truth (or probability) of theories, one first has to establish their meaning. Thus the problem of *inductive definition*, 'constitution' or 'reduction' of theoretical to observational terms, came to be a crucial one for logical empiricism, and the successive failures of its solutions led to the so-called 'liberalization' of the verifiability criterion of meaning and to further failures.[3]

[1] Characteristic of classical epistemology is the sceptic–dogmatist controversy. The sceptical trend in classical empiricism invokes the restriction of the sources of knowledge to sense experience only to show that there is no authoritative source of knowledge whatsoever: even sense experience is deceptive, and therefore there is no such thing as knowledge. In the context of the present discussion I neglect the sceptical pole of the classical justificationist dialectic.

This analysis of the classical or justificationist theory of knowledge is one of the main pillars of Karl Popper's philosophy; cf. the Introduction of his [1963*a*]. For a further discussion cf. volume 1, chapter 1.

[2] Informal mathematical proof and inductive generalization are essentially analogous from this point of view. Cf. my [1976*c*], especially n. 2 on p. 81.

[3] Popper had criticized this trend as early as 1934, in his *Logic of Scientific Discovery*, end of § 25; later he gave an interesting critical exposition of the problem in his [1963*a*] esp. pp. 258–79. The criticism was either ignored or misrepresented; but, on the whole, the process of 'liberalization of logical empiricism' has been nothing but a piecemeal and incomplete, independent and not so independent, rediscovery of Popper's 1934 arguments. The historian of thought cannot help but see a general pattern: a school of thought is established; it receives crushing criticism from outside; this external

The *methodological implications of this logic of justification* were clear. Classical method in general demands that the path of knowledge should be cautious, slow progress from proven truth to proven truth, avoiding self-perpetuating error. For empiricism in particular this meant that one had to start from indubitable factual propositions from which, by gradual valid induction, one could arrive at theories of ever higher order. The growth of knowledge was an accumulation of eternal truths: of facts and 'inductive generalizations'. This theory of 'inductive ascent' was the methodological message of Bacon, Newton and – in a modified form – even of Whewell.

Critical practice demolished the classical idea of valid content-increasing inferences in both mathematics and science, and separated valid 'deduction' from invalid 'informal proof' and 'induction'. Only inferences which did not increase logical content came to be regarded as valid.[1] This was the end of the logic of justification of classical empiricism.[2] Its logic of discovery was first shaken by Kant and Whewell, then crushed by Duhem,[3] and finally replaced by a new theory of the growth of knowledge by Popper.

2 THE ONE MAIN PROBLEM OF NEOCLASSICAL EMPIRICISM: WEAK INDUCTIVE JUSTIFICATION (DEGREE OF CONFIRMATION)

Following the defeat of classical empiricism most empiricists refused to draw the sceptical conclusion that theoretical science – undefinable from observational terms, unprovable from observational statements – is nothing but sophistry and illusion. They thought that a good

criticism is ignored; internal difficulties set in; 'revisionists' and 'orthodox' fight about them, with the orthodox turning the original doctrine into a 'fairly dried-up petty-foggery' by criticism-reducing stratagems and with the revisionists slowly and incompletely discovering and digesting the critical arguments which have been there for decades. From the outside, where these critical arguments have become commonplaces, the 'heroic' struggle of revisionists – whether within marxism, freudianism, catholicism or logical empiricism – looks trivial and occasionally even comical. ('Fairly dried-up petty-foggery' was Einstein's description of later logical empiricism; cf. Schilpp [1959–60], p. 491.)

[1] A reconstruction of this historical process is one of the main topics of my [1963–4].

[2] Of course, classical rationalists may claim that inductive inferences are enthymematic deductive inferences with synthetic *a priori* 'inductive principles' as hidden premises. Also cf. *below*, p. 163.

[3] One of the most important arguments in the history of philosophy of science was Duhem's crushing argument against the inductive logic of discovery which showed that some of the deepest explanatory theories are *fact-correcting*, that they are *inconsistent* with the 'observational laws' on which, according to Newtonian inductive method, they were allegedly 'founded' (cf. his [1906], pp. 190–5 of the 1954 English translation). Popper revived and improved Duhem's exposition in his [1948], and his [1957]. Feyerabend elaborated the theme in his [1962]. I have shown that a similar argument applies in the logic of mathematical discovery: as in physics one may not explain what one has set out to explain, so in mathematics one may not prove what one has set out to prove (cf. my [1976c]).

empiricist could not give up science. But how then could one be a good empiricist – retaining both science and some basic core of empiricism?

Some thought that the breakdown of induction destroyed science as knowledge but not as a socially useful instrument. This was one source of modern instrumentalism.

Others shrank back from this degradation of science to the level of 'glorified plumbing' (as Popper put it) and set out to save science as knowledge. Not as knowledge in the classical sense, for that had to be restricted to mathematical and logical knowledge; but as knowledge in some weaker sense, as fallible, conjectural knowledge. There could have been no more radical departure from classical epistemology: according to classical epistemology conjectural knowledge was a contradiction in terms.[1]

But then two new problems arose. The *first* problem was *the appraisal of conjectural knowledge*. This new appraisal could not possibly be a black and white appraisal like the classical one. It was not even clear whether such an appraisal was possible; whether it would have to be conjectural itself; whether there was a possibility of a quantitative appraisal; and so on. The *second* problem was *the growth of conjectural knowledge*. The theories of the inductive growth of (certain) knowledge (or of 'inductive ascent') – from Bacon and Newton to Whewell – had collapsed: they had urgently to be replaced.

In this situation *two schools of thought emerged*. One school – *neoclassical empiricism* – started with the first problem and never arrived at the second.[2] The other school – *critical empiricism* – started by

[1] This switch from the classical epistemology to fallibilism was one of the great turning points in Carnap's intellectual biography. In 1929 Carnap still thought that only indubitably true statements could be admitted to the body of science; Reichenbach's probabilities, Neurath's dialectic, Popper's conjectures, made him give up his original idea that 'there was a certain rock bottom of knowledge...which was indubitable. Every other kind of knowledge was supposed to be firmly supported by this basis and therefore likewise decidable with certainty.' (Cf. Carnap's Autobiography in Schilpp (*ed.*) [1963], esp. p. 57. Also cf. Reichenbach's amusing recollections in his [1936].) It is interesting that the same switch was one of the great turning points also in Russell's intellectual biography (cf. *above*, chapter 1, p. 11ff.).

[2] Most 'neoclassicists' were – and some still are – blissfully unaware of the second problem. They thought that even if science does not produce certainty, it produces near-certainty. Ignoring Duhem's master-argument against induction, they insisted that the main pattern of scientific progress is 'non-demonstrative inference' from factual premises to theoretical conclusions. According to Broad the unsolved problem of *justification of induction* was a 'scandal of philosophy' *because* inductive method was the 'glory of science': scientists proceeded successfully from truths to richer truths (or, at least, to very probable truths) while philosophers toiled unsuccessfully to justify this procedure (cf. his [1952], pp. 142–3). Russell held the same view. Other neoclassicists occasionally admit that at least *some* of creative science may consist of irrational jumps which then have to be closely followed by severe assessment of the degree of evidential support. (Carnap's position will be analysed in detail *below*, in section 4.) But whether they assume that science proceeds by induction, by irrational insight, or by conjectures and refutations, they assume it unthinkingly: for most neoclassicists have a distinct aversion to taking the problem of the growth of science seriously. Cf. *below*, pp. 135ff.

solving the second problem and went on to show that this solution solves the most important aspects of the first too.

The first school – culminating in Carnap's neoclassical empiricism – approached the problem from the classical point of view of the logic of justification. Since it was clear that theories could not be classified as provably true or false, they had (according to this school) to be classified as at least 'partially proved', or in other words, as 'confirmed (by facts) to a certain degree'. It was thought that this 'degree of evidential support' or 'degree of confirmation' should somehow be equated with probability in the sense of the probability calculus.[1] The acceptance of this identity suggested a vast programme;[2] to define a – possibly computable – countably additive measure function over the field of the sentences of the complete language of science, satisfying also some further adequacy requirements inspired by the intuitive idea of 'confirmation'. Once such a function is defined, the degree to which a theory h is confirmed by the evidence e can be calculated simply by taking $p(h, e) = p(h \cdot e)/p(e)$. If there are several different possible functions, further secondary axioms have to be added to Kolmogorov's primary axioms until the function is uniquely determined.

Thus Carnap – following the Cambridge school (Johnson, Broad, Keynes, Nicod, Ramsey, Jeffreys), Reichenbach, and others – set out to solve the following problems: (1) to justify his claim that the degree of confirmation satisfies Kolmogorov's axioms of probability; (2) to find and justify further secondary adequacy requirements for the determination of the sought-for measure function; (3) to construct – piecemeal – a complete, perfect language of science in which all propositions of science can be expressed; and (4) to offer a definition of a measure function which would satisfy the conditions laid down in (1) and (2).

Carnap thought that while science was conjectural, the theory of probabilistic confirmation would be *a priori* and infallible: the axioms, whether primary or secondary, would be seen to be true in the light of *inductive intuition*, and the language (the third ingredient) would of course be irrefutable – for how can one refute a language? (At first, he may also have hoped that the measure function would be computable: that once a machine is programmed with the perfect language and the axioms, it will churn out probabilities for any

[1] Throughout the paper the terms 'probability' and 'probabilistic' will be used in this sense.

[2] In fact, probabilistic inductive logic was a Cambridge invention. It stemmed from W. E. Johnson. Broad and Keynes attended his lectures and then developed his ideas. Their approach rested on a simple logical howler (going back to Bernoulli and Laplace). As Broad put it: 'induction cannot hope to arrive at anything more than probable conclusions and *therefore* the logical principles of induction must be the laws of probability' ([1932], p. 81, my italics). The premise refers to likelihood or verisimilitude, the conclusion to the mathematical calculus of probability. (It is interesting that before Popper's criticism of Keynes and Reichenbach in 1934 nobody had pointed out this conflation.)

hypothesis relative to the evidence that is fed in. Science is fallible, but the degree of its fallibility is precisely and infallibly measurable by a machine. But, of course, he realized that Church's theorem shows that in general this is impossible.[1]) Now, since according to logical empiricism – I prefer to call it neoclassical empiricism[2] – only analytic statements can be infallible, Carnap took his 'inductive logic' to be *analytic*.[3]

He also found that the construction of a complete language for science is a piecemeal and possibly never-ending process – but then one had better make sure that the gradual construction of the confirmation function follows closely the gradual construction of this language and that the already established values of the function are not altered in the process of completion. This was, I think, the ideal that Carnap tried to approximate first with his *requirement of* (relative) *completeness* of the language,[4] then with his axiom *C6* in his 1952 system,[5] and later with *Axiom 11* of his 1955 system.[6] The underlying ideal seems to be a principle that one could call the *principle of minimal language*, that is, the principle according to which the degree of confirmation of a proposition *depends only on the minimal language in which the proposition can be expressed.* Thus the degree of confirmation would remain invariant while the language is being enriched.

It soon turned out that the difficulties involved in the construction of the confirmation function increase steeply with the increasing complexity of the language. Despite the tremendous work done by Carnap and his collaborators during the last twenty years, the research

[1] Cf. Carnap [1950], p. 196. Also cf. Hintikka [1965], p. 283, n. 22. *But the function may be computable for finite languages or for languages without universal statements.*

[2] The main feature of classical empiricism was the domination of the logic of justification over the logic of discovery. This feature essentially survived in logical empiricism: partial justification or appraisal, but not discovery, of theories was its primary interest. On the other hand, in Popper's treatment of growth without foundations the logic of discovery dominates the scene: I call this approach 'critical empiricism'; but 'critical rationalism' (or 'rational empiricism'?) may be even more suitable.

[3] This later led to philosophical troubles, cf. *below*, p. 160, n. 2, pp. 188ff, 196ff. Also, strictly speaking, throughout this paragraph, 'infallible' should be replaced by 'practically infallible'; for Carnap, since Gödel's results, occasionally warns that neither deductive nor inductive logic is perfectly infallible. But he still seems to think that inductive logic is no *more* fallible than deductive logic, which for him includes arithmetic (cf. his [1968*b*], p. 266). The fallibility of 'analytic' statements, of course, knocks out a major pillar of logical empiricism (cf. *above*, chapter 2).

[4] This lays down that the system of predicates should be 'sufficiently comprehensive for expressing all qualitative attributes exhibited by the individuals in the given universe' ([1950], p. 75); the requirement also stipulated that 'any two individuals differ only in a finite number of respects' (p. 74).

[5] 'We may assume that it has no influence on the value of $c(h, e)$, where h and e contain no variables, whether in addition to the individuals mentioned in h and e there are still other individuals in the universe of discourse or not' (Carnap [1952], p. 13).

[6] 'The value of $c(h, e)$ remains unchanged if further families of predicates are added to the language.' Cf. Schilpp (*ed.*) [1963], p. 975, and also the Preface to the Second Edition of Carnap [1950], 1962, pp. xxi–xxii.

programme of 'inductive logic' still has not produced measure functions on languages including analysis, or physical probability, which, of course, are needed in the formulation of the most important scientific theories. But the work is going on.[1]

Submerged in this difficult programme, Carnap and his school completely ignored the problem of scientific *method. The classical twin problems of induction were the justification of theories, and the discovery of theories from facts. Carnap's neoclassical solution provides at best a solution of the problem of weak justification. It leaves the problem of discovery, the problem of the growth of knowledge, untouched.* Logical empiricists thus made a very significant *cut* in the empiricist programme. There can be little doubt that neither Carnap nor his associates paid any serious attention to genuine methodological problems. Indeed, Carnap – and Carnapians – do not even seem to have a word for what used to be called 'methodology', 'heuristic', or 'logic of discovery': the rational reconstruction of patterns of growth (or, as Bar-Hillel would put it, 'diachronics'). Carnap says that the concept of confirmation is 'basic in the *methodology* of empirical science'.[2] This expropriation of 'methodology' for the method of justification, for the study of fully fledged theories – and, moreover, the implicit or explicit exclusion from rational investigation of the study of their growth – is widespread.[3]

Carnap uses the term 'methodology' primarily to denote the discipline of *applying* inductive logic. Thus 'inductive logic' concerns itself with the construction of the *c*-function while 'methodology of induction' offers advice about *applying* the *c*-function. 'Methodology of induction' is then a chapter within the logic of justification.[4]

An interesting example of the expropriation of the natural terminology of growth-concepts for use as confirmation-concepts is the Carnapian usage of the term 'inference-making'[5]

[1] I wonder, will the next interesting development in inductive logic produce impossibility results which may prove that certain elementary adequacy requirements regarding the construction of *c*-functions cannot possibly be fulfilled for rich – and even for not so rich – languages? (But I am very sceptical whether such results would, as Gödel's did, pave the way to new patterns of growth.)

[2] Cf. Schilpp (*ed.*) [1963], p. 72, my italics.

[3] Carnap says, characteristically, in the first sentence of his [1928]: 'The aim of epistemology is the foundation of a method for the justification of cognitions.' Method, *qua* logic of discovery, disappears – we have nothing but a 'method' of justification.

A similar situation has arisen in the philosophy of mathematics, where 'methodology', 'proof', etc. are all expropriated for concepts in the logic of justification. Cf. my [1963–4]. In this paper my deliberate *mixed* usage of the justificationist term 'proof' and of the heuristic term 'proof' creates – as I intended – a paradoxical impression.

[4] Cf. Carnap [1950], §44*A*: 'Methodological problems'. He remarks that he applies here the term 'methodology' only 'for lack of a better term'; but undeniably his usage is made possible only by the fact that in his philosophy he does not need the term in its original meaning. (I understand from Bar-Hillel that now Carnap uses instead of the pair 'inductive logic' and 'methodology of induction' the pair '*pure inductive logic*' and '*applied inductive logic*'.)

[5] For another example (Carnap's expropriation of the methodological term 'improvement of guesses') cf. *below*, p. 150, n. 5; also, for a discussion of his use of 'universal inference', cf. *below*, p. 144.

To understand the problem we have to go back to classical empiricism. According to *classical empiricism* science proceeds by inductive inference: first one collects some facts and then one 'makes an inductive inference' to a theory which the inductivist perfectly naturally calls a 'generalization'. According to Popper's *critical empiricism*, one starts with speculative theories which one then tests severely. Here there is only deductive inference and no 'generalizations'. Carnap time and again seems to agree with Popper that the facts are not the necessary starting points of discovery. When he uses 'inductive inference', he uses it as a technical term in the sense of the logic of justification but not in the sense of the logic of discovery. For him 'inference-making' is nothing but the assignment of a probability value to an ordered pair $\langle h, e \rangle$. This is certainly a strange terminology, for 'inference-*making*' is the characteristic term of old-fashioned inductivist logic of *discovery*.

This example is not meant so much to demonstrate the excusable expropriation of the terms of the logic of discovery for use in a logic of justification but also to show the specific dangers inherent in the *non-conscious* character of this expropriation. Occasionally a faint ghost of the logic of discovery appears with confusing consequences. For instance, in his present paper Carnap writes: 'I would not say that it is wrong to regard inference-making as the aim. But from my point of view it seems *preferable* to take as the *essential* point in inductive reasoning the determination of probability values' (my italics).[1] But what is 'inductive inference-making' if not 'determination of probability values'? Why does not Carnap say that, indeed, it *is* wrong to regard inference-making as the aim of induction unless it is taken to be determination of the degree to which a hypothesis follows from factual evidence?[2]

Why did logical empiricists have no interest in the logic of discovery? The historian of thought may explain this in the following way. Neoclassical empiricism replaced the old idol of classical empiricism – *certainty* – by the new idol of *exactness*.[3] But one cannot describe the

[1] Cf. his [1968*b*], p. 258. Also cf. his [1968*c*], 'in my conception probabilistic ("inductive") reasoning consists *essentially* not in making inferences but *rather* in assigning probabilities' (p. 311, my italics).

[2] I have little doubt that at least part of the answer is that Carnap is frequently too generous to his adversaries, and that he almost always lets them get away as long as they are willing to *share* the disputed area. While being addicted to his own system of ideas, he never follows up the enemy in hostile terrain. It is symptomatic of the Carnap school that while Popper and Popperians have written several essays in the critical history of Carnapian ideas, Carnapians have never even tried to do the same with Popperian ideas. ('*Live and let live*' is not a good rule for the dialectic of intellectual progress. If one does not follow up a critical clash to the bitter end, one may leave uncriticized not only the adversary but also oneself: for the best way to understand critically one's own position is through the relentless criticism of contrary positions.)

[3] 'In this post-rationalist age of ours, more and more books are written in symbolic languages, and it becomes more and more difficult to see why: what it is all about,

growth of knowledge, the logic of discovery, in 'exact' terms, one cannot put it in formulae: it has therefore been branded a largely 'irrational' process; only its completed (and 'formalized') product can be *judged* rationally. But these 'irrational' processes are a matter for history or psychology; *there is no such thing as 'scientific' logic of discovery*. Or, to put it in a slightly different way: classical empiricists thought that there are *rules of discovery*; neoclassical empiricists learned (many of them from Popper[1]) that there are no such rules; so they thought there was nothing to learn about it. But there are, according to Carnap, rules of *confirmation*; therefore confirmation *is* a subject suitable for 'scientific inquiry'.[2] This also explains why they felt free to expropriate the terminology of the logic of discovery.

All this has never been formulated in such sharp terms; most Carnapians graciously agree that one can say a few informal (and therefore neither very serious nor very significant) things about the logic of discovery, and that Popper has indeed said them. Bar-Hillel, generously, even goes further and suggests a 'division of labour, the Carnapians concentrating mostly on a rational "*synchronic*" reconstruction of science and the Popperians remaining mostly interested in the "*diachronic*" growth of science'.[3]

This division of labour seems to imply that the two problems are somehow independent. But they are not. I think the lack of recognition of this interdependence is an important shortcoming of logical empiricism in general and of Carnap's confirmation theory in particular.

The most interesting phenomenon in this proposed 'division of

and why it should be necessary, or advantageous, to allow oneself to be bored by volumes of symbolic trivialities. It almost seems as if the symbolism were becoming a value in itself, to be revered for its sublime "exactness": a new expression of the old quest for certainty, a new symbolic ritual, a new substitute for religion' (Popper [1959], p. 394).

[1] Cf. Carnap [1950], p. 193.

[2] For a further discussion cf. *below*, p. 169.

[3] Cf. his [1968*a*], p. 66. Bar-Hillel's generosity will be still more appreciated if one remembers that for orthodox logical empiricism even the meaningfulness of 'sloppy' Popperian logic of discovery could be called into question. It is not psychology or history, so it is not empirical. But it would be stretching the concept of analyticity a bit too far to accommodate it. However, Carnap was prepared to do so in order to rescue Popper (although he seemed to think that Popper's usage of the term 'methodology' was a bit idiosyncratic): 'Popper calls the field of his inquiries *methodology*. However, the logical character of the methodological statements and rules is left open. According to Popper (as opposed to the positivist view) there is a third, not precisely characterized field besides the analytical statements of logic and the empirical statements of science; and methodological statements and rules belong to this field. He does not elaborate his position which I regard as rather doubtful. But this does not seem to be essential to Popper's general philosophy. Just the other way round: Popper himself says that methodology rests on conventions and its rules should be compared with the rules of chess; but this clearly implies that they are analytic' (Carnap [1935], p. 293).

Indeed, Popper himself, in §82 of his book, says that his appraisal of theories is analytic. While distrusting the logical empiricists' dogma that all meaningful statements are either analytic or synthetic (§§10–11), he never elaborated an alternative position.

labour' is the thesis, implicit in Carnap's work (at least between 1945 and 1956[1]), that theories certainly play an important role in the growth of science but none in the logic of confirmation. This is why the whole problem can be best approached through a discussion of Carnap's elimination of universal hypotheses in his 1950 theory of confirmation and of Popper's criticism of it. This discussion will also try to explain why and how the central problem of classical empiricism, the (strong or weak) justification of *theories*, has disappeared from Carnap's programme.

3 THE WEAK AND STRONG ATHEORETICAL THESES

(a) *Carnap abandons the Jeffreys–Keynes postulate. Qualified instance confirmation versus confirmation*

The original problem of confirmation was no doubt the confirmation of *theories* rather than that of *particular predictions*. It was the burning epistemological problem of empiricism to 'prove' theories at least partially from 'the facts' in order to save theoretical science from the sceptics. It was agreed that a crucial adequacy requirement for confirmation theory was that it should *grade* theories according to their evidential support. Broad, Keynes, Jeffreys – indeed everybody in Cambridge – saw clearly that, if confirmation is probability, and if the *a priori* confirmation of a hypothesis is zero, no finite amount of observational evidence can lift its confirmation above this level: this is why they assigned zero prior probability only to impossible propositions. According to Wrinch and Jeffreys: 'However sceptical we may be about the finality of a particular law, we should say that its probability is finite'. We shall refer to this as the Jeffreys–Keynes postulate.[2]

It is clear from Carnap's *Testability and Meaning* that in the 1930s he had similar ideas about degree of confirmation[3] (which he had not yet definitely identified with probability[4]). But in the early 1940s Carnap found that in his newly-developed theory the degree of confirmation of all *genuinely* universal propositions (i.e. those which refer to infinitely many individuals) was zero. This was clearly inconsistent with the adequacy requirements of his two main forerunners, Jeffreys and Keynes, and also with his own original adequacy requirements. Now he had several possible ways of solving this inconsistency:

[1] For Carnap's present position cf. *below*, p. 142, n. 6 and pp. 159–60; also p. 166, n. 2.

[2] See Wrinch and Jeffreys [1921], especially pp. 381–2. This is the problem background of the well-known Wrinch–Jeffreys simplicity ordering; this served to solve the problem of how one can learn 'inductively' about theories. (For a critical discussion cf. Popper [1959], appendix *viii.)

[3] 'Instead of verification we may speak...of gradually increasing confirmation of the law' (Carnap [1936], p. 425).

[4] *Ibid.*, pp. 426–7.

(1) By regarding $c(\boldsymbol{l}) = 0$[1] as absurd since it clashes with the Jeffreys–Keynes thesis. But since $c(\boldsymbol{l}) = 0$ follows from $p(\boldsymbol{l}) = 0$ and $c = p$, at least one of these two theses has to be given up.[2] One has either (1*a*) to design a probability function which takes positive values on universal propositions or (1*b*) to abandon $c = p$.

(2) By regarding $c(\boldsymbol{l}) = 0$ as acceptable and abandoning the Jeffreys–Keynes postulate, and with it the idea that a theory can ever be made even probable. But this may seem to many a counter-intuitive solution and certainly shows the need also for some *other* degree of confirmation.

(3) By reducing the domain of the confirmation function to particular propositions, and claiming that this function solves all the problems of confirmation theory: theories are dispensable for confirmation theory. In a Pickwickian sense this solution may be said to leave the Jeffreys–Keynes thesis valid in the restricted domain.[3] But here again one needs a serious and persuasive argument before one accepts that all confirmation theory is in fact about particular propositions.

Carnap tried possibilities (1*a*), (2), and (3), *but never* (1*b*): to give up $c = p$ was for him inconceivable. First he chose (2) and put forward an interesting argument in its defence. This was his theory of 'qualified instance confirmation'. He then seemed to be inclined to entertain (3). And now, one understands, he is working on a solution on the lines of (1*a*).

Popper, on the other hand, thought that (1*a*), (2), and (3) had to be ruled out: (1*b*) *was the only solution*. Both (2) and (3) were inconceivable for him because he thought that no worthwhile confirmation theory could fail to explain how we can learn about theories and how we can grade them according to their empirical support. As for (1*a*) he claimed to have barred that by proving that $p(\boldsymbol{l}) = 0$ for *all* 'acceptable' probability functions.[4] But if $c(\boldsymbol{l})$

[1] Following Carnap throughout the rest of this paper I use $\boldsymbol{l}$ for universal, *h* primarily for particular positions. (*My term 'particular' corresponds to his 'molecular'*.)

[2] It is to be remembered that throughout this paper *p* refers to countably additive measure functions (cf. p. 133, n. 1).

[3] In a Pickwickian sense, since the Jeffreys–Keynes thesis refers, primarily if not exclusively, to *genuinely* universal hypotheses. (Also, one has to remember that not only universal but also precise particular numerical predictions would have zero measure in Carnap's 1950 approach; but, of course, one could again argue that such predictions can never be confirmed in a strict sense, and that confirmation theory should be further restricted to predictions within some finite interval of error of measurement: precise predictions also are dispensable for confirmation theory.) For Popper (and for me) without *genuinely universal* propositions there can be no scientific theories. Of course, if one postulates that the universe can be described in a finite language then the theories themselves are ('L-')equivalent to particular propositions. In defence of Carnap's 1950 theory one may set up a philosophy of science based on the idea of such a 'finite' universe. For a critical discussion of such an idea cf. Nagel [1963], pp. 799–800.

[4] Cf. Popper [1959], appendices *vii and *viii. (According to Popper a probability function in this context is 'acceptable' if, roughly, (*a*) it is defined on a language

must *not* be uniformly zero and $p(\boldsymbol{l})$ *is* uniformly zero, then $c \neq p$.

But is $c(\boldsymbol{l}) = 0$ so absurd? Carnap, following a hint by Keynes and Ramsey, gave a spirited defence of it.

The main point of Carnap's defence is that naive intuition was misguided into $c(\boldsymbol{l}) \neq 0$. He indicates that 'degree of confirmation' is a vague concept, but if we substitute for it the concept of 'rational betting quotient', the mistake transpires at once.[1] For – Carnap argues – when a scientist or engineer says that a scientific theory is 'well-founded' or 'reliable' he does not mean that he would bet that the theory was true for all its instances in the whole universe and for eternity. (In fact, he would always bet against such a preposterously daring proposal, exactly according to $c(\boldsymbol{l}) = 0$.) What he means is that he would bet that the next instance of the theory will comply with it. But this second bet is really on a particular proposition. The scientist's and, still more, the engineer's phrase 'confirmation of a law', according to Carnap, refers to the next instance. He called this 'confirmation of the next instance' the 'qualified instance confirmation' of the law. Of course the qualified instance confirmation of a law is not zero; but, on the other hand, qualified instance confirmation or 'reliability' of a law is *not* probability.[2]

Carnap's argument is very interesting. It had already been proposed by Keynes, who had grave doubts whether any reasonable grounds could be found for his $p(\boldsymbol{l}) > 0$. This led him to this aside: 'Perhaps our generalisation should always run: "It is probable that any given ϕ is f", rather than, "It is probable that all ϕ are f". Certainly, what we commonly seem to hold with conviction is the belief that the sun will rise *tomorrow*, rather than the belief that the sun will *always* rise.'[3] Keynes' doubts about his postulate were soon followed by Ritchie's disproof. Ritchie offered a proof that the probability of any inductive generalization – as such, in the absence of *a priori considerations* – is zero.[4] This, of course, did not disturb Broad and Keynes who did not

containing infinitely many individual constants (say, names of moments of time) and (*b*) it is defined for particular statements as well as for universal statements.) Good claimed (in his review of Popper's book, [1960], p. 1173) that the proof was incorrect, but this interesting problem has unfortunately not been discussed since, in spite of some recently proposed systems with positive probabilities for universal propositions. Also, one should here point out that for Popper $p(\boldsymbol{l}) = 0$ was important because it purported to show that the neoclassical rule '*Aim at highly probable theories*' was hopelessly utopian – as had been the classical rule: '*Aim at indubitably true theories*'. (For Popper's charge that Carnap adopts this neoclassical rule cf. *below*, pp. 145–6).

Also cf. Ritchie's 1926 proof that inductive generalizations, 'as such', have zero probability, *below*, p. 141.

[1] Carnap does not put it as clearly as this. But his argument is an interesting anticipation of his later problemshift from degree of confirmation to betting quotients (cf. *below*, section 4).

[2] Carnap, of course, must have realized this, but did not care to say so. It was Popper who first pointed this out in his [1955–6], p. 160. Also cf. Popper [1968], p. 289.

[3] Keynes [1921], p. 259.

[4] Ritchie [1926], esp. pp. 309–10 and 318.

mind metaphysical speculation; but it seems to have disturbed Ramsey. Replying to Ritchie, he pointed out that

we can agree that inductive generalisations need have no finite probability, but particular expectations entertained on inductive grounds undoubtedly do have a high numerical probability in the minds of all of us... If induction ever needs a logical justification it is in connection with [such probabilities].[1]

However, the idea of qualified instance confirmation creates an awkward difficulty. It ruins the programme of a *unified* theory of probabilistic confirmation of laws. It looks as if there are *two* important and utterly different measures of confirmation at least for ('empirical') laws (i.e. 'low-level' theories):[2] $c_1(\boldsymbol{l}) = 0$ for any theory whatsoever – and a non-probabilistic measure of confirmation oriented towards 'reliability' ($c_2(\boldsymbol{l}) \approx 1$ if $\boldsymbol{l}$ has overwhelming verifying evidence).

Popper's main criticism of Carnap's qualified instance confirmation was that the explanation was '*ad hoc*', introduced by Carnap only 'in order to escape from an unintended result', namely that '[Carnap's] theory has not supplied us with an adequate definition of "degree of confirmation" [for theories]'.[3] In particular, he pointed out that according to Carnap's 'reliability measure' a refuted theory scarcely loses its reliability by refutation; indeed, there is no guarantee that a *refuted* law obtains a lower qualified instance confirmation than any of those which have stood up to tests.

More generally, if a theory is again and again falsified, on the average, in every n-th instance, then its (qualified) 'instance confirmation' approached $1-1/n$ instead of 0, as it ought to do, so that the law 'All tossed pennies always show heads' has the instance confirmation ½ instead of 0. In discussing in my *L. Sc. D.* a theory of Reichenbach's which leads to mathematically equivalent results, I described this unintended consequence of his theory as 'devastating'. After 20 years, I still think it is.[4]

But Popper's argument would only hold if Carnap maintained – as Reichenbach indeed did – that 'qualified instance confirmation' had anything to do with 'confirmation'. But according to Carnap it had *not*; and he indeed agrees with Popper that the law 'All tossed pennies always show heads' – while having ½ reliability – has *zero* confirmation. Carnap introduced his 'reliability measure' only to explain why engineers think, mistakenly, that $c(\boldsymbol{l}) \neq 0$: because they conflate confirmation measure with reliability measure, because they confuse betting on a law with betting on its next instance.

[1] Ramsey [1926], pp. 183–4.

[2] In Carnap's 1950 system the only theories that figured were 'empirical generalizations' which could be expressed in his 'observational language' and contained only monadic predicates. The generalization of the concept of qualified instance confirmation to theories in general could be extremely difficult if not impossible.

[3] Popper [1963*a*], p. 282.

[4] *Ibid.*, p. 283. He repeats this statement in his [1968*b*], p. 290.

(*b*) *The weak atheoretical thesis: confirmation theory without theories*

Since the term 'qualified instance confirmation of *theories*' was only a *manner of speech* for the confirmation of certain particular propositions (of the 'next instances'), it was, strictly speaking, redundant. In 1950 Carnap still kept 'qualified instance confirmation' for the sake of those who found it difficult to get rid of the old-fashioned idea that the main problem of inductive logic is the partial justification of *theories* by evidential support; but after 1950 he abandoned qualified instance confirmation, 'reliability', altogether – he did not mention it either in his 'Replies' in the Schilpp volume or in his present paper. (This elimination of the theory of reliability also solved the awkward problem of having *two* confirmation theories, especially since one of the two was not probabilistic.) In addition he decided to omit the 'ambiguous' term 'degree of confirmation' (and 'evidential support') and use exclusively 'rational betting quotient'.[1]

But this decision was more than terminological. It was partly motivated by Popper's 1954 criticism which showed that Carnap's intuitive idea of confirmation was inconsistent,[2] and partly by the reconstruction and strengthening in 1955 of an earlier result of Ramsay and De Finetti by Shimony (and following him by Lehmann and Kemeny).[3]

According to the Ramsey–De Finetti theorem a betting system is 'strictly fair' or as Carnap puts it, a system of beliefs is 'strictly coherent', if and only if it is probabilistic.[4] Thus, at the time when Popper showed that there is something wrong with Carnap's idea of evidential support and when Carnap himself felt that his original arguments even for equating rational betting quotients and logical probability were 'weak',[5] this result seemed to provide solid foundations for his inductive logic: at least rational betting quotients and degrees of rational belief were proved to be probabilistic. The final solution of the problem of evidential support in terms of rational betting quotients could then be left for later. But Carnap had to pay a price for the support offered by the Ramsey–De Finetti theorem: he had to abandon *any* reference in his theory to universal propositions, for the proof of this theorem hinges on the lemma that $p(h) \neq 0$ for all contingent propositions.[6]

[1] Cf. his 'Replies' in Schilpp (*ed.*) [1963], p. 998 (written about 1957); and the Preface to the second edition of his [1950], 1962, p. xv.

[2] Cf. *below*, pp. 353–6.

[3] Cf. Shimony [1955]; Lehman [1955]; Kemeny [1955].

[4] For a clear explanation of the terms 'strict fairness' or 'strict coherence' see Carnap [1968*a*], pp. 260–2.

[5] *Ibid.*, p. 266.

[6] In his [1968*a*] he bases his theory on the Ramsey–De Finetti theorem (p. 260), but does not mention that it does not apply to universal propositions at all as long as $p(\boldsymbol{l}) = 0$. (Shimony, in his important paper (*loc. cit.*, pp. 18–20), indicates that it may be impossible to extend the field of applicability of the Ramsey–De Finetti theorem to countably additive fields at all even if one experiments with other probability metrics.)

So first $p(l) = 0$ led to a uniform, trivial appraisal of universal propositions; then $p(h) \neq 0$, as a condition of the Ramsey–De Finetti theorem, led to their total exclusion. What finally emerged was a 'confirmation theory' which (1) was *essentially concerned with betting on particular predictions.* But this theory had also another important feature: (2) *the rational betting quotient for any particular prediction was to be independent of the available scientific theories.*[1]

I shall refer to these two theses – which make *theories dispensable in the logic of confirmation* – jointly as the '*weak atheoretical thesis*'; I shall refer to the thesis that *theories are dispensable both in confirmation theory and in the logic of discovery,* as the '*strong atheoretical thesis*'.

The shift from the original problem about confirmability and confirmation of *theories* to the weak atheoretical thesis is not a minor shift. Carnap seems to have taken a long time to make it. This is indicated by his original hesitation among three views about confirmation theory: should the confirmation of theories or of predictions play the primary role, or should they be on a par?

Some believe that, with respect to the evidence at hand, our primary judgments concern the reliability of theories, and that judgments as to the reliability of predictions of single events are derivative in the sense that they depend upon the reliability of the theories used in making the predictions. *Others believe that judgments about predictions are primary, and that the reliability of a theory cannot mean anything else than the reliability of the predictions to which the theory leads.* And according to a third view, there is a general concept of the reliability of a hypothesis of any form with respect to given evidence. Theories and molecular predictions are in this case regarded as merely two special kinds of hypotheses.[2]

He then decided to opt for the second view.[3] But we find some trace of hesitation even in his *Logical Foundations of Probability,* for the explicit announcement of his final decision comes rather unexpectedly at the very end of the book, where he introduced $c(l) = p(l) = 0$ and its far-reaching consequences. Some subtle formulations in the book now, of course, with hindsight, appear significant.[4] But nowhere does he indicate before p. 571 in the *Appendix* that probabilistic appraisal applies only to particular predictions, and not to theories. But he knew the result all along, since he had already published it in 1945 in his paper in *Philosophy of Science,* 'On Inductive Logic'. The clearest account of what happened is already contained in this early publication:

[1] 'Scientific' is meant here in Popper's sense.

[2] Carnap [1946], p. 520, my italics.

[3] A particularly clear statement of this can be found in his [1966], p. 252: 'Once we see clearly which features of predictions are desirable, then we may say that a given theory is preferable to another one if the predictions yielded by the first theory possess on the average more of the desirable features than the predictions yielded by the other theory.'

[4] E.g.: 'The *two* arguments [of logical probability] in general refer to facts' (p. 30, my italics).

The universal inductive inference is the inference from a report on an observed sample to a hypothesis of universal form. Sometimes the term 'induction' has been applied to this kind of inference alone, while we use it in a much wider sense for all non-deductive kinds of inferences.[1] The universal inference is not even the most important one; it seems to me now that the role of universal sentences in the inductive procedures of science has generally been overestimated... The predictive inference is the most important inductive inference.[2]

So Carnap first 'widens' the classical problem of inductive justification and then omits the original part.

One cannot help wondering what persuaded Carnap to resign himself to this radical problemshift. Why did he not at least *try*, perhaps following the Wrinch–Jeffreys idea of simplicity ordering (expounded in 1921),[3] to introduce *immediately* a system with positive measures for universal propositions? A tentative answer is of course that he would have encountered many more technical difficulties and he first wanted to try a relatively easier approach. There is, of course, nothing wrong with such a consideration: it is understandable that one tries to simplify the technical difficulties of one's research programme. But one still might be more cautious and not shift one's philosophical position *too* hastily under such temptation.[4] (This is of course not to say that problem-cuts and problemshifts – and feedbacks from technical difficulties to basic philosophical assumptions – are *not* inevitable companions of any major research programme.)

Calling attention to a problemshift, even critically, does not imply that the shifted problem may not be very interesting and correctly solved. Therefore the critic's next step should be to appraise Carnap's solution of the problem of rational betting quotients on particular propositions. But before we do this (in section 5), it will be worthwhile to make a few comments on the two main lines of Popper's attack on Carnap's programme: (*a*) his criticism of Carnap's alleged strong atheoretical thesis (in the remaining parts of section 3) and (*b*) his criticism of Carnap's identification of evidential support and logical probability (in section 4).

[1] Of course this is a rather misleading way of putting it; for Carnap's 'universal inductive inference' is not an 'inference *from* a sample *to* universal hypothesis', but a metalinguistic statement of the form $c(\boldsymbol{l}, e) = q$. Cf. *above*, pp. 136ff.

[2] Also cf. Carnap [1950], p. 208: 'The term "induction" was in the past often restricted to universal induction. Our later discussion will show that actually the predictive inference is more important not only from the point of view of practical decisions but also from that of theoretical science.' Again, this problemshift goes back to Keynes: 'Our conclusions should be in the form of inductive correlations rather than of universal generalisations' (Keynes [1921], p. 259).

[3] Cf. *above*, p. 138, n. 1.

[4] However, according to Popper, Carnap's 'antitheoretical turn' was rather a *return* to his old antitheoretical position of the late 1920s. Cf. *below*, pp. 145–6.

(c) *The conflation of the weak and the strong atheoretical theses*

In the *Appendix* of his *Logical Foundations* Carnap concludes the section 'Are Laws Needed for Making Predictions?' with this statement: 'We see that the use of laws is not indispensable for making predictions. Nevertheless it is expedient, of course, to state universal laws in books on physics, biology, psychology, etc.'

This is certainly a sharp statement of an unusual position. As we shall see, it was intended to express nothing more than Carnap's 'weak atheoretical thesis'. But the unfortunate formulation made some readers think that it did express more, that it expressed the 'strong thesis': theories are *altogether* dispensable in science. They thought that if Carnap had meant only the weak thesis he would have said: 'Nevertheless universal laws are vital ingredients in the growth of science', instead of referring only to their (mnemonical?) expediency in textbooks.[1] And very few people could realize that for Carnap 'making predictions' is not predicting unknown facts from known facts but rather assigning probability values to such predictions already made.[2]

It was this passage that provoked Popper into his onslaught on Carnap's alleged strong atheoretical thesis and into neglecting the criticism of the 'weak thesis'. Popper, of course, remembered his old heroic Vienna days when he fought the Vienna Circle in order to prevent them from banning theories on the grounds that they could not be strictly justified ('*verified*'). He thought he had won that battle. Now, to his horror, he thought that Carnap was going to ban them again because they could not be at least probabilistically justified ('*confirmed*'):

> With his doctrine that laws may be dispensed with in science, Carnap in effect returns to a position very similar to the one he had held in the heyday of verificationism...and which he had given up in the *Syntax* and in *Testability*. Wittgenstein and Schlick, finding that natural laws are non-verifiable, concluded from this that they are not genuine sentences...Not unlike Mill they described them as rules for the derivation of genuine (singular) sentences – the *instances* of the law – from other genuine sentences (the initial conditions). I criticized this doctrine in my *L. Sc. D.*; and when Carnap accepted my criticism in the *Syntax* and in *Testability* I thought that the doctrine was dead.[3]

Carnap, in his reply, unfortunately, ignored Popper's desperate protest and did not clear up the misunderstanding. But, on many other occasions, Carnap *did* try to avoid any impression that he should have thought that theories were dispensable in the logic of discovery.

At least after the *Logik der Forschung* Carnap agreed with Popper's

[1] Even some of his closest collaborators misunderstood this passage. Bar-Hillel – agreeing with Watkins's interpretation of it – describes it as expressing an 'overly instrumentalistic attitude' (Bar-Hillel [1968*c*], p. 284).

[2] Cf. *above*, pp. 136ff.

[3] Popper [1963*b*], pp. 283–4.

logic of discovery and with its emphasis on the central role played by theories in the growth of science. For instance, in 1946 he wrote:

From the purely theoretical point of view, the construction of a *theory* is the end and goal of science...the theory is not *discovered* by a wholly rational or regulated procedure; in addition to knowledge of the relevant facts and to experience in working with other theories, such nonrational factors as intuition or the inspiration of genius play a decisive role. Of course, once a theory is proposed, there may be a rational procedure for *examining* it. Thus it becomes clear that the relation between a theory and the observational evidence available is, strictly speaking, not that of *inferring* the one from the other but rather that of *judging* the theory on the basis of the evidence when both are given.[1]

Carnap kept stressing that all that he was interested in was how to *judge* ready theories, not how to *discover* them, that even if judging theories could be reduced to judging particular predictions, *discovering* theories could not be reduced to discovering particular predictions:

The task of inductive logic is not to find a law for the explanation of given phenomena. This task cannot be solved by any mechanical procedure or by fixed rules; it is rather solved through the intuition, the inspiration, and the good luck of the scientist. The function of inductive logic begins *after* a hypothesis is offered for examination. Its task is to measure the support which the given evidence supplies for the tentatively assumed hypothesis.[2]

A recent passage in Carnap's Intellectual Autobiography shows interestingly Carnap's reluctant, restricted, but undeniable appreciation of the role of theories in the *growth* of science:

the interpretation of theoretical terms is always incomplete, and the theoretical sentences are in general not translatable into the observation language. These disadvantages are more than balanced by the great advantages of the theoretical language, *viz.* the great freedom of concept formation and theory formation, and the explanatory and predictive power of a theory. These advantages have so far been used chiefly in the field of physics; the prodigious growth of physics since the last century depended essentially upon the possibility of referring to unobservable entities like atoms and fields. In our century other branches of science such as biology, psychology, and economics have begun to apply the method of theoretical concepts to some extent.[3]

Why then Carnap's misleading formulation in the *Appendix*? *The explanation, I think, lies in the conflation of the conceptual and terminological frameworks of the logics of justification and discovery, caused by Carnap's neglect of the latter. This led then to the subsequent – unintended – conflation of the weak and strong atheoretical theses.*

[1] Carnap [1946], p. 520.

[2] Carnap [1953], p. 195. Carnap does not draw the attention of the unsuspecting reader to the fact that according to his theory (*anno* 1953) the measure of support which any evidence can supply for a tentatively assumed *universal* hypothesis is zero.

[3] Carnap [1963], p. 80.

(*d*) *The interconnection between the weak and strong atheoretical theses*

But why was Popper misled by Carnap's slip? I think because he could not imagine how one could possibly combine a Popperian logic of discovery with Carnap's logic of justification. *For him the weak and strong theses were inseparable.* He, mistakenly, thought that 'those who identify confirmation with probability must believe that a high degree of probability is desirable. They implicitly accept the rule: "Always choose the most probable hypothesis".'[1] Why *must*? Why must Carnap *implicitly accept* the rule which he *explicitly rejects*? (He even says – following Popper – that scientists devise 'daring guesses on slender evidence'.[2])

To be fair to Popper, one has to point out that his assertion that 'those who identify confirmation with probability. . .implicitly accept the rule: "Always choose the most probable hypothesis"' may have applied to Jeffreys, Keynes, Russell, and Reichenbach – to those whom he criticized in 1934. And this is no coincidence: there is, indeed, a deep connection between confirmation theory and heuristic. In spite of his mistake about Carnap's actual beliefs, *Popper touched here on a basic weakness of Carnap's philosophy: the loose, and even paradoxical, connection between his elaborate logic of confirmation (or reliability) and neglected logic of discovery.*

What sort of connection is there between a theory of confirmation and a logic of discovery?

A theory of confirmation assigns marks – directly or indirectly[3] – to theories, it gives a *value-judgment*, an *appraisal* of theories. Now the appraisal of any finished product is bound to have decisive pragmatic consequences for the method of its production. Moral standards, by which one judges people, have grave pragmatic implications for education, that is, for the method of their production. Similarly, scientific standards by which one judges theories, have grave pragmatic implications for scientific method, the method of their production. An important pattern of criticism of moral standards is to show that they lead to absurd educational consequences (for instance utopian moral standards may be criticized by pointing to the hypocrisy to which they lead in education). There should be an analogous pattern of criticism for confirmation theory.

The methodological implications of Popperian appraisals are relatively easily discernible.[4] Popper wants the scientist to *aim* at highly falsifiable bold theories.[5] He wants him to *aim* at very severe tests of

[1] Popper [1963*a*], p. 287. [2] Carnap [1953], p. 128.

[3] Indirectly with the help of qualified instance confirmation.

[4] Cf. *below*, section 6.

[5] By the way, Popper's occasional slogan: '*Always choose the most* improbable *hypothesis*' (e.g. [1959], p. 419; or [1963*a*], p. 218), is a careless formulation, since according to Popper *all* universal hypotheses have the same improbability, namely 1; this gives no guidance on which to choose; the guidance is only provided by his non-quantitative theory of 'fine structure of content' in appendix *vii of his *Logic of Scientific Discovery*.

his theories. But would Carnap want the scientist to *aim* at theories of, say, high qualified instance confirmation? Or should he only *rely on* them but not *aim at them*? It can be shown by a simple example that he not only *should not* aim at them, but *must not* aim at them.

Let us take Carnap's principle of 'positive instantial relevance', that is, in the language of qualified instance confirmation,

$$c_{qi}(\boldsymbol{l}, e) < c_{qi}(\boldsymbol{l}, e \,\&\, e')$$

where e' is the 'next instance' of $\boldsymbol{l}$. According to Carnapians, this principle is a precise version of an axiom of informal inductive intuition.[1]

But to Nagel it is not:

According to the formulae Carnap obtains for his system, the degree of confirmation for a hypothesis is in general increased if the confirming instances for the hypothesis are multiplied – even when the individuals mentioned in the evidence cannot be distinguished from each other by any property expressible in the language for which the inductive logic is constructed.[2]

Indeed, Carnap's reliability theory puts a premium on the completely mechanical repetition of the same experiment – indeed, a decisive premium, for such mechanical repetitions may make the qualified instance confirmation of any statement of the form 'All *A*'s are *B*'s' not merely constantly increase but actually converge to unity.[3]

Now this inductive judgment seems to lead to strange pragmatic consequences. Let us have two rival theories such that both 'work' in certain well-defined experiments. We programme two machines to perform and record the two experiments respectively. Will the victory go to the theory whose machine works faster in producing confirming evidence?

This is connected with what Keynes called the problem of the 'weight of evidence'. Indeed, it is a simple *paradox of weight of evidence*. Keynes noticed (as some of his predecessors already had) that the *reliability* and the *probability* of a hypothesis may differ. On our paradox, no doubt, he would have simply commented: 'The *probabilities* of the two theories, of course, differ, but the *weight of evidence* in their favour is the same.' Keynes emphasized that 'weight cannot, then, be explained in terms of probability'[4] and that 'the conclusion that the "weight" and the "probability" of an argument are independent properties, may possibly introduce a difficulty into the discussion of the application of probability to practice'.[5]

[1] Carnap and Stegmüller [1959], p. 244.

[2] Nagel [1964], p. 807. Nagel had already put this argument forward against Reichenbach (Nagel [1939], §8).

[3] Cf. formula 17 in Carnap [1950], p. 573.

[4] Keynes [1921], p. 76.

[5] *Ibid.* The intuitive discrepancy between the 'weight of evidence' and probability does not only introduce a mere 'difficulty', as Keynes would have it, but is alone 'difficult' enough to destroy the very foundations of inductive logic. For another paradox of the weight of evidence cf. *below*, p. 163, n. 2.

Of course this criticism does not hit Popper who admits as evidence only results of sincere attempts at refutation.[1] Carnap, however, can escape this criticism only by insisting that his appraisal of theories in respect of supporting evidence must have no methodological implications as to how to collect such evidence. *But can one completely sever the appraisal of theories from its methodological implications? Or perhaps several different appraisals are needed, some appraising theories from the point of view of methodology, others from the point of view of confirmation?*

(*e*) *A Carnapian logic of discovery*

Is there then a 'Carnapian' logic of discovery which would be married as naturally to Carnap's inductive logic as Popper's logic of discovery is to his theory of empirical content and corroboration (or as the classical logics of discovery are married to their corresponding logics of justification)?

It so happens that Kemeny *did* offer such a Carnapian heuristic.

Kemeny's heuristic is not simply 'Aim at *theories* with high Carnapian marks' – for he does not seem to regard theory-construction as a task of science. According to Kemeny,[2] the task of the theoretical scientist is to explain 'certain data collected through careful observations' with the help of 'scientifically acceptable' hypotheses. 'The selection of such hypotheses can be analysed into three stages: (1) The choice of a language in terms of which the hypothesis is to be expressed... (2) The choice of a given statement from this language, which is to serve as the hypothesis. (3) The determination of whether we are scientifically justified to accept the hypothesis on the given evidence.' Then Kemeny continues: 'It is the last step that Carnap is interested in' (and which he solves by his *c*-functions). It can be seen that if Carnap succeeds in solving (3), he makes (2) superfluous:

> Given the language, we can consider any meaningful statement of it as a potential theory. Then the 'best confirmed hypothesis relative to the given evidence' is well defined and may be selected. (Uniqueness is assumed for convenience only; it is easy to modify the argument by the addition of an arbitrary selection among equally confirmed hypotheses.)

Kemeny says these are the three stages of 'selecting an acceptable hypothesis'. But could not these three stages represent a full *account of scientific method*? There would then be three stages in the growth of science: (1′) the construction of languages (and the determination of λ), (2′) the calculation of the *c*-values for non-universal hypotheses, and (3′) the application (interpretation) of the *c*-functions.[3] The second stage, since *h* and *e* are not universal statements, could be programmed

[1] Cf. e.g. Popper [1959], p. 414.

[2] Kemeny [1963], p. 711ff.

[3] According to Carnap, in the construction of languages one has to follow an inductive path, starting with observation-language. Cf. e.g. his [1960], p. 312.

on an inductive machine.[1] The third seems to be trivial. But, Kemeny consoles us, this 'would not put the scientist out of business'; he would be occupied with the first stage, devising languages which 'would remain as the truly creative step'.

Let us now look more closely at Kemeny's methodology. First we devise a language. Then we define a probability distribution over the Boolean algebra of its (possibly only particular) sentences. Then we perform experiments and calculate, according to the Bayes formula, $p(h, e^k)$ where e^k is the conjunction of the outcomes of k experiments. Our 'improved' distribution resulting from the Bayesian learning process will be $p_k(h) = p(h, e^k)$. So all that we do is to feed $p_k(h)$ into a machine and fit it with a data-recorder: then we read off the 'improved' guesses $p_k(h)$ each evening. This 'learning process', this way of 'improvement of our guesses', is known as 'Bayesian conditionalization'.

What is wrong with 'Bayesian conditionalization'? Not only that it is '*atheoretical*' but that it is *acritical*. There is no way to discard the Initial Creative Act: the learning process is strictly confined to the original prison of the language. Explanations that break languages[2] and criticisms that break languages[3] are impossible in this set-up. The strongest criticism *within* a language – refutation in the hard sense in which one can refute a deterministic theory – is also ruled out, for in this approach *science becomes statistics writ large. But statistics becomes Bayesian conditionalization writ large,* for refutation by statistical rejection methods is ruled out too: no finite sample can ever prevent a 'possible world' from exerting eternal influence on our estimates.

In this method there is no place of honour accorded any more to *theories or laws.* Any sentence is as good as any other, and if there is a preferred class, then – at least in Carnap's present systems – it is the class of particular sentences. The concept of *explanation* (again[4]) disappears; though we may retain the term as a manner of speech for those sentences whose instantiations have high confirmation. *Testability* disappears too, for there are no potential falsifiers. No state of affairs is ever excluded. The recipe is: guesses, with different and changing degrees of probability, but without criticism. Estimation replaces testing and rejecting. (It is curious how difficult it is for many people to understand that Popper's guesswork idea of science means not only a – trivial – admission of fallibility but also a demand for criticizability.)[5]

[1] Carnap [1950], p. 196.

[2] The deepest explanations are exactly the 'fact-correcting' explanations: those which radically reformulate and reshape the explanandum, and change its 'naive' language into 'theoretical' language. Cf. *above*, p. 131, n. 3.

[3] A paradigm of language-breaking criticism is concept-stretching refutation; cf. my [1963–4].

[4] As it had done once already in the early days of logical empiricism.

[5] Carnap, in a recent paper [1966], again stresses his agreement with Popper that 'all knowledge is basically guessing' and that the aim of inductive logic is 'precisely to improve our guesses and, what is even more of fundamental importance, to improve

One can scarcely deny that Kemeny's inductive method (of discovery) is naturally associated with Carnap's inductive method (of confirmation). Carnap's 'weak atheoretical thesis' – no theories in the logic of confirmation – strongly suggests Kemeny's 'strong atheoretical thesis' – no theories in the logic of discovery either. But Carnap himself never followed up this suggestion – even at the price of a stark contrast between his occasionally expressed, almost Popperian views on the method of discovery and his own method of confirmation.[1]

Kemeny's heuristic, of course, in a sense vindicates Popper's fears: the 'weak atheoretical thesis' strongly suggests the 'strong atheoretical thesis'. But while the historian of thought must point out the strong connection between the two, he must not condemn the 'weak thesis' for 'guilt by association'. Popper's conflation of the strong methodological thesis and the weak justificationist thesis led him on many occasions to flog the dead horse and spare the live. But the criticism of the weak thesis has to be direct. However, before embarking on this criticism (in section 5), let us see how Carnap's programme shifted, not only from theories to particular propositions but also from evidential support to rational betting quotients.

4 PROBABILITY, EVIDENTIAL SUPPORT, RATIONAL BELIEF AND BETTING QUOTIENTS

Carnap's shift of the problem of inductive logic from universal to particular propositions was accompanied by a parallel shift from the

our general methods for making guesses'. The deceptive similarity in the terminology covers up very different meanings. '*Improvement of a guess*', for Popper, means refuting a theory and replacing it by an unrefuted one with higher empirical content and, preferably, in a new conceptual framework. A Popperian improvement of a guess is then part of his logic of discovery and, incidentally, a critical, creative and purely theoretical affair. '*Improvement of a guess*', for Carnap, means taking all the 'alternatives' to a particular hypothesis available in some given language *L*, estimating their probabilities relative to the total (or 'relevant') evidence and then choosing among them the one which seems the most rational to choose according to the purpose of one's action. 'Improvement of the general *methods* for making guesses' is then an improvement of the methods of choosing a *c*-function and possibly also an improvement of the pragmatic rules of application discussed in §50 of his [1950] and in his just quoted [1966]. A Carnapian improvement of a guess is then a mechanical (or almost mechanical) and essentially pragmatic affair – creativity is shifted to the 'methods' of making guesses, which then of course are 'of more fundamental importance'. While in Popper's 'improvement of a guess', refutation – critically throwing a theory overboard – plays a crucial role, it plays no role in Carnap's 'improvement of a guess'. (Of course, one may ask whether Carnap's 'improvement of a guess' is part of the logic of discovery or part of the logic of confirmation. It certainly fits well in Kemeny's heuristic framework – with a pragmatic flavour added to it.)

[1] Carnap in fact praised Kemeny's paper as having been 'very successful in presenting ... the aims and methods of inductive logic' (Carnap [1963], p. 979). But I do not think he paid serious attention to that part of Kemeny's (by the way, in many respects excellent) paper which I analysed. I should like to stress once again Carnap's *unawareness* of problems concerning the logic of discovery.

interpretation of inductive logic as providing primarily degrees of evidential support to its interpretation as providing primarily rational betting quotients. In order to appreciate this important parallel problemshift, let me give another piece of potted history.

Neoclassical empiricism had a central dogma: *the dogma of the identity of*: (1) *probabilities*, (2) *degrees of evidential support* (*or confirmation*), (3) *degrees of rational belief, and* (4) *rational betting quotients.*

This '*neoclassical chain of identities*' is not implausible. For a true empiricist the only source of rational belief is evidential support: thus he will equate the degree of rationality of a belief with the degree of its evidential support. But rational belief is plausibly measured by rational betting quotients. And it was, after all, to determine rational betting quotients that the probability calculus was invented.

This chain was the basic hidden assumption underlying Carnap's whole programme. At first he was, as it transpires from *Testability and Meaning*, primarily interested in evidential support. But in 1941, when he embarked on his research programme, he saw his basic task primarily as that of finding a satisfactory 'explication' of the concept of logical probability. He wanted to perfect the work initiated by Bernoulli, Laplace, and Keynes.

But Bernoulli, Laplace and Keynes developed their theory of logical probability not for its own sake but only because they took logical probability to be identical with rational betting quotients, degrees of rationality of belief and of evidential support.

And so did Carnap. A brief glance at the *order* of his problems (confirmation, induction, probability) on page 1 of his *Logical Foundations of Probability* shows this. Thus his theory of probability was to solve the time-honoured problem of induction, which, according to Carnap, was to judge laws and theories on the basis of evidence. But as long as evidential support = probability, *Logical Foundations of Probability* = *Logical Foundations of Evidential Support* = *Logical Theory of Confirmation.* Carnap, after some hesitation, decided to call his *explicatum* for logical probability 'degree of confirmation' – a choice which later turned out to be something of an embarrassment.

(*a*) *Are degrees of evidential support probabilities?*

Already at an early stage of his work Carnap came to *feel* that evidential support is the weak point in the chain of neoclassical empiricism. Indeed, the discrepancy between rational betting quotients and degrees of evidential support was so glaring in the case of *theories* that he had to split the two already in his 1950 exposition. For the rational betting quotient for any theory is zero, but its 'reliability' (that is, its evidential support) varies. Therefore he split his concept of confirmation for theories into two: their 'degree of confirmation', he claimed, was zero, but their degree of confirmation (i.e. qualified instance confirmation) was positive.[1]

This throws new light on Carnap's first step in his 'atheoretical' problemshift: the first atheoretical move was due to the first crack in the neoclassical chain.

But very soon he found that even by formulating his philosophy of science completely in terms of particular propositions he could not prevent further cracks. The identity of degrees of evidential support and rational betting quotients for particular propositions is not self-evident either: the probabilistic character of the second may seem to be clear, but the probabilistic character of the first is anything but clear. This is what he had in mind when he wrote: 'Although this explanation [i.e. explanation of logical probability as evidential support] may be said to outline the primary and simplest meaning of probability$_1$, it alone is hardly sufficient for the clarification of probability$_1$ as a quantitative concept.'[2] Since Carnap had, at this point, already realized that his argument that evidential support = logical probability, is based on 'entirely arbitrary' assumptions,[3] he shifted the emphasis to betting intuition. But he did *not* realize that not only is his argument for the thesis concerning the identity of evidential support and logical probability based on unsatisfactory assumptions but the thesis may be altogether false – even in the case of particular propositions.

Without realizing it, he introduced two different concepts in his *Logical Foundations of Probability* for rational betting quotients and for degrees of evidential support. For rational betting quotients he used $p(h, e)$; for degrees of evidential support he used $p(h, e)-p(h)$. But he conflated the two: in most of his book (in his quantitative and comparative theory) he claimed that *both* rational betting quotients and degrees of evidential support are $p(h, e)$; in §§86, 87, 88 (in his classificatory theory), however, he slipped into the thesis that *both* are $p(h, e)-p(h)$.

It is the irony of the story that in these sections Carnap criticized Hempel for having two different *explicanda* for evidential support in mind,[4] and for having, in the main, opted for the wrong, probabilistic, betting approach.

The two conflated notions are, of course, radically different. The Carnapian bettor's $p(h, e)$ is maximal when h is a tautology: the probability of a tautology, on any evidence, is 1. The Carnapian scientist's $p(h, e)-p(h)$ is minimal when h is a tautology: the evidential

[1] Cf. *above*, p. 140.

[2] Carnap [1950], p. 164. This is the first mention in the book of this insufficiency: indeed, earlier there is a marked confidence that there will be no such insufficiency. But great books are usually characterized by a certain inconsistency – at least in emphases. One modifies, self-critically, one's position when elaborating it, but one seldom rewrites – if only for lack of time – the whole book on each such occasion.

[3] *Ibid.*, p. 165.

[4] *Ibid.*, p. 475.

support of a tautology is always zero. For $p(h, e)$ the following 'consequence condition' holds: $p(h, e)$ can never decrease when transmitted through deductive channels, that is, if $h \rightarrow h'$, then $p(h', e) \geqslant p(h, e)$. But for $p(h, e) - p(h)$ this condition, in general, does not hold. The differences are due to the fact that two rival and mutually inconsistent intuitions are at play. According to our *betting intuition*, any conjunction of hypotheses, whatever the evidence, is at least as risky as any of the conjuncts. (That is, $(h)(h')(e)(p(h, e)) \geqslant p(h \,\&\, h', e)$.) According to our *intuition of evidential support*, this cannot be the case: it would be absurd to maintain that the evidential support for a more powerful theory (which is, in the Carnapian projection of the language of science onto the distorting mirror of particular hypotheses, a conjunction of hypotheses) *must not* be more than the evidential support for a weaker consequence of it (in the case of the Carnapian projection, for any of the conjuncts). Indeed, intuition of evidential support says that the more a proposition says, the more evidential support it may acquire. (That is, within Carnap's framework, it should be the case that $(\exists h)(\exists h')(\exists e)(c(h, e)) < c(h \,\&\, h', e)$.) But then degrees of evidential support cannot be the same as degrees of probability in the sense of the probability calculus.

All this would be trivial if not for the powerful time-honoured dogma of what I called the 'neoclassical chain' identifying, among other things, rational betting quotients with degrees of evidential support. This dogma confused generations of mathematicians and of philosophers.[1]

The first philosopher to challenge the dogma was Popper.[2] He set out to break the neoclassical chain by proving that degrees of evidential support cannot possibly be probabilities – whatever the interpretation of the latter. That is, he set out to prove that the function $C(h, e)$, evidential support, confirmation, or corroboration of h by the evidence e, does not obey the formal calculus of probability.

Popper proposed two different critical arguments: one in 1934, and one in 1954. (In 1954 he also proposed a 'rival formula'.)

Popper's 1934 argument was that

[1] Now we can see that Hempel fell prey to the *same* confusion. He realized that there are two rival subcurrents in the theory of confirmation: one can be characterized primarily by the consequence condition, the other by the condition that if e confirms h, it also confirms any hypothesis that entails h: the entailment or 'converse consequence condition'. He showed that, given some simple, generally accepted assumptions, the two conditions are inconsistent. (Cf. his [1945], p. 104.) After some hesitation, and indeed, confusion, he rather arbitrarily chose in 1945 the former, in 1965 the latter. (Cf. his Postscript to his 1945 paper in his [1965], p. 50.) Incidentally, his famous 'paradox of confirmation' looks dramatically different according to which condition one adopts: this is the keynote of the discussion of his paradox between Popperians and Carnapians. (For this point, cf. Mackie [1962–3].)

[2] The first statistician to challenge the dogma was Fisher. He equated 'degree of rational belief' with his non-additive likelihood function (cf. his [1922], p. 327, n. *). But he could not argue his position sufficiently clearly since he did not have Popper's idea of empirical content or his theoretical outlook.

the corroborability of a theory and also the degree of corroboration of a theory...stand both, as it were, in inverse ratio to its logical probability...But the view implied by probability logic is the precise opposite of this. Its upholders let the probability of a hypothesis increase in *direct proportion* to its logical probability – although there is no doubt that they intend their 'probability of a hypothesis' to stand for much the same thing that I try to indicate by 'degree of corroboration'.[1]

Or: *degree of evidential support is proportional not to probability but to improbability.*

According to Popper's new footnote, added in 1959 to this passage, these lines 'contain the crucial point of [his] criticism of the probability theory of induction'.[2]

But the supporting argument wobbles. It hinges on two points:

(1) The first crucial point of the argument is that corroboration varies inversely with probability, that is, if $p(h) \geqslant p(h')$, then for all e, $C(h, e) \leqslant C(h', e)$. But this unqualified assertion about the direct proportionality of degree of corroboration and of corroborability (or about the inverse proportionality of degree of corroboration and probability) is absurd;[3] so much so that Popper himself, when in 1954 he gave a detailed list of his adequacy requirements, postulated that, at least when h and h' imply e, the degree of corroboration must vary *directly, not inversely,* with (prior) probability, that is, if $p(h) \geqslant p(h')$, then for all e, $C(h, e) \geqslant C(h', e)$.[4]

(2) The second crucial point of the argument is that 'posterior probability', unlike corroboration, varies directly with 'prior probability', that is, if $p(h) \geqslant p(h')$, then for all e, $p(h, e) \geqslant p(h', e)$.[5] But this, as Bar-Hillel pointed out to me, is false, and counterexamples can easily be constructed.

Nevertheless the heart of Popper's argument is sound and can easily be corrected in the light of his later work. First, while one has to abandon his first thesis about the direct ratio between degree of corroboration and empirical content, one may keep his weaker thesis about the direct ratio between degree of corroborability and empirical content: if we increase the content of a theory, its corroborability also increases. This can be achieved for instance by fixing the upper bound

[1] Popper [1934], §83.

[2] Popper [1959], p. 270, n. *3.

[3] One wonders whether it was this assertion that misled Carnap into believing that Popper seems to use the term 'degree of corroborability' as synonymous with the term 'degree of corroboration'. (Cf. his Reply to Popper in Schilpp (*ed.*) [1963], p. 996).

[4] Cf. his [1954], desideratum *viii* (*c*), republished in his [1959], p. 401. Incidentally, Popper used to accuse Carnap of 'choosing' the most probable theory. But *then* he himself could be accused, on the ground of his *viii* (*c*), of 'choosing' the most probable theory from among those which can explain a given piece of evidence.

[5] For another formulation of this same point see his [1959], p. 363: 'The laws of the probability calculus demand that, of two hypotheses, the one that is logically stronger, or more informative, or better testable, and thus the one which can be *better corroborated,* is always *less probable* – on any given evidence – than the other.'

of degree of corroboration of a hypothesis at its empirical content,[1] and by allowing evidence to raise the corroboration of the more informative theory above the maximum level of the corroboration of the less informative one. That is, if h implies h' and the empirical content of h is greater than of h', then $C(h, e) > C(h', e)$ should be possible for some e. This indeed *is* excluded by Carnap's inductive logic, according to which probability can only increase when transmitted through deductive channels: if h implies h', then for all e, $p(h, e) \leqslant p(h', e)$. But then corroboration (or confirmation or evidential support) cannot be probability.

Popper's 1954 argument, like Popper's 1934 argument, was an important one. But, as in the previous case, his formulation suggested a stronger thesis than the one he had actually proved; and thereby, as in the previous case, he weakened and delayed its impact.

What he claimed to have established was once more 'a mathematical refutation of all those theories of induction which identify the degree to which a statement is supported or confirmed or corroborated by empirical tests with its degree of probability in the sense of the calculus of probability'.[2] But what he in fact proved was that Carnap's 1950 'grand theory' was inconsistent. I call Carnap's 'grand theory' the trinity of classificatory, comparative, and quantitative concepts of confirmation soldered into one 'grand theory' by the requirement that they should be related as the concepts, Warm, Warmer, and Temperature.[3] Popper's argument showed that the inconsistency was due to the fact that Carnap had inadvertently two different '*explicanda*' in mind, namely, evidential support (something like Popper's degree of corroboration) and logical probability.[4] Popper claimed that Carnap fell prey to the historical 'tendency to confuse measures *of* increase or decrease with the measures *that* increase and decrease (as shown by the history of the concepts of velocity, acceleration and force)'.[5]

By now Carnap and most Carnapians have accepted the gist of Popper's criticism. Carnap, in 1962, in the Preface to the second edition of his *Logical Foundations of Probability*, separated the two explicanda and decided that, in the future, he would call $p(h, e)$ not 'degree of confirmation' but 'rational betting quotient' or, simply, 'probability'. But together with the *term* 'confirmation', went his *theory* of confirmation, that is, his theory of evidential support. Popper rightly stated in 1955: 'there is no "current [Carnapian] theory of confirmation"'.[6] Bar-Hillel was the first Carnapian who proposed a new theory of confirmation and suggested what I would call a 'vectorial' instead of a 'scalar' appraisal of hypotheses, consisting of an ordered pair: ⟨'initial informative content', 'degree of confirmation'⟩.[7] In 1962

[1] Cf. the third of Popper's 1954 *desiderata*, [1959], p. 400.
[2] *Ibid.*, pp. 389–90. Also cf. pp. 396–8.
[3] Cf. Carnap [1950], p. 15.
[4] Popper [1959], p. 393.
[5] *Ibid.*, p. 399.
[6] Cf. his [1955–6], p. 158.
[7] Bar-Hillel [1955–6].

Carnap decided to take Bar-Hillel's advice.[1] But now he seems to have changed his mind again and returned to his old idea of a 'scalar' rather than a 'vectorial' confirmation theory; he now suggests $p(h, e) \cdot (1 - p(h))$ for degree of confirmation. Bar-Hillel interprets this as a symptom of Carnap's incurable 'acceptance syndrome'.[2]

However, Carnap – and his followers – certainly do not panic at this disarray. As they see it, inductive logic is primarily concerned with the explication of logical probability, and not with the problem of evidential support, which will eventually be solved with its help. It was a mistake, they assert, but only a slight and primarily *terminological* mistake to call the *explicatum* of logical probability 'degree of confirmation'.

No doubt Carnap's research programme of inductive logic had sufficient tenacity to survive the shattering blow to its *direct* interpretation as a theory of evidential support.[3] But while he may claim that his 'theory of confirmation' does not collapse with his original theory of confirmation (mistakenly christened 'theory of confirmation'); while Bar-Hillel and he also may correctly challenge Popper and his followers to criticize his inductive logic interpreted as a theory of rational betting quotients rather than as a theory of evidential support,[4] there is not much point in retorting to Popper's demolition of Carnap's theory of confirmation that Popper's interpretation of 'degree of confirmation' as degree of confirmation is a 'misinterpretation'.[5]

(*b*) *Are 'degrees of rational belief' degrees of evidential support or are they rational betting quotients?*

Even if Carnapians found a new, satisfactory theory of evidential support, they would face a new problem. Since Popper broke the chain between degrees of evidential support and probabilities (and hence, according to Carnap, rational betting quotients), to which side, if any, should 'degrees of rational belief' belong? Or should rational belief be split into two? Carnap seems to take it for granted that degrees of rational belief are betting quotients. Popper seems to take it for granted that degrees of rationality of belief be equated with his degrees of evidential support.[6]

It has been a cornerstone of empiricism that the only justification,

[1] Carnap [1966].

[2] Cf. Bar-Hillel [1968*b*], p. 153.

[3] When Carnap finally understood that Popper's criticism contains a valid point, he saved his *programme*, if not his 1950 *theory*, on a mere two pages of the Preface of the 1962 edition of his *Logical Foundations*. Popper underestimated the tenacity of research programmes when he thought that the 'contradictoriness [of Carnap's 1950 theory] is not a minor matter which can be easily repaired' (Popper [1959], p. 393).

[4] Bar-Hillel [1956–7], p. 248; and Carnap's Reply to Popper, in Schilpp (*ed.*) [1963], p. 998.

[5] *Ibid.*

[6] Cf. Popper [1959], p. 415. But see *below*, p. 196, n. 10.

total or partial, of a hypothesis – and therefore the only rational ground for total or partial belief in it – is its evidential support. And there has also been a long-standing dogma about degrees of belief, namely, that their best touchstone is how much one is willing to bet on them. (Carnap attributed this idea to Ramsey,[1] but Ramsey himself refers to it as to an 'old-established' thesis.[2]) But if evidential support was to determine degrees of belief, and degrees of belief were to be measured by betting quotients, then these three concepts naturally were merged into one. But now this old-established trinity is split. And this rings the death knell of the *one* concept of 'rational belief', 'credence', 'credibility', etc., in any *objective* sense of these terms.

Thus the breaking up of the chain of neoclassical empiricism implies the collapse also of its theory of rational belief. Already in 1953 Kemeny and Oppenheim distinguished 'degrees of inductive support' (which were identical with Carnap's rational betting quotients or degrees of 'firmness') and 'degrees of factual support' (which were related to Carnap's degrees of 'increase of firmness').[3] Which should measure the rationality of belief?

There are some obvious arguments for $p(h, e)$. But those philosophers who still take logical empiricism seriously but have become convinced that Carnap's inductive logic contains aprioristic metaphysical assumptions,[4] are bound to ask whether it is not a betrayal of true empiricism to claim that $p(h, e)$ should determine the rational degree of belief? For the true empiricist surely there must be no other source of rational belief but empirical evidence (and, of course, genuinely tautological logic). But why should the true empiricist take $p(h, e)$ for evidential support, rather than $p(h, e) - p(h)$, when it is obvious that a large ingredient of the value of $p(h, e)$ is simply the putative probability of h in the light of no evidence whatever?

Proliferation of more or less different formulae with more or less different name-tags does not solve the problem. In one paper we read:

> The upshot is essentially as follows: (*A*) Carnap is concerned to analyze the measure inherent in the question: 'How sure are we of p if we are given q as evidence?' (*B*) Popper and Kemeny–Oppenheim deal with the question: 'How much surer are we of p given q than without q?' (*C*) The present measure of evidential relevance deals with the question: 'How much is our confidence in the truth of p increased or decreased if q is given?'[5]

But there is no critical discussion of the rights or wrongs of any of these different measures: instead, we are told that it is 'impolite' to deny that each of them measures *something*.[6]

[1] Cf. his [1968*b*], p. 259.
[2] Cf. his [1926], p. 172.
[3] Cf. Kemeny and Oppenheim [1953], pp. 307–24.
[4] Cf. *below*, p. 160, n. 2.
[5] Rescher [1958], p. 87.
[6] *Ibid.*, p. 94.

(c) *Are rational betting quotients probabilities?*

The safest link in the neoclassical chain seemed to be the one between probabilities and rational betting quotients, supported by the Ramsey–De Finetti theorem. But several arguments undermine this support. For instance Putnam pointed out that in scientific predictions 'we are not playing against a malevolent opponent but against nature, and nature does not exploit "incoherencies"'.[1] Indeed, if we assume, for the sake of the argument, that betting quotients do measure degrees of rational belief, but also that the only rational source of belief is evidential support, and finally that evidential support is not probabilistic, then what is the correct conclusion from the Ramsey–De Finetti theorem? The correct conclusion is that it is irrational to base our theory of rationality on the manichean assumption that if we do not arrange our bets (or degrees of belief) probabilistically an evil power will catch us out by a shrewdly arranged system of bets. If once this unrealistic assumption is abandoned, we may just as well have another look at non-probabilistic theories of rational betting quotients, like Wald's minimax method or even possibly formulae akin to Popper's degree of corroboration,[2] etc., which are now being regarded by Carnapians as conclusively refuted by the Ramsey–De Finetti theorem alone.

Putnam's argument in itself is enough to shake the universal validity of Carnap's theory of rational belief and of rational betting quotients. But I shall propose also a different, independent argument. This argument will not question that rational betting quotients should be probabilistic; it will not question that rational betting quotients be restricted to particular hypotheses; but it will question the second clause of the weak atheoretical thesis underlying Carnap's theory of rational betting quotients (or of degrees of rational belief) for particular propositions, that is, the thesis that restricts even the domain of the second argument of his *c*-functions to particular propositions and forbids taking into account the appraisal of theories altogether.[3] However, I shall show that in calculating rational betting quotients of particular hypotheses one cannot escape appraising (genuinely universal) *theories*. Now Carnap's inductive logic cannot appraise theories, because theories, the vehicles of scientific growth, cannot be satisfactorily appraised without a theory of scientific growth. But if so, Carnap's inductive logic fails not only as a theory of evidential support, but also as a theory of rational betting quotients.

(In his 'Replies' ([1963*b*]), Carnap writes that he has now constructed new, probabilistic, *c*-functions in which theories may have positive measures (p. 977). But since these new *c*-functions are, as I understand,

[1] Putnam [1967], p. 113.

[2] Popper's *desideratum viii* (*c*) incorporates a crucial element of betting intuition into his theory of evidential support (cf. his [1959], p. 401).

[3] Cf. *above*, p. 141.

still probabilistic and therefore invariably obey the consequence condition, they are scarcely suitable for measuring evidential support; since they assign positive values to universal hypotheses, they cannot possibly mean rational betting quotients in view of Keynes's, Ramsey's and Carnap's arguments to the effect that only a fool would bet on the universal truth of a scientific theory. (No wonder that in Hintikka's recent papers, in which he too develops metrics with $c(\boldsymbol{l}) > 0$, the term 'betting' never occurs.) But then a strange situation may arise: Carnap may have to leave the definition *both* of rational betting quotients and of degrees of evidential support (in terms of his *c*-function) for later and elaborate an inductive logic hanging completely in the air. Of course, as already mentioned, there is no *logical* limit to the tenacity of a research programme – but one may come to doubt whether its shift was 'progressive'.)

5 THE COLLAPSE OF THE WEAK ATHEORETICAL THESIS

(*a*) '*Correctness of language*' *and confirmation theory*[1]

Carnap's atheoretical confirmation theory rests on a number of theoretical assumptions, some of which, in turn, will be shown to depend on the available scientific theories. The epistemological status of these theoretical assumptions – the axioms governing the *c*-functions, *L*, the value of λ – has not been sufficiently clarified; Carnap's original claim that they were 'analytic' may have now been abandoned but has not yet been replaced. For instance λ may be interpreted either as a measure of the degree of complexity of the world,[2] or as the speed at which one is willing to modify one's *a priori* assumptions under the influence of empirical evidence.[3] I shall concentrate on *L*.

[1] Throughout this section 'confirmation theory' stands for 'inductive logic' as Carnap *now* interprets it: as a theory of rational belief, of rational betting quotients. Of course, once we have shown that there can be no atheoretical 'confirmation theory', that is, no atheoretical inductive logic, we also have shown that there can be no atheoretical confirmation theory, that is, no definition of degrees of evidential support in terms of atheoretical betting quotients.

[2] Cf. Popper's and Nagel's papers in Schilpp (*ed.*) [1963], especially pp. 224–6 and pp. 816–25. Carnap did not comment, although in his *Autobiography* he seems to say that λ depends on the 'world structure', which, however, the 'observer is free to choose' according to his degree of 'caution' (*ibid.*, p. 75). But already in an earlier paper he had said that λ 'somehow corresponds to the complexity of the universe' [1953], p. 376). In his present paper he argues that 'for fruitful work in a new field it is not necessary to be in possession of a well-founded epistemological theory about the sources of knowledge in the field' ([1968*c*], p. 258). No doubt, he is right. But then the programme of inductive logic, which originally was deeply embedded in austere empiricism, in fact presupposed, if Popper's and Nagel's arguments are correct, an apriorist epistemology.

[3] Kemeny seems to favour the latter interpretation. He calls λ an 'index of caution' which puts a brake on modifying one's position too fast under the influence of empirical evidence (cf. his [1952–3], p. 373 and his [1963], p. 728). But which is the *rational* index of caution? With infinite caution one never learns, so this is irrational. Zero caution may come under criticism. But otherwise the inductive judge seems to refuse to pronounce on λ – he leaves its choice to the scientist's instinct.

The choice of a language for science implies a conjecture as to what is relevant evidence for what, or what is connected, by natural necessity, with what. For instance, in a language separating celestial from terrestrial phenomena, data about terrestrial projectiles may seem irrelevant to hypotheses about planetary motion. In the language of Newtonian dynamics they become relevant and change our betting quotients for planetary predictions.

Now how should one find the 'correct' language which would state correctly what evidence is relevant for a hypothesis? Although Carnap never asked this question, he implicitly answered it: in his 1950 and 1952 systems 'the observational language' was expected to fulfil this requirement. But Putnam's and Nagel's arguments imply that the 'observational' language cannot be 'correct' in my sense.[1]

However, both Putnam and Nagel discussed this problem as if there were some unique theoretical language opposed to *the* observation language, so that this theoretical language would be the correct one. Carnap countered this objection by promising to take *the* theoretical language into account.[2] But this does not solve the general problem, the problem of 'indirect evidence' (I call '*indirect evidence relative to L in L**' an event which does not raise the probability of another event when both are described in *L*, but does so if they are expressed in a language *L**). In the examples given by Putnam and Nagel *L* was Carnap's 'observational language' and *L** the superseding theoretical language. But a situation of the same kind may occur whenever a theory is superseded by a new theory couched in a new language. Indirect evidence – a common phenomenon in the growth of knowledge – makes the degree of confirmation a function of *L* which, in turn, changes as science progresses. Although growth of the evidence *within* a fixed theoretical framework (the language *L*) leaves the chosen *c*-function unaltered, growth of the theoretical framework (introduction of a *new* language *L**) may change it radically.

Carnap tried his best to avoid any 'language-dependence' of inductive logic. But he always assumed that the growth of science is in a sense cumulative: he held that one could stipulate that once the degree of confirmation of *h* given *e* has been established in a suitable 'minimal language', no further argument can ever alter this value. But scientific change frequently implies change of language and change of language implies change in the corresponding *c*-values.[3]

This simple argument shows that Carnap's (implicit) 'principle of minimal language'[4] does not work. This principle of gradual construction of the *c*-function was meant to save the fascinating ideal of an eternal, absolutely valid, *a priori* inductive logic, the ideal of an

[1] Schilpp (*ed.*) [1963], pp. 779 and 804ff.

[2] *Ibid.*, pp. 987–9.

[3] Where theories are, there is fallibility. Where scientific fallibility is, there is refutability. Where refutability is, refutation is nearby. Where refutation is, there is change. How many philosophers go to the end of the chain?

[4] Cf. *above*, p. 133.

inductive machine that, once programmed, may need an *extension of the original programming* but *no reprogramming*. Yet this ideal breaks down. The growth of science may destroy any particular confirmation theory: the inductive machine may have to be reprogrammed with each new major theoretical advance.

Carnapians may retort that the revolutionary growth of science will produce a revolutionary growth of inductive logic. But how can inductive logic grow? How can we change our whole betting policy with respect to hypotheses expressed in a language L whenever a new theory couched in a new language L^* is proposed? Or should we do so only if the new theory has been – in Popper's sense – corroborated?

Obviously we do not always want to change our c-function on the appearance of a fancy new theory (in a fancy new language) which has no better empirical support than the extant ones. We certainly would change it if the new theory has stood up to severe tests, so that it could be said that 'the new language has empirical support'.[1] But *in this case we have reduced the Carnapian problem of confirmation of languages (or, if you prefer, of the choice of a language) to the Popperian problem of corroboration of theories.*

This consideration shows that *the essential part of 'language planning' far from being the function of the inductive logician is a mere by-product of scientific theorizing.*[2] The inductive logician can, at best, say to the scientist: 'if *you* choose to accept the language L, then I can inform you that, in L, $c(h, e) = q$.' This, of course, is a complete retreat from the original position in which the inductive judge was to tell the scientist, on the sole basis of h and c, how much h was confirmed on the basis of e: 'All we need [for calculating $c(h, e)$] is a logical analysis of the meanings of the two sentences'.[3] But if the inductive judge needs to get from the scientist in addition to h and e also the language of the most advanced, best-corroborated theory, what then does the scientist need the inductive judge for?

Yet the situation of the Carnapian inductive judge becomes still more precarious if there are two or more rival theories formulated in completely different languages. In this case there does not seem to be any hope of the inductive judge deciding between them – unless he asks some super-judge to set up a secondary c-function in order to appraise languages. But how should that be done?[4] Perhaps the 'best' they can do is – instead of getting into an infinite regress of meta-confirmation functions – to ask *the scientist* for those degrees of belief

[1] In this case we might even talk about the 'refutation of a language'.

[2] It is interesting that some inductive logicians, still unaware of this fact, think that language planning is only 'formalization' and therefore a mere 'routine' (though quite possibly laborious) job for the inductive logician.

[3] Carnap [1950], p. 21.

[4] Bar-Hillel, pointing out that 'there exist no generally accepted criteria for the comparison of two language-systems', says that 'here lies an important task for present-day methodology' ([1963], p. 536). Also cf. L. J. Cohen [1968], pp. 247ff.

which he cares to attach to his rival theories, and weight the values of the rival functions accordingly.

Inductive justice is perhaps at its weakest in the prescientific stage of 'empirical generalizations' or 'naive conjectures'.[1] Here inductive judgments are bound to be highly unreliable, and the language of such conjectures will in most cases soon be replaced by some radically different new language. However, present-day inductive judges may give very high *c*-values to predictions couched in a naive language which the scientist's instinctive 'hunch' may rate very low.[2] The inductive judge has only two ways out: either by appealing to some super-judge with a request for an appraisal of languages, or by appealing to the scientist's instinctive rating. Both courses are fraught with difficulty.

To sum up, it seems that rational betting quotients can best be provided to customers who – at their own risk and left to their own devices – specify the language in which the quotients are to be calculated.

All this shows that there is here something that makes inductive logic dramatically different from deductive logic: if A is a consequence of B, no matter how our empirical knowledge develops, A will remain a consequence of B. But with the increase of our empirical knowledge, $c(A, B)$ may change radically. Since we opt for a new language usually in the wake of a victorious theory *corroborated by empirical facts*, Carnap's claim that 'the principal characteristic of the statements in both fields [i.e. in deductive and inductive logic] is their independence of the contingency of facts'[3] is shattered, and thereby also the justification of applying the common term 'logic' to both fields.

A historical consideration of the notorious 'principles of induction' may throw some more light on this situation. The main problem of classical epistemology was to *prove* scientific theories; the main problem of neoclassical empiricism is to *prove* degrees of confirmation of scientific hypotheses. One way of trying to solve the classical problem was to reduce the problem of induction to deduction, to claim that inductive inferences are enthymematic and that in each of them there is a hidden major premise, a synthetic *a priori* '[*classical*] *principle of induction*'. Classical inductive principles would then turn scientific theories from mere conjectures into proven theorems (given, of course, the certainty of the minor premise, expressing the factual evidence). This solution, of course, was incisively criticized by empiricists. But neoclassical empiricism wants to *prove* statements of the

[1] Cf. my [1976*c*], chapter 1, §7.

[2] This is, in fact, a further paradox of the 'weight of evidence' (cf. *above*, p. 148). Curiously, according to Carnapians, it is exactly this prescientific domain where inductive logic can be most successfully applied. This mistake stems from the fact that probability measures have been constructed only for such 'empirical' (or, as I prefer putting it, 'naive') languages. But unfortunately such languages express only very accidental, superficial features of the world and therefore yield particularly uninteresting confirmation estimates.

[3] Carnap [1950], p. 200.

form $p(h, e) = q$; therefore neoclassical empiricism too needs some indubitably true premise (or premises), that is, a '[*neoclassical*] *principle of induction*'. Keynes, for instance, refers to this neoclassical principle when he assumes 'some valid principle darkly present to our minds, even if it still eludes the peering eyes of philosophy'.[1] Unfortunately in the literature the two different kinds of inductive principles have been persistently conflated.[2]

That inductive logic must depend on metaphysical principles was a commonplace for Broad and Keynes; they had no doubt that for deriving a probability metric they needed such principles. But in trying to work out these principles they became very sceptical about whether they could find them and if they did find them whether they could justify them. In fact, both, but especially Broad, gave up hope of the latter; Broad thought that all such principles could achieve would be to *explain* rather than to *prove* some probability metric which one would have to take for granted.[3] But he must have thought that this was a scandalous retreat: we may call it the neoclassical scandal of induction. The '*classical scandal of induction*' was that the inductive principles, needed to *prove* theories (from facts), could not be justified.[4] The '*neoclassical scandal of induction*' is that the inductive principles, needed to *prove* at least the degree of confirmation of hypotheses, could not be justified either.[5] The neoclassical scandal of induction meant that since the inductive principles could not serve as proving but only as explanatory premises, *induction could not be part of logic but only of speculative metaphysics.* Carnap, of course, could not admit *any* metaphysics, whether 'proven' or speculative: so he solved the problem by hiding the rattling metaphysical skeleton in the cupboard of Carnapian 'analyticity'. This is how the diffident metaphysical speculation of the Cambridge school turned into confident Carnapian language-construction.

[1] Keynes [1921], p. 264.

[2] An interesting example is the difference between the classical and neoclassical Principle of Limited Variety. Its main classical version is the Principle of Eliminative Induction: 'there are only n possible alternative explanatory theories'; if factual evidence refutes $n-1$ of them, the nth is *proved.* Its main neoclassical version is Keynes's principle 'that the amount of variety in the universe is limited in such a way that there is no one object so complex that its qualities fall into an infinite number of independent groups' – unless we assume this, no empirical proposition whatsoever can ever become highly probable (*loc. cit.*, p. 258). Also while classical principles of induction may be formulated in the object language (for instance, by postulating the disjunction of the 'limited variety' of theories), neoclassical principles of induction can be formulated *only* in the meta-language (cf. *below*, p. 188, n. 1).

[3] Cf. his [1959], p. 751.

[4] Cf. above, p. 321, n. 1. In order to appreciate the 'scandal' one should read Keynes (or Russell) rather than Broad. In setting out his programme Keynes had two uncompromisable, absolute basic requirements: the certainty of evidence and the logically proven certainty of statements of the form $p(h, e) = q$. (*A Treatise on Probability*, 1921, p. 11.) If these cannot be satisfied, inductive logic cannot fulfil its original aim: to save scientific knowledge from scepticism.

[5] These two 'scandals' have also been conflated in the literature.

Now Carnap's 'analytical' inductive principles consist partly of his explicit axioms, partly of his implicit meta-axioms about the correctness of L and λ. We showed that as far as the correctness of a chosen L is concerned, it is not only an unprovable but a refutable premise. But then confirmation theory becomes no less fallible than science itself.

(b) The abdication of the inductive judge

We saw that since the inductive judge cannot appraise languages, he cannot pass judgment without asking the scientist to perform this task for him. But on further scrutiny it turns out that the scientist is asked to participate still more actively in fixing degrees of confirmation. For even after a language (or several languages and their respective weights) has been agreed on, difficult situations may still arise.

What if there is a very attractive, very plausible theory in the field, which so far has little or no evidence to support it? How much should we bet on its next instance? The sober advice of the inductive judge is that no theory should be trusted beyond its evidential support. For instance, if the Balmer formula is proposed after only three lines of the hydrogen spectrum have been discovered, the sober inductive judge would keep us from becoming over-enthusiastic and prevent us – against our hunch – from betting on a fourth line as predicted by the formula.

The inductive judge is then unimpressed by striking theories without sufficient factual backing. But he may be unimpressed also by dramatic refutations. I shall show this by another example. Let us imagine that we already have c-functions defined on very complex languages. For instance, let us postulate a language in which all Newtonian and Einsteinian mechanics and theories of gravitation are expressible. Let us imagine that hitherto all the (billions of) predictions from both theories have been confirmed, except for the predictions concerning the perihelion of Mercury where only Einstein's theory was confirmed while Newton's theory failed. The numerical predictions from the two theories differ in general but because of the smallness of their difference or the imprecision of our instruments the difference was measurable only in the case of Mercury. Now new methods have been devised, precise enough to decide the issue in a series of crucial experiments. How much should we then bet on the Einsteinian predictions h_E^i versus the corresponding Newtonian h_N^i? A scientist would take into account that Newton's theory had after all been refuted (in the case of Mercury) while Einstein's theory survived, and would therefore suggest a very daring bet. A Carnapian however, cannot, with his weak atheoretical thesis, take Newton's or Einstein's theories (and the refutation of the former) into account. Thus he would find little difference between $c(h_N^i, e)$ and $c(h_E^i, e)$, and might therefore suggest a very cautious bet: he would regard a high bet on Einstein's theory against Newton's, under these circumstances, as a bet on a mere hunch.

Now the interesting point is that although a Carnapian might be prevented by the weak atheoretical thesis from taking theories (and their refutations) into account in calculating *c*-values, he may suggest that it would be wise in such cases to ignore the calculated 'rational betting quotient' and rather bet on the 'hunch'. Carnap himself explains that

it is true that many non-rational factors affect the scientist's choice, and I believe that this will always be the case. The influence of some of these factors may be undesirable, for instance a bias in favour of a hypothesis previously maintained publicly or, in the case of a hypothesis in social science, a bias caused by moral or political preferences. But there are also non-rational factors whose effect is important and fruitful; for example, the influence of the 'scientific instinct or hunch'. Inductive logic does not intend to eliminate factors of *this* kind. Its function is merely to give the scientist a clearer picture of the situation by demonstrating to what degree the various hypotheses considered are confirmed by the evidence: This logical picture supplied by inductive logic will (or should) influence the scientist, but it does not uniquely determine his decision of the choice of hypothesis. He will be helped in this decision in the same way a tourist is helped by a good map. If he uses inductive logic, the decision still remains his; it will, however, be an enlightened decision rather than a more or less blind one.[1]

But if inductive logicians agree that 'hunches' may frequently overrule the exact rules of inductive logic, is it not misleading in the extreme to stress the rule-like, exact, quantitative character of inductive logic?

Of course, the inductive judge, instead of abdicating his responsibilities to the scientist's hunches, may try to strengthen his legal code. He may construct an inductive logic that throws the atheoretical thesis overboard and makes his judgments dependent on the theories proposed in the field. But then he would have to grade these theories according to their trustworthiness prior to constructing his main *c*-function.[2] If this falls through, he again will have no alternative but

[1] [1953*a*], pp. 195–6. Of course, some scientists may protest against calling even an untested speculation a 'non-rational factor' in prediction-appraisal; though perhaps they would not mind taking it to be a 'non-empirical factor'.

[2] Indeed, Carnap, in one of his new and as yet unpublished systems, has already made a first step in this direction. He is referring to this system in his reply to Putnam: 'In inductive logic, we might consider treating the postulates although they are generally synthetic, as "almost analytic" ([Logical Foundations of Probability, 1950], D58–1a), i.e., assigning to them the *m*-value 1. In this connection it is to be noted that only the fundamental principles of theoretical physics would be taken as postulates, no other physical laws even if they are "well established". What about those laws which are not logical consequences of the postulates, but are "proposed" in Putnam's sense? In my form of inductive logic I would assign to them the *m*-value 0 (for another alternative see my comments on (10), §26 III); but their instance confirmation may be positive. As mentioned earlier, we could alternatively consider here, in analogy to Putnam's idea, making the rules such that the d.c. of a singular prediction would be influenced not only by the form of the language and thereby by the postulates,

to ask the scientist to provide him with these grades. If the inductive judge, in addition, recognizes that one cannot take (as Carnap may have hoped originally) evidence statements as at least practically certain,[1] then he will have to ask the scientist for still more information about his hunches: he will have to ask also for the scientist's (changing) degrees of belief attached to each evidence statement.

Indeed, this further step was inevitable. After Popper's 1934 argument concerning the theoretical character of 'basic statements' there could be no doubt that any evidence statement was theoretical.[2] To insist on their 'molecular' character is misleading: in a 'correct' language all 'observation statements' appear as logically universal statements, whose 'trustworthiness' would have to be measured by the 'trustworthiness' of the background theories to which they belong.[3] And this trustworthiness may occasionally be remarkably low: whole bodies of accepted evidence statements may be overthrown when a new theory throws new light on the facts. The growth of the body of 'evidence statements' is no more cumulative and peaceful than the growth of explanatory theories.

But then, in the light of these considerations, the following picture emerges. The weak atheoretical thesis has collapsed. Either (1) the atheoretical *c*-functions have to be overruled in each single case by the scientist's hunches or else (2) new 'theoretical' *c*-functions must be constructed. The latter case again gives rise to two possibilities: either (2*a*) a method can be designed to provide both a 'prior' *c*-function (that is a function assessing the trustworthiness of theories prior to the construction of the final language) and the 'proper' *c*-function for calculating the degree of confirmation of the 'particular' statements of the final language: or, (2*b*) if such a method cannot be designed, the new 'theoretical' *c*-function will have to depend on the scientist's hunches about the trustworthiness of his languages, theories, evidence and about the rational speed of the learning process. But in this case (2*b*), it is no business of the inductive judge to criticize these 'hunches'; for him these are data about the scientist's *beliefs*. His judgment will be: if *your* beliefs (rational or irrational) about languages, theories, evidence, etc., are such and such, then *my* inductive intuition provides

but also by the class of proposed laws. At the present moment, however, I am not yet certain whether this would be necessary.' [1963], pp. 988–9; for his reference to '(10), §26 III' cf. *ibid.*, p. 977).

[1] Not that Carnap would not have preferred to find a way to deal with uncertain evidence: 'For many years I have tried in vain to correct this flaw in the customary procedure of inductive logic.' (Cf. his [1968*b*], p. 146.)

[2] Keynes still thought that inductive logic *had* to be based on the *certainty* of evidence statements; if evidence statements are inevitably uncertain, inductive logic is pointless (cf. his [1921], p. 11). The need for taking into account the uncertainty of evidence statements in inductive logic was first mooted by Hempel and Oppenheim [1945], pp. 114–15, but only Jeffrey made the first concrete step in this direction (cf. his [1968], pp. 166–80).

[3] For further discussion of this point cf. my [1968*c*].

you with a calculus of 'coherent' beliefs about all the other hypotheses within the given framework.

There is really not much to choose between an 'atheoretical' inductive logic whose verdicts are to be freely overruled by the scientist's theoretical considerations (which fall beyond its legislation) on the one hand, and, on the other, a 'theoretical' inductive logic which is based on theoretical considerations fed in from outside. In both cases the inductive judge abdicates his historic responsibilities.[1] All that he can do is to keep coherent the scientist's beliefs; so that if the scientist should bet on his beliefs against a shrewd and inimical divine bookmaker, he would not lose merely because his betting system was incoherent. The abdication of the inductive judge is complete. He promised to hand down judgment on the rationality of beliefs; now he is ending up by trying to supply a calculus of coherent beliefs on whose rationality he cannot pronounce. The inductive judge cannot claim any longer to be a 'guide of life' in any relevant sense.

(Carnapians, while accepting the gist of this evaluation, may not regard all this as a failure. They may argue that what is happening is perfectly analogous to what happened in the history of deductive logic: originally it was devised for *proving* propositions and only later had to confine itself to *deducing* propositions from others. Deductive logic was originally intended to establish both firm truth-values and also safe channels for their transmission; the fact that it had to abdicate proving and concentrate on deducing liberated its energy rather than clipping its wings and reducing it to irrelevance. Inductive logic was originally intended to establish both objective, rational degrees of beliefs and their rational coordination; now, forsaking the former and concentrating wholly on the 'rational' coordination of subjective beliefs, it can still claim to be *a* guide of life. I do not think this argument holds water but I have little doubt that it will be taken up and eagerly and ingeniously defended.)

With the abdication of the inductive judge even the last tenuous threads are severed between inductive logic and the problem of induction. A mere calculus of coherent beliefs can at best have marginal significance relative to the central problem of the philosophy of science. Thus, in the course of the evolution of the research programme of inductive logic its problem has become much less interesting than the original one: the historian of thought may have to record a 'degenerating problemshift'.

This is not to say that there are no interesting problems to which Carnap's calculus might apply. In some betting situations the basic beliefs may be specified as the 'rules of the game'. Thus (1) a language *L* may be laid down which is 'correct' by decree: indirect evidence is excluded as violating the rules of the game; (2) nothing may be permitted to influence the events from unspecified, not previously

[1] Nobody has yet tried (2*a*); and I bet nobody will.

agreed, sources – such influence constitutes a breach of the rules; (3) there may be an absolutely mechanical decision procedure for establishing the truth-values of the evidence statements. If conditions such as these are fulfilled, inductive logic may provide useful service. And, indeed, these three conditions are essentially fulfilled in some standard betting situations. We could call such betting situations '*closed games*'. (It can easily be checked that the Ramsey–De Finetti–Shimony theorem applies only to such 'closed games'.) But if science is a game, it is an *open game*. Carnap was mistaken in claiming that the difference between the usual (closed) games of chance and science was only in degree of complexity.[1]

Carnap's theory may also apply to 'closed statistical problems', where for practical purposes both the 'correctness' of the language and the certainty of the evidence may be taken for granted.

The programme of inductive logic was overambitious. The gap between rationality and formal rationality has not narrowed quite as much since Leibniz as Carnapians seem to have thought. Leibniz dreamt of a machine to decide whether a hypothesis was true or false. Carnap would have been content with a machine to decide whether the choice of a hypothesis was rational or irrational. But there is no Turing machine to decide either on the truth of our conjectures or on the rationality of our preference.

There is then an urgent need to look again at those fields in which the inductive judge has abdicated his responsibilities: first of all, at problems concerning the appraisal of theories. Solutions to these problems had been offered by Popper – even before Carnap started his programme. These solutions are methodologically oriented and unformalizable. But if relevance is not to be sacrificed on the altar of precision, the time has come to pay more attention to them.

Moreover, it will be shown with the help of Popper's approach that the appraisal of theories has a fine structure unnoticed by Carnap; and that, using this fine structure, one can offer a rival even to Carnap's theory of rational betting quotients for particular hypotheses. But the difference between the Popperian and the Carnapian approach cannot be put simply as a difference between different solutions of the same problem. Solving the problem interestingly always involves reformulating it, putting it in a fresh light. In other words: an interesting solution always *shifts* the problem. *Rival solutions of a problem frequently imply rival problemshifts.* The discussion of the rival Popperian problemshift will also throw further light on Carnap's problemshift. As Carnap and his school shifted the original centre of gravity of the problem of induction away from informality to formality, away from methodology to justification, away from genuine theories to particular propositions, away from evidential support to betting quotients, so Popper and his school shifted it in exactly the opposite direction.

[1] Cf. his [1950], p. 247.

6 THE ONE MAIN PROBLEM OF CRITICAL EMPIRICISM: METHOD

While neoclassical empiricism inherited from classical empiricism only the problem of a monolithic, all-purpose appraisal of hypotheses, Popper's critical empiricism focussed attention on the problem of their discovery. The Popperian scientist makes *separate* appraisals corresponding to the separate stages of discovery. I shall use these methodological appraisals ('acceptability$_1$', 'acceptability$_2$', etc.) to construct an appraisal even of the *trustworthiness* of a theory ('acceptability$_3$'). This 'acceptability$_3$' comes closest to Carnap's degree of confirmation. But since it is based on Popper's *methodological* appraisals, I shall first have to discuss these at some length.

(a) '*Acceptability*$_1$'

The first, prior appraisal of a theory immediately follows its proposal: we appraise its '*boldness*'. One of the most important aspects of 'boldness' may be characterized in terms of '*excess empirical content*',[1] or, briefly, '*excess content*' (or '*excess information*' or '*excess falsifiability*'): a bold theory should have some novel potential falsifiers which none of the theories in the extant body of science has had; in particular, it should have excess content over its '*background theory*' (or its '*touchstone theory*'), that is, over the theory it challenges.

The background theory may not have been articulated at the time when the new theory is proposed; but in such cases it can easily be reconstructed. Also, the background theory may be a double or even a multiple theory in the following sense: if the relevant background knowledge consists of a theory T_1 and of a falsifying hypothesis T'_1 of it, then a challenging theory T_2 is bold only *if it entails some novel factual hypothesis* which had not been entailed either by T_1 or T'_1.[2]

A theory is the bolder the more it revolutionizes our previous picture of the world: for instance, the more surprisingly it unites fields of knowledge previously regarded as distant and unconnected; and even possibly the 'more inconsistent' it is with the 'data' or with the 'laws' it set out to explain (so that if Newton's theory had not been

[1] For the all-important concept of 'empirical content' cf. Popper [1934], §35: the empirical content of a theory is its set of 'potential falsifiers'. Incidentally, Popper, with his two 1934 theses that (1) the empirical information a sentence conveys is the set of states of affairs which it forbids and that (2) this information can be measured by improbability rather than probability, founded the semantic theory of information.

[2] My 'background theory' and 'background knowledge' should not be confused with Popper's 'background knowledge'. Popper's 'background knowledge' denotes 'all those things which we accept (tentatively) as *unproblematic* while we are testing the theory', such as initial conditions, auxiliary theories, etc. etc. (Popper [1963*a*], p. 390, my italics). My 'background theory' is *inconsistent* with the tested theory, Popper's is *consistent* with it.

inconsistent with Kepler's and Galileo's laws, which it set out to explain, Popper would give it a lower grade for boldness[1]).

If a theory is judged 'bold', scientists '*accept*$_1$' it as a part of the 'body of science' of the day. On acceptance into the body of science several things may happen to the theory. Some may try to criticize it, test it; some may try to explain it; others may use it in the determination of the truth-values of potential falsifiers of other theories; and even technologists may take it into consideration. But above all, this *acceptance*$_1$ is acceptance for serious criticism, and in particular for testing: it is a certificate of testworthiness. If a 'testworthy' theory is explained by a new 'bold' higher level theory, or if some other theory is falsified with its help, or again, if it becomes shortlisted for technological use, then its testing becomes still more urgent.

We may call acceptance$_1$ '*prior acceptance*' since it is *prior* to testing. But it is *usually not prior to evidence*: most scientific theories are designed, at least partly, to solve an explanatory problem.

One may be tempted to characterize the boldness of a theory merely by its 'degree of falsifiability' or 'empirical content', that is, by the set if its potential falsifiers. But if a new theory T_2 explains some available evidence already explained by some extant theory T_1, the 'boldness' of T_2 is gauged not simply by the set of potential falsifiers of T_2 but by the set of those potential falsifiers of T_2 which had not been potential falsifiers also of T_1. A theory which has no more potential falsifiers than its background theory has then at most zero 'excess falsifiability'. Newton's theory of gravitation has very high excess falsifiability over the conjunction of its background theories (Galileo's theory of terrestrial projectiles and Kepler's theory of planetary motions): therefore it was bold, it was scientific. However, a theory akin to Newton's theory, but applying to all gravitational phenomena except the orbit of Mercury has negative excess empirical content over Newton's original, unrestricted theory: therefore this theory, proposed after the refutation of Newton's theory, is not bold, is not scientific. *It is excess empirical content, that is, the increase in empirical content, rather than empirical content as such, that measures the boldness of the theory.* Clearly, one cannot decide whether a theory is bold by examining the theory in isolation, but only by examining it in its historico-methodological context, against the background of its available rivals.

Popper proposed in 1957 that theories should not only be 'testable' (that is, falsifiable) but also 'independently testable', that is, they should have 'testable consequences which are different from the *explicandum*'.[2] This requirement, of course, corresponds to 'boldness' as here

[1] Popper calls such 'fact-correcting' explanations '*deep*' – which is another word for a certain type of outstanding boldness (cf. his [1957], p. 29). It goes without saying that it is impossible to compare at all, let alone numerically, 'degrees' of 'depth'.

[2] Popper [1957], p. 25; also cf. his [1963*a*], p. 241.

defined: it suggests that in the prior appraisal of theories falsifiability should be supplemented by excess falsifiability.

But it transpires already from his *Logic of Scientific Discovery*, 1934, that falsifiability without excess falsifiability is of little importance. His demarcation criterion draws a line between 'refutable' (scientific) and 'irrefutable' (metaphysical) theories. But he uses 'refutable' in a rather Pickwickian sense: he calls a theory T 'refutable' if *two* conditions are satisfied: (1) if it has some 'potential falsifiers', that is, one can specify statements conflicting with it whose truth-value can be established by some generally accepted experimental technique of the time,[1] and (2) if the discussion partners agree not to use 'conventionalist stratagems',[2] that is, not to replace T, following its refutation, by a theory T' which has less content than T. This usage, although explained very clearly in several sections of Popper's *Logic of Scientific Discovery*, is responsible for many misunderstandings.[3]

My terminology is different, and nearer to ordinary usage. I call a theory 'refutable' if it satisfies Popper's *first* condition. Correspondingly, I call a theory *acceptable*$_1$ *if it has excess refutability over its background theory*. This criterion stresses the Popperian idea that it is growth that is the crucial characteristic of science. *The crucial characteristics of growing science are excess content rather than content, and, as we shall see, excess corroboration rather than corroboration.*

We must distinguish carefully between the concept of boldness, based on Popper's 1957 concept of independent testability, and Popper's 1934 concept of testability or scientificness. According to his 1934 criterion a theory, arrived at, after the refutation of its background theory, by a 'content-decreasing stratagem'[4] may still be 'scientific', if it is agreed that this procedure will not be repeated. But such a theory is not independently testable, not bold: according to my model it cannot be accepted (accepted$_1$) into the body of science.

It should be mentioned that Popper's Pickwickian usage of 'refutability' leads to some queer formulations. For instance, according to Popper, 'once a theory is refuted, its empirical character is secure and shines without blemish'.[5] But what about Marxism? Popper, correctly, says that it is *refuted*.[6] He also holds, in his usage correctly, that it is *irrefutable*,[7] that it has lost its empirical character because its defenders, after each refutation, produce a new version of it with *reduced empirical content*. Thus in Popper's usage irrefutable theories may be refuted. But of course, content-reducing stratagems do not, in the ordinary sense of the term, make a *theory* 'irrefutable': they make rather *a series*

[1] For further discusssion of this condition cf. my [1968*c*].

[2] Cf. his [1934], §§ 19–20.

[3] For the clearest discussion of Popper's demarcation criterion cf. Musgrave [1968], pp. 78 ff.

[4] I prefer this term to Popper's term 'conventionalist stratagem'.

[5] Popper [1963*a*], p. 240.

[6] *Ibid.*, pp. 37 and 333.

[7] *Ibid.*, pp. 34–5 and 333.

of theories (or a 'research programme') irrefutable. The series, of course, consists of non-bold theories and represents a degenerative problemshift; but it still contains *conjectures and refutations* – in the 'logical' sense of 'refutations'.

Finally, let me say that I also tried to avoid the possibly misleading expression 'degree' of excess falsifiability or of falsifiability. In general, as Popper frequently stresses, the (absolute) empirical contents of two theories are not comparable. One can state in the case of rival theories that one has excess content as compared with the other, but again, the excess is in no sense measurable. If one grants, with Popper, that all theories have equal logical probability, namely zero, logical probability cannot show the difference between the empirical content of theories. Moreover, a theory T_2, of course, may have excess content relative to T_1 while its (absolute) content may be much less than that of T_1; also, T_2 may be bold relative to T_1 and T_1 at the same time bold relative to T_2. Thus boldness is a binary, transitive but not anti-symmetrical relation (a 'pre-ordering') among theories. However, we might agree to say that 'T_2 has higher empirical content than T_1' if and only if T_2 has excess empirical content over T_1 but not *vice versa*. Then '*has higher empirical content than*' is a partial ordering relation, although '*bold*' is not.

(*b*) '*Acceptability*$_2$'

Bold theories undergo *severe tests*. The 'severity' of tests is assessed by the difference between the likeliness of the positive outcome of the test in the light of our theory and the likeliness of its positive outcome in the light of some rival 'touchstone theory' (already articulated in the extant body of science, or only articulated on the proposal of the new theory). Then there are only two types of 'severe' tests: (1) those which refute the theory under test by 'corroborating'[1] a falsifying hypothesis of it and (2) those which corroborate it while refuting the falsifying hypothesis. If all the hundreds of millions of ravens on earth had been observed and all had been found to be black, these observations would still not add up to a single severe test of the theory *A*: 'All ravens are black'; the 'degree of corroboration' of *A* would still be zero. In order to test *A* severely one must use some 'bold' (and, still better, some already corroborated) touchstone theory, say, *A'*: 'a specified substance *a* injected into the liver of birds always turns their feathers white without changing any other genetic characteristic'. Now if *a*, injected in ravens, turns them white, *A* is refuted (but *A'* corroborated). If it does not, *A* is corroborated (but *A'* refuted).[2] Indeed, in Popper's ruthless society of theories, where the law is the (shortlived) survival of the fittest, a theory can become a hero only through murder. A theory becomes testworthy on presenting a threat

[1] The definition of 'corroboration' follows *below*, p. 174.

[2] It is clear from this that there is nothing psychologistic about 'severity'.

to some extant theory; it becomes 'well-tested' when it has proved its mettle by producing a new fact that realizes the threat and liquidates a rival.

This Popperian jungle contrasts starkly with Carnap's civilized society of theories. The latter is a peaceful welfare state of fallible but respectably aging theories, reliable (according to their qualified instance confirmation) to different but always positive degrees, which are registered daily with pedantic precision in the office of the inductive judge. Murders are unknown – theories may be undermined but never refuted.[1]

Bold theories, after having been severely tested, undergo a second 'posterior' appraisal. A theory is '*corroborated*' if it has defeated *some* falsifying hypothesis, that is, if *some* consequence of the theory survives a severe test. It then becomes 'accepted$_2$' in the body of science. A theory is '*strictly corroborated at time t*' if it has been severely tested and not refuted in any test up to time t.[2]

A severe test of T relative to a touchstone theory T' tests, by definition, the *excess* content of T over T'. But then *a theory* T *is corroborated relative to* T' *if its excess* content *over* T' *is corroborated, or if it has 'excess corroboration' over its touchstone theory.* We may also say that *a theory is corroborated or accepted$_2$ if it is shown to entail some novel facts.*[3] Thus, just as 'acceptability$_1$' is related to excess content, 'acceptability$_2$' is related to excess corroboration. This, of course, is in keeping with the (Popperian) idea that it is the progressing problematic frontiers of knowledge, and not its relatively solid core, which give science its scientific character; according to justificationism it is just the other way round.

Just as T_2 may have excess content over T_1 and *vice versa*, T_2 may have excess corroboration over T_1 and *vice versa*. Excess corroboration, like boldness, is a transitive but not an anti-symmetrical relation. However, just as we defined a partial ordering on the basis of the pre-ordering 'has higher content than', we may define a partial ordering on the basis of the pre-ordering 'has excess corroboration over': 'T_2 *has a higher degree of corroboration than* T_1' if T_2 has excess corroboration over T_1 but not *vice versa*.[4]

[1] It is another characteristic example of the 'dramatic' air of Popper's theory that while he would regard the first observation of relativistic light deflection in 1919 as a severe test of Einstein's theory, he does not regard repetitions of the 'same' test as severe (since the result of the first test has now become background knowledge), and does not allow them to contribute to the corroboration achieved by the first test (cf. his [1934], §83). Popper rejects slow inductive 'learning$_{ind}$', which is based on long chains of evidence: in a dramatic 'learning$_{popp}$' process one learns in a flash.

[2] The two concepts are somewhat different. Popper's 'well-tested' is closer to my 'strictly corroborated'. Cf. *below*, p. 177.

[3] Or, more precisely, some novel factual hypotheses which stand up to severe tests.

[4] According to these definitions, T_2 may have 'higher degree of corroboration than' T_1 but T_1 may be 'bold' with respect to T_2 – although T_1 cannot have 'higher content than' T_2. I indulge in these pedantries if only to show how absurd it is to dream of an all-purpose monolithic appraisal of theories.

Popper's 1934 concept of the severity of a test of T was related to the concept of a sincere effort to overthrow T. In 1959 he still wrote: 'The requirement of sincerity cannot be formalised'.[1] But in 1963 he said that 'the severity of our tests can be objectively compared; and, if we like, we can define a measure of their severity'.[2] He defined severity as the difference between the likelihood of the predicted effect in the light of the theory under test together with background knowledge *and* the likelihood of the predicted effect in the light of background knowledge only, where Popper's 'background knowledge' is, unlike my 'background knowledge', the *unproblematic* knowledge we assume while testing.[3] That is, while my degree of severity of a test-evidence e of a theory T relative to a touchstone theory T' might be symbolized by $p(e, T) \dot{-} p(e, T')$, Popper's degree of severity of e of T relative to unproblematic background knowledge b might be symbolized by $p(e, T \& b) - p(e, b)$. In my interpretation initial conditions are *part* of a theory; indeed, Popper's b is part both of my T and of my T'. The difference is only very slight; my definition, I think, gives an additional stress to the Popperian idea that methodological concepts should be related to competitive growth. Thus for Popper a novel test *of Einstein's theory* may be 'severe' even if its result also corroborates Newton's theory. In my framework such a test is a 'severe' test *of Newton's* rather than Einstein's theory. But in both formulations the degree of severity of a test depends on the extant body of science, on some conception of the available background knowledge.

The scientist '*accepts*$_2$' a theory for the *same* purposes as he *accepted*$_1$ the bold theory before the tests. But *acceptance*$_2$ lends an added distinction to the theory. An *accepted*$_2$ theory is then regarded as a supreme challenge to the critical ingenuity of the best scientists: Popper's starting point and his paradigm of scientific achievement was the overthrow and replacement of Newtonian physics. (Indeed, it would be very much in the spirit of Popper's philosophy to issue only temporary certificates of acceptance: if a theory is accepted$_1$ but does not become accepted$_2$ within n years, it is eliminated; if a theory is accepted$_2$ but has had no lethal duels for a period of m years, it is also eliminated.)

It also follows from the definitions of acceptance$_1$ and acceptance$_2$ that a theory may be accepted$_1$ but already, at the time of its acceptance, be known to be false. Again, a theory may be accepted$_2$ but may have failed *some* severe tests. That is, one should accept bold theories into the body of science whether or not they have been refuted. One should accept them for further criticism, for further testing, for the purpose of explaining them etc., at least *as long as* there is no new bold superseding theory. This methodology allows the 'body of science' to be inconsistent, since some theories may be 'accepted$_1$', together

[1] Popper [1959], p. 418.
[2] Popper [1963*a*], p. 388.
[3] Cf. *above*, p. 170, n. 2.

with their falsifying hypotheses. Thus consistency (and, in view of 'acceptance$_0$', refutability[1]) should be construed as a regulative principle rather than a precondition of acceptance. (An important consequence of this consideration is that since the body of science may be inconsistent, it cannot be an object of rational belief. This is yet another argument for Popper's thesis that 'belief-philosophy' has nothing to do with the philosophy of science.)

The rule – 'keep testing and explaining even a refuted theory until it is superseded by a better one' – suggests a counterpart to our criteria of acceptance: a theory T_1 is '*superseded*' and *eliminated* from the body of science ('*rejected*$_1$') on the appearance of a new theory which has corroborated excess content over T_1 while T_1 has no corroborated excess content over T_2.[2]

A bold theory always challenges some theory in the extant body of science; but the *supreme* challenge is when it not only claims that the challenged theory is false but that it can explain all the truth-content of the challenged theory. Thus *refutation alone is not sufficient reason to eliminate the challenged theory.*

Because of deeply ingrained justificationist prejudices against refuted theories, scientists frequently play down refuting instances and do not take a falsifying hypothesis seriously before the latter gets embedded into a higher-order rival theory which explains also the partial success of the refuted theory. Until then falsifying hypotheses are usually kept out of the *public* body of science. But it also happens that a theory is publicly refuted though not yet replaced: the theory, known to be false, continues to be explained and tested. In such cases the theory is officially recorded as one which, in its extant version, applies only to 'ideal' or 'normal' etc. cases, and the falsifying hypotheses – if mentioned are recorded as 'anomalies'.[3]

But such 'ideal', 'normal' cases, of course, frequently do not even exist. For instance it was always known that 'ideal' hydrogen 'models'

[1] Theories like the conservation principles, which are not 'bold' in our technical sense, and may not even be testable, may also be accepted ('*accepted*$_0$') into the body of science as regulative principles or scientific research programmes. For detailed discussion and literature on this '*acceptance*$_0$' cf. volume 1, chapter 1. (Three similar 'acceptances' are discussed by J. Giedymin, in his [1968], pp. 70ff.)

[2] The 'rules of elimination' may be varied; their actual form does not matter much. But *it is vital that there be some procedures for eliminating theories from the body of science* in order to save it from following for too long a track that may lead nowhere.

[3] A detailed heuristic analysis of such situations is to be found in my [1976c] and in chapter 1, volume 1. One of the *leitmotivs* of my [1963–4] is that a refuted hypothesis may not be expelled from the body of science: that, for instance, *one may bravely – and profitably – go on to 'explain' a hypothesis known to be false.* My original case study was taken from informal mathematics but only to show that such patterns are characteristic not only of the growth of science but also of the growth of mathematics.

One may use the expression 'explain T_1 with the help of T_2' in the sense of 'explain with the help of T_2 what looks like the truth-content of T_1 in the light of T_2'. The semantics of ordinary language is unsuitable in discussing these matters, since it is based on false theories of scientific growth according to which one only explains what are alleged to be true reports of facts. Agassi's [1966] also discusses this problem.

like the ones described by Bohr's first theory do not exist – not to mention 'models' in economic theories. But this shows that theories rarely pass severe tests of their new content with flying colours; even some of the best theories may never get 'strictly corroborated'.[1] But such theories, even if in a strict sense they fail all their tests, may have some of their excess empirical content, namely some weaker but still interesting consequences, corroborated; thus they may still entail novel facts and thus be accepted$_2$. For instance, Bohr's first theories of the hydrogen atom were immediately falsified by the fact that spectral lines were multiplets;[2] but the subsequent discovery of Lyman's, Brackett's, and Pfund's series corroborated, and indeed, strictly corroborated, the weaker but still novel, previously undreamt of, consequence of Bohr's theory that there are spectral lines in the immediate neighbourhood of predicted wavelengths. Theories, while failing all their tests quantitatively, frequently pass some of them 'qualitatively': and if they lead to novel facts, then according to our definition they can still be 'accepted$_2$'.

(According to Popper's definition of corroboration a theory is *either* corroborated *or* refuted. But even some of the best theories have failed to be corroborated by Popper's austere 'strict' standards; indeed, most theories are born refuted.)

These considerations make us revise our previous criterion of *elimination* of a theory from the body of science.[3] Refutation is definitely not *sufficient* for the elimination of a theory – but its becoming 'superseded' by a more powerful theory is not *necessary* either. For if a theory, however bold, has no excess corroboration, that is, if it entails no novel *facts*,[4] we may eliminate it from the body of science without its becoming superseded by a challenger. But, of course, it may be much more difficult to establish that a theory is in this sense 'fully refuted' (in the sense that *all* its novel consequences are refuted) than to establish that it is corroborated, that is, that at least *one* of its novel consequences is corroborated. It is easier to establish acceptance$_2$ than this (full) rejection$_2$.

Acceptance$_2$ then draws a methodologically important demarcation between bold theories: while a first-class (accepted$_2$) theory is eliminated from the body of science only when it is superseded by a bold new theory, a second-class theory (with no excess corroboration) is eliminated on mere refutation, there being nothing in it that had not

[1] A theory is 'strictly corroborated' if it is corroborated but has not been refuted (cf. *above*, p. 174).

[2] The 'fine structure' of the hydrogen spectrum – amounting to a refutation both of Balmer's series and of Bohr's first models – was discovered by Michelson in 1891, twenty-two years before the publication of Bohr's theory.

[3] Cf. *above*, p. 175.

[4] It should be stressed again that we never *know* that a theory entails novel facts. All we may know is that it entails novel corroborated hypotheses. Corroborated hypotheses are the fallible (epistemological–methodological) counterparts of (ontological) facts; corroborated hypotheses may not be 'truly corroborated'.

been explained before: the (putative) truth-content of the body of science will not be decreased by such 'rejections$_2$', just as it will not be decreased by 'rejections$_1$'. A theory without excess corroboration has no excess explanatory power.[1]

All this serves as a good illustration of the Popperian problemshift: the decisive difference between the corroboration and the refutation of a theory is primarily a matter for the logic of discovery and not for the logic of justification. *Thus one may accept$_1$ and accept$_2$ theories even if they are known to be false, but reject (reject$_1$) theories even if there is no evidence against them.*[2] This conception of acceptance and rejection is a feature totally alien to the classical outlook. We accept theories if they indicate *growth* in truth-content ('progressive problemshift'); we reject them if they do not ('degenerative problemshift'). This provides us with rules for acceptance and rejection even if we assume that all the theories we shall ever put forward will be false.

One of the most important features of the two methodological appraisals of theories is their '*historical character*'.[3] They depend on the state of background knowledge: the prior appraisal on the background knowledge at the time of the proposal of the theory and the posterior appraisal also on the background knowledge at the time of each test.

The 'historical character' of these appraisals has interesting implications. For instance, a theory explaining three hitherto seemingly unrelated lower level well-corroborated theories but nothing more would have no excess empirical content over the conjunction of its predecessors, and therefore both its prior and posterior appraisal may turn out to be unfavourable. But if the theory had followed the first two lower level theories but preceded the third, it would have had 'excess falsifiability' and would also have had excess corroboration; it would have ranked high on both counts. The Bohr–Kramers–Slater theory ranked high on the first appraisal but it failed its first test and

[1] Incidentally, it is interesting that for universal propositions Popper's formulae for explanatory power and for degree of corroboration coincide: one gets the formula for degree of corroboration by multiplying the formula for explanatory power by $1+p(h)p(h,e)$. But $p(h)$ for universal propositions, according to Popper, is zero. Therefore if e is interpreted as the total *test*-evidence, explanatory power and degree of corroboration will become synonymous. (Cf. Popper [1959], p. 401.)

[2] Perhaps at this point we should mention that the 'body of science' as a body of deductively perfectly organized theories with crystal clear rules of acceptance$_1$ and acceptance$_2$, rejection$_1$ and rejection$_2$, is an *abstraction*. I neither assume that such a 'body of science' has existed at any time nor that even as an abstraction it is useful for *all* purposes. For many purposes, it is better to construe science – as Popper stresses in his later philosophy – rather as a *body of problems* than a body of theories.

[3] After having stated that prior and posterior appraisal are mere corollaries to methodology, this should be no surprise. Methodology is wedded to history, since methodology is nothing but a rational reconstruction of history, of the growth of knowledge. Because of the imperfection of the scientists, some of the actual history is a caricature of its rational reconstruction; because of the imperfection of the methodologists, some methodologies are caricatures of actual history. (And, one may add, because of the imperfection of historians, some histories of science are caricatures both of actual history and of its rational reconstruction.)

never became corroborated. But had it been put forward earlier, it could have survived a few first tests and thus it would have gained also high posterior ranking before its death.[1] Agassi points out that such examples indicate that the second appraisal – at least in some cases – may remunerate a 'prior' virtue of the theory, namely that it was proposed boldly and *quickly*, before its whole true 'factual' content was discovered without its stimulus.[2] One might have thought that while a positive prior appraisal indirectly applauds the inventor of the theory, a positive posterior appraisal can show nothing but his luck; one can aim at bold theories but one cannot *aim* at well-corroborated theories. It is up to us to devise bold theories; it is up to Nature whether to corroborate or to refute them. But Agassi's analysis shows that this is not *entirely* correct: the bolder the theory, the more chance it has of excess corroboration. Thus *prior and posterior appraisal appraise conjointly the growth of our knowledge produced by the theories rather than the theories in themselves.*

Popper never actually introduced the requirement of 'acceptability$_2$'; nevertheless, 'acceptability$_2$' is only an improved version of his new '*third requirement*' that a satisfactory theory should pass some independent tests *and* should not fail on the *first* one.[3] It was careless of Popper to attach importance to the *first* test being failed.

Popper's supporting arguments can also, I think, be improved. In order to do so, let me first contrast *two models of scientific growth.*

In the '*Popperian model*' the growth of science is regulated by the rules of acceptance and rejection as outlined above. The '*Agassite model*' differs from this in one single aspect: complete lack of excess corroboration is no reason for rejection$_2$, and if a theory explains all the truth-content of its predecessor, it may 'supersede' it even without having excess corroboration.

Now let us take the following sequence of theories and refutations:

(1) A major theory T_0, accepted$_2$, is refuted by a minor falsifying hypothesis f_1, which is also accepted$_2$.[4] The (relevant part of the) body of science in both models consists of T_0 and f_1.

(2) T_1 is proposed. T_1 is bold, explains all the truth-content of T_0 as well as f_1; its excess content is e_1. But e_1 is 'fully refuted', T_1 is rejected$_2$. The refuting hypothesis is f_2 and it is accepted$_2$.

[1] Similarly, the theory T_0, that at dusk flying saucers fly over Hampstead Heath is, I guess, 'fully refuted' if put forward in 1967. But let us imagine that we have never before seen any flying objects – animate or inanimate; and that according to our theories it is *impossible* that objects should fly. Now if T_0 is put forward in such historical circumstances and is carefully tested, it might well be corroborated by the observation of some flying owls. The theory then has led to the discovery of a revolutionary new fact: that there exist (well specifiable) flying objects. Flying saucers would have entered the history of science rather as Newton's forces acting at a distance did.

[2] Cf. his most interesting [1961], pp. 87–8.

[3] Popper [1963*a*], pp. 240–8.

[4] 'Major' means comprehensive, having a huge content; 'minor' means a low-level, factual hypothesis.

In the Popperian model the body of science now consists of T_0, f_1, f_2. In the Agassite model it consists of T_1 and f_2.

(3) T_2 is proposed. T_2 is bold, explains all the truth-content of T_1 as well as f_2; its excess content is e_2. But e_2 is 'fully refuted', T_2 is rejected$_2$. The refuting hypothesis is f_3 and it is accepted$_2$.

In the Popperian model the body of science now consists of T_0, f_1, f_2, f_3. In the Agassite model it consists of T_2 and f_3. And so on.

This shows that Popper rejects theories T_1 and T_2 as *ad hoc*,[1] while Agassi does not. For Popper such growth is *pseudo-growth* which does not live up to his *ideal of growth*. He would agree that T_1 and T_2 are heuristically stimulating since they 'led' to f_2 and f_3; but according to him, in such growth theories are mere stimulants, 'mere instruments of exploration'.[2] For Popper there is no 'growth of knowledge' without at least a chance of growth in verisimilitude;[3] in the Agassite model verisimilitude seems to stagnate or even decrease in the sequence $\{T_0, T_1, T_2\}$, which, therefore, for Popper, represents a degenerating shift. The verisimilitude of the sum of the increasing number of falsifying hypotheses may of course increase; but, for Popper, this is an 'inductivist' disintegration of science into a collection of isolated phenomena.

But, following the line of Agassi's argument, let us imagine that after T_0 and f_1, T_2 is immediately proposed. T_2 will then be accepted$_1$ and also accepted$_2$, for f_2 is part of its excess content. Now why should $\{T_0, T_1, T_2\}$ represent a degenerative shift when $\{T_0, T_2\}$ represents a progressive shift?

The argument is interesting. But instead of being an argument *against* the 'Popperian model', it gives a final touch to its clarification. According to Popper the essence of science is growth: *fast potential growth* (acceptability$_1$) and *fast actual growth* (acceptability$_2$). Slow growth is not good enough to live up to Popper's ideal image of science. If imagination does not fly fast enough ahead of the discovery of facts, science degenerates.[4] The Popperian model exposes this degeneration, the Agassite model covers it up.

[1] Popper uses his pejorative term '*ad hoc*' in two clearly distinguishable senses. A theory without excess content is '*ad hoc*' (or rather '*ad hoc*$_1$'): cf. his [1934], § 19 and his [1963*a*]. p. 241. But since 1963, he has also called a theory without excess corroboration '*ad hoc*' (or rather '*ad hoc*$_2$'): cf. his [1963*a*], p. 244.

[2] Cf. his [1963*a*], p. 248, n. 31.

[3] In Popper's philosophy 'verisimiltude' denotes the difference between the truth-content and falsity-content of a theory. Cf. chapter 10 of his [1963*a*].

[4] Mary Hesse was then wrong when she claimed that Popper's 'third requirement' was 'not consistent with the main planks of Popper's anti-inductivist position' [1964], p. 118). So was, in this respect, Agassi, who, as Popper tells us, regarded Popper's third requirement as 'a residue of verificationist modes of thought'; but so was Popper, who, in spite of his right instinct, 'admitted' that 'there may be a whiff of verificationism here' [1963*a*], p. 248, n. 31). In fact, this requirement can be regarded as a main plank of Popper's anti-inductivist position, since, without it, by Popper's standards, science would degenerate into a collection of facts. Moreover this is one of the most typical examples of how independent Popper's methodology is of 'inductive' considerations.

Whether science will be able to live up to these wonderful standards is, of course, a different matter. If it does, then it will have proceeded not simply through conjectures and refutations, but through (bold) *conjectures, verifications and refutations.*[1]

(c) '*Acceptability$_3$*'

Popper's methodological appraisals contrast sharply with the classical and neoclassical tradition. Since 'ordinary' thinking and 'ordinary' language are saturated with this tradition, laymen (and also 'ordinary' philosophers and scientists) find Popper's ideas difficult to digest. No doubt, they will find my use of the term 'acceptance' for his concepts of acceptance$_1$ and acceptance$_2$ particularly idiosyncratic. What is so unusual for most people is the underlying idea that one may 'accept' a statement into the body of science before even looking at the evidence; that the degree to which it has transcended (or even negated!) the accepted evidence should count in its favour instead of to its disadvantage. Acceptance$_1$ runs counter to the classical dogma that 'to discover is to prove', and to the neoclassical dogma that the degree of scientific 'acceptance' of a hypothesis increases when the gap between hypothesis and evidence decreases. Moreover, the idea that hypotheses which are false and even known to be false may, under certain strange conditions, be 'accepted', sounds totally incredible to traditional empiricists. Also they may find it difficult to comprehend that for acceptance$_2$ the facts which the theory was devised to explain (i.e. which had been discovered before testing began) are irrelevant; and so are all further observations unless they represent severe tests between the theory and some competitor. All this runs counter to the classical and neoclassical dogma according to which each confirming instance counts, however little.[2]

But, apart from the details, Popper's spectrum of appraisals confuses the 'justificationists'. For them – whether orthodox or revisionist – there is only one unique, monolithic, all-purpose '*scientific acceptance*': acceptance of a theory into the body of science *to the degree to which it has been proved.*[3]

Such an idea has always been rejected by Popper. Nonetheless, many philosophers, even if they agree that Popper *did* open up a gamut of important appraisals, would still contend that there are vital problems

[1] 'Verification' stands here for 'excess corroboration'.

[2] Recently, under the influence of Popper's arguments, Hintikka constructed an inductive logic, which, in this respect – and also in some others – deviates from neoclassical dogma (cf. his [1968], p. 191ff.).

[3] This is the historical explanation of what Bar-Hillel very aptly calls the '*acceptance syndrome*' which results from the assumption that '"to accept" has one unique meaning in all contexts'. The idea of degree of provenness may have been tacitly discarded by many philosophers, but the syndrome of all-purpose acceptance – 'truly amazing in its lack of sophistication' – still lingers on. (Cf. Bar-Hillel [1968*b*], p. 150ff.)

for the solution of which even he needs *some* concept of '*acceptability*$_3$' ('inductive acceptability', 'trustworthiness', 'reliability', 'evidential support', 'credibility', etc.). This 'acceptability$_3$' – in whatever particular version – is required for an appraisal of the *future performance* of the theory, and it is alleged that it cannot be estimated without *some* inductive principle.

Acceptability$_3$ was originally the dominating aspect both of the classical and of the neoclassical all-purpose appraisal. A theory was 'accepted' primarily if it was judged to yield reliable predictions.

The reason for the paradoxical air of Popper's methodological appraisals is that for all philosophers before him there had been one single conception of acceptability: acceptability$_3$. But for Popper acceptability meant primarily acceptability$_1$ and/or acceptability$_2$.

One example of this confusion is the widespread argument purporting to show that Popper's scientific method rests on inductive considerations concerning acceptability$_3$. The argument proceeds as follows:

(1) Let a theory T be falsified in 1967. Nobody would regard its falsification as a sign of good future prospects for a theory, as a sort of infantile disease which even the healthiest hypotheses could be depended upon to catch, but which then rendered them immune to future attack. Hence, we reject the 'counterinductive policy' of replacing a theory T_1, refuted in 1967, by a modified version T_2 of it which restricts its validity to the period after 1967.

(2) But the *only* possible rationale for rejecting such a 'counterinductive policy' is the tacit inductive principle that theories refuted in the past will continue to be refuted in future, that is, that T_2 is untrustworthy, that it is not *acceptable*$_3$.

(3) Since Popper too would reject T_2, he *must* hold this inductive principle: he must regard T_2 as *not acceptable*$_3$. But then his methodology rests on an inductive principle, contrary to his claims. Q.E.D.[1]

But (2) is false; and so is the concluding part of (3). Popper does not bar T_2 because it is unacceptable$_3$, but because it has no excess empirical content over T_1: because it is unacceptable$_1$. In formulating Popper's methodology there is no need to refer to acceptability$_3$.

(i) '*Total corroboration*' *as a measure of* '*acceptability*$_3$' *of theories.* It seems obvious to me that the *basis* for a definition (or 'explication', as Carnap would put it) of the intuitive idea of acceptability$_3$ should be Popper's 'verisimilitude': the difference between the truth-content and falsity-content of a theory.[2] For surely a theory is the more acceptable$_3$, the nearer it is to the truth, that is, the greater its verisimilitude.

[1] Cf. Ayer [1956], pp. 73–4. Also cf. Wisdom [1952], p. 225.

[2] Popper [1963*a*], chapter 10, esp. pp. 233–4; also cf. Watkins [1968], p. 271ff. *But see Miller [1974] and Tichý [1974]. (*Eds.*)

Verisimilitude is Popper's reconstruction of non-probabilistic 'probability':[1] but while Carnap claims to be able to calculate *his* probability infallibly, Popperian 'probability' – verisimilitude – cannot be infallibly known, for in Popper's philosophy there is no way of discovering with certainty the truth-values of propositions.

But which are the most 'verisimilar' theories? I think that these can (tentatively) be constructed in the following way: we take the extant 'body of science' and replace each refuted theory in it by a weaker unrefuted version. Thus we increase the putative verisimilitude of each theory, and turn the inconsistent body of scientific theories (accepted$_1$ and accepted$_2$) into a consistent body of accepted$_3$ theories, which we may call, since they can be recommended for use in technology, the 'body of technological theories'.[2] Of course, some accepted$_3$ theories will not be acceptable$_1$ or acceptable$_2$ since we arrived at them by content-reducing stratagems; but here we do not aim at scientific growth but at reliability.

This simple model is a rational reconstruction of the actual practice of choosing the most reliable theory. Technological choice follows scientific choice: acceptable$_3$ theories are modified versions of acceptable$_1$ and acceptable$_2$ theories: the way to the acceptable$_3$ theories leads through acceptable$_1$ and acceptable$_2$ theories. *For the appraisal of trustworthiness the methodological appraisals are indispensable.*

We may try to assess acceptability$_3$ also in terms of 'degree of corroboration'. Severity and corroboration (or rather 'excess corroboration') as we defined them, are binary relations between the tested theory T and some touchstone theory T' (or possibly even ternary relations between T, T' and some test-evidence e). Because of this, corroboration turned out to have 'historical character'. But it may seem that the verisimilitude of a theory in the light of evidence must be independent of its prehistory. Indeed, it is a deeply entrenched dogma of the logic of justification that evidential support depends only on the theory and the evidence and certainly not on the growth that they represent in relation to former knowledge.[3] As Keynes put it: 'The peculiar virtue of prediction...is altogether imaginary...the question as to whether a particular hypothesis happens to be propounded before or after [the] examination of [its instances] is quite irrelevant.'[4] Or, as a recent critic of Popper put it: '[To enquiries concerning evidential support] it is quite irrelevant whether in fact

[1] Popper [1963*a*], esp. pp. 236–7 and second edition, 1965, pp. 399–401.

[2] If we have two rival, inconsistent theories in the body of science such that neither of them has superseded the other, then their 'trimmed', 'technological' versions may still be inconsistent. In such cases we may either cautiously choose the maximal consistent subset of the propositions of the two theories, or daringly choose the theory with the more empirical content.

[3] There have been notable exceptions, e.g. Whewell: see Agassi [1961], pp. 84 and 87. Popper's requirement for 'independent testability' has a long – and interesting – prehistory.

[4] Keynes [1921], p. 305.

scientists always or usually or never make their observations before conceiving their theories or *vice versa*.'[1] However, the dogma of independence of evidential support from prehistory is false. It is false because the problem of the *weight of evidence* cannot be solved without historico-methodological criteria for 'collecting' theories and evidence.[2] Both the truth-content and the falsity-content of any theory contains infinitely many propositions. How can we minimize the bias of the sample? Most of those people who were aware of this problem suggested the *tabula rasa* solution: evidence must be collected without theoretical bias. The final destruction of this solution was achieved by Popper. *His* solution was that only evidence which was the result of severe tests, 'test-evidence', should count: the only admissible positive evidence for a theory are the corpses of its rivals. Evidential support *is* a historico-methodological concept.

But here one has to be careful. Evidential support for a theory obviously depends not just on the *number* of the corpses of its rivals. It also depends on the *strength* of the killed. That is, evidential support is, as it were, a hereditary concept: it depends on the total number of rival theories killed by the killed rivals. Their set, *in toto*, determines the '*total corroboration*' of a theory.[3] In the assessment of the reliability of a theory *all* the corpses on the long road leading from the most naive expectations to the theory should be taken into consideration.[4] (The conflation of excess corroboration (the tentative estimate of growth) and total corroboration (the tentative estimate of reliability) is a confusing element in Popper's – and Agassi's – presentation.[5])

This argument shows that Popper's 'best-corroborated' theories (in the 'total' sense) almost exactly coincide with our accepted_3 theories.

But whichever criterion of acceptability_3 we choose, it will have *two very serious shortcomings*. The first is that *it gives us very limited guidance*. While it offers us a body of 'most reliable' theories we cannot compare with its help the reliability of any two theories among these 'most reliable' ones. One cannot compare Popper's (total) 'degrees of corroboration' for two unrefuted theories which have stood up to severe tests. All that we can know is that the theories in our latest body of accepted_3 theories have higher degrees of corroboration than their 'predecessors' in any past, discarded body of accepted_3 theories. A theory T_2 that supersedes T_1, inherits from T_1, the set of theories which

[1] Cf. Stove [1960], p. 179.

[2] Also cf. *above*, p. 148.

[3] *Throughout the rest of the paper, 'corroboration' stands for 'total corroboration'; 'best-corroborated' stands for 'with maximal total corroboration'.*

[4] We may, alternatively, articulate a 'most naive expectation' as touchstone theory and assess the severity of tests relative to that reconstructed theory. If the theory under test is a statistical theory, we may use some prior Laplacean distribution as a touchstone theory. But this approach may lead to misleading results. (Cf. *below*, p. 198ff.)

[5] It is, incidentally, connected with Popper's failure to distinguish sharply between the merits of excess content and content, discussed *above*, p. 172ff.

T_1 had defeated: the corroboration of T_2 will clearly be higher than the corroboration of T_1. But the corroborations of two theories T_1 and T_2 can only be compared when the set of defeated theories in T_1's past is a subset of the set of defeated theories in T_2's past: that is, when T_1 and T_2 represent different stages of the same research programme. This circumstance reduces drastically the practical, technological use of corroboration as an estimate of reliability for competing technological designs. For each such design may be based on some theory which, in its own field, is the most advanced; therefore each such theory belongs, in its own right, to the 'body of technologically recommendable theories', of theories accepted$_3$; and therefore their degrees of corroboration will not be comparable. There is not, and cannot be, any *metric* of 'degree of corroboration' – indeed the expression 'degree of corroboration', in so far as it suggests the existence of such a metric, is misleading.[1]

But where corroborations of two theories are incomparable, so are their reliabilities. This is entirely plausible. We can only judge the reliability of eliminated theories from the vantage point of our present theories. For instance, we can give a detailed estimate of the reliability, or verisimilitude, of Newton's theory from the point of view of Einstein's theory: we may issue the warning that it is particularly unreliable for high velocities, etc. Even so, the estimate will be fallible, since Einstein's theory is fallible. But we cannot give even a fallible absolute estimate of Einstein's theory itself before it, in turn, is superseded by another theory. *Thus we cannot grade our best available theories for reliability even tentatively, for they are our ultimate standards of the moment.* Only God could give us a correct, detailed estimate of the absolute reliability of *all* theories by checking them against *his* blueprint of the universe. Inductive logicians, of course, do offer such an estimate: but their estimate depends upon an *a priori* superscientific inductive knowledge.[2]

The *second serious shortcoming* of our criterion of reliability is that it is unreliable. Even where comparisons are possible, one can easily conceive of conditions which would make the estimate of verisimilitude by corroboration false. The successive scientific theories may be such that each increase of truth-content could be coupled with an even larger increase in hidden falsity-content, so that the growth of science would be characterized by increasing corroboration and decreasing verisimilitude. Let us imagine that we hit on a true theory T_1 (or on one with very high verisimilitude); in spite of this we manage to 'refute' it with the help of a corroborated falsifying hypothesis f,[3] replace it

[1] For a criticism of Popper's metric for degree of corroboration of statistical theories see *below*, p. 197ff.

[2] Cf. *below*, p. 188.

[3] If a theory is refuted, it is not necessarily false. If God refutes a theory, it is 'truly refuted'; if a man refutes a theory, it is not necessarily 'truly refuted'. Ordinary language does not distinguish sufficiently between truth and alleged truth, between

by a bold new theory T_2 which again gets corroborated, etc., etc. Here we would be following, unknowingly, the twists and turns of a disastrous problemshift, moving even further from the truth – while assuming that we are soaring victoriously towards it. Each theory in such a chain has higher corroboration and lower verisimilitude than its successor: such is the result of having 'killed' a true theory.

Alternatively, let us imagine that in 1800 somebody proposed a stochastic law that in natural processes entropy decreases. This was corroborated by a few interesting facts. Then somebody else discovered that the facts were only due to fluctuations, each with zero probability, and set up the second law of thermodynamics. But what if the decrease of entropy is indeed a natural law, and only our small space-time corner of the universe is characterized by such a major, very unlikely fluctuation?[1] The most rigorous observance of Popperian method may lead us away from truth, accepting false and refuting true laws.

Thus the estimates of reliability or verisimilitude by Popper's 'degree of corroboration' may be false – and therefore, of course, they are unprovable. Certainly, *if science, as it progresses, approximated to truth* (in the sense that its verisimilitude increased with increasing corroboration) *then our estimates would be correct.* The question immediately arises, is *this* assumption the inductive principle on which our philosophy of technology hinges? But in my view, whether or not a proposition is an 'inductive principle' depends not only on the proposition in itself, but also on its epistemic status and function: the truth of an inductive principle must be established *a priori*, because its function is to be the premise in a *proof* or justification. It is certainly interesting to ask what metaphysical conditions would make our estimates of verisimilitude correct. But the metaphysical statements (*not inductive principles*) specifying these conditions will not *prove* the thesis that the ordering generated by degree of corroboration necessarily equals the ordering generated by degree of verisimilitude: they will rather call attention to the possibility that they might not be satisfied and thereby *undermine* its universal validity. *There is nothing wrong with fallible*

methodological concepts and their metaphysical counterparts. The time is ripe for purifying it of such sacrilegious usages.

Corroborated falsifying hypotheses (or 'falsifying facts') are widely believed to be particularly hard facts; nevertheless they too are frequent casualties of scientific growth. However, it so happens that even if a corroborated falsifying hypothesis of T is refuted, it always has enough strength left to keep T refuted. If not for this incomprehensible feature of the growth of knowledge, Popper could not have ruled that falsification is (methodologically) 'final', that 'a corroborative appraisal made at a later date...can replace a positive degree of corroboration by a negative one, but not *vice versa*' ([1934], §82).

[1] This again may seem to many people very unlikely. Boltzmann actually thought it likely, as it transpires from his [1896–8], §90. (I owe this reference to Popper.)

speculative metaphysics, but only with interpreting some such metaphysical statements as infallible inductive principles.[1]

For instance, there is nothing wrong with speculating about the conditions for a body – and in particular, a growing body – of scientific (or technological) theories to come into existence – and stay in existence (one of the many possible conditions would be that there be natural laws), or about the conditions necessary for our survival while acting upon our best theories. One possibility is that *our best-corroborated theories happen to have relatively large verisimilitude in the set of those of their consequences which are related to the small spatio-temporal corner of the universe which is our 'home', and their verisimilitude grows with the advancement of science.* This simple but crucial metaphysical assumption would explain mankind's technological success, but it may be false. But since it is irrefutable, we can never discover that it *is* false, if it is: the biggest disasters cannot disprove it (just as the biggest successes cannot prove it). We may 'accept_0' this assumption into our body of 'influential' metaphysical theories[2] *without believing it,* just as we keep accepting_1 and accepting_2 false and even mutually inconsistent theories into our body of science.

These considerations show that even technology can do, or *rather, must do* without 'inductive principles' although it may 'rely' on some (technologically) influential metaphysics.

But there is an immense difference between the ordinary (classical) and probabilistic (neoclassical) conceptions of reliability on the one hand, and our Popperian conception of 'reliability' on the other.

The *classical conception* regards a theory as reliable if it is true, unreliable if it is false. Rationality is acting on true theories; and rational action is unfailingly rewarded by success. The ultra-dogmatist wing of classical empiricism maintains that it can recognize – like God – the truth or falsehood of theories; the ultra-sceptical wing that knowledge, and therefore, rational action, is impossible.

The *neoclassical conception* regards a theory as reliable to a certain degree – in Carnap's 1950 view according to its 'qualified instance confirmation'. To each theory, at any time, there belongs a number-between 0 and 1, indicating, with the certainty of logic, its reliability. Thus *reliability has a precise and absolutely reliable proven metric.* The most reliable theory may, however, let one down: the cleavage between rational action and success is larger than in the classical conception. *But we can still know what risk we take and foresee the possible sorts of disaster,* together with their respective probabilities. Each proposition has a precise quantitative measure of reliability. In Carnap's approach the super-scientific inductive knowledge needed to determine this

[1] It should be stressed that in my usage 'inductive principle' is not restricted to principles which imply a *probabilistic* confirmation function, but is any principle claimed to be *a priori* true which implies a confirmation function – whether the latter is probabilistic or not.

[2] For the idea of (scientifically) 'influential metaphysics' cf. Watkins's important [1958].

metric stems from his (hidden and, indeed, disavowed) claim that he knows the probability distribution of the possible worlds before the blindfolded Creator selected one and turned it into the real world.[1]

The *Popperian conception* of 'reliability', as explained here, differs in being a conception of reliability, which, in turn, is itself unreliable. In it we distinguish between '*true reliability*' – unknown to us – and '*estimated reliability*'. This is a direct consequence of the lack of inductive principles in this approach.

It is also evident that in this Popperian approach reliability has nothing to do with 'rational belief': why should it be 'rational' to believe that the universe satisfies all those conditions which would make corroboration a correct estimate of verisimilitude?

The Popperian approach offers no metric, no absolute degrees of reliability, but, at best, a very weak (and, of course, in addition, unreliable) partial ordering among theories. The monadic predicate 'reliable' is replaced by the binary predicate: 'more reliable than'.

Moreover, not only may the most reliable theory let one down, but the theory of reliability is itself unreliable. *We cannot know what risk we are taking and we cannot foresee the possible shapes of disasters and still less their precise probability.* According to Carnap, for instance, even if you rationally predict that you will pull out a blue ball from an urn, you must be prepared (to a well-definable degree) to pull out a white or a red one (according to his metaphysical theory of possible universes, as reflected in his language). But for Popper the possible variety of the universe is unlimited: you may equally well pull out a rabbit, or your hand may be caught in the urn, or the urn may explode, or, rather, you may pull out something dramatically unexpected that you cannot possibly understand or even describe. Urn games are poor models of science.

The cleavage between rationality and 'success' is then much wider in the Popperian approach than it was in the previous approaches: so much so that Popperian 'reliability' should always be in quotes.

(ii) *Popper's opposition to 'acceptability$_3$'*. Popper has never cared to devote much attention to the problem of acceptability$_3$. He regards the problem as 'comparatively unimportant'.[2] Indeed, he was right: what one can say about it is not very much. But his casual remarks about the subject are confusing.

On the one hand he stresses time and again that 'the best we can say of a hypothesis is that *up to now* it has been able to show its

[1] Incidentally, Carnap's programme has at least one superficial similarity to Hilbert's: both gave up intrinsic certainty for propositions in the object-language, but both wanted to re-establish it for propositions in the meta-language. For Hilbert indubitable meta-mathematics was to establish – by a feedback-effect, as it were – at least the consistency of mathematics. For Carnap indubitable meta-science (inductive logic) was to establish at least the reliability metric of science. (Also cf. *above*, chapter 1).

[2] Popper [1968*a*], p. 139.

worth';[1] one cannot infer from 'degree of corroboration' to trustworthiness. Anyone, then, who interprets his 'degree of corroboration' as having anything to do with degree of corroboration as generally understood, misinterprets him. As Watkins put it: 'A Popperian corroboration-appraisal *is* analytic and does *not* have predictive implications'.[2] Popper's theory of corroboration then maintains a stony silence about the future prospects of the theory. But if all yet unchecked states of affairs were, say, equally possible, the degree of rationality of belief in any yet unchecked particular proposition would be zero. Indeed, sometimes one wonders whether Popper and Watkins would regard *any* consideration of acceptability$_3$ as a criminal act of 'induction'.

Salmon rightly points out that if Popper's appraisal of scientific theories is analytic, then Popper cannot explain how science can be a guide of life.[3] If degree of corroboration does not serve as an estimate, however fallible, of verisimilitude, then Popper cannot explain the rationality of our practical actions, cannot have any practical philosophy and especially, any philosophy of technology which is *based on science.*

One reaction of Popper and some of his colleagues adds up to a curious doctrine that practical rationality is independent of scientific rationality. Popper stresses that for practical purposes 'false theories often serve well enough: most formulae used in engineering or navigation are known to be false'.[4] As Watkins put it: 'Our method of hypothesis-selection in practical life should be well-suited to our practical aims, just as our methods of hypothesis-selection in theoretical science should be well-suited to our theoretical aims; and the two kinds of method may very well yield different answers.'[5] Moreover, Watkins claims that a theory 'may very well be both *better* corroborated by past tests and *less* likely to survive future tests'.[6] So reliability may even be inversely proportional to corroboration! Down with applied science?

On the other hand, we also find strong hints that, even for Popperians, science *is* a guide of life. For instance, Popper writes: 'Admittedly it is perfectly reasonable to believe that...well-tested laws will continue to hold (since we have no better assumption to act upon), but it is also reasonable to believe that such a course of action will lead us at times into severe trouble.'[7] Moreover, he even seems to hint that degree of corroboration might be a reasonable estimate of verisimilitude.[8] And I have already mentioned his one statement that 'degree

[1] The quotation is from a note published in *Erkenntnis* 1935 and reprinted in Popper [1959], p. 315. The italics are mine.

[2] Watkins [1968*a*], p. 63.

[3] Cf. his [1968], pp. 95–7.

[4] Popper [1963*a*], p. 57.

[5] Watkins [1968*a*], p. 65.

[6] *Ibid.*, p. 63.

[7] Popper [1963*a*], p. 56. Also cf. Watkins' concluding statement, in similar vein, of his [1968*a*], p. 66.

[8] *Ibid.*, p. 235.

of corroboration' can be interpreted as 'degree of rationality of belief'.[1]

Now where does Popper stand? Is *any* sort of 'acceptability$_3$' an 'inductivist concept' or not? Unfortunately Popper does not define (contemporary) '*inductivism*' with sufficient clarity. As far as I can see he interprets it as a combination of *three doctrines*.

The first doctrine of inductivism is *the doctrine of inductive method*: it postulates the primacy of 'facts' in the logic of discovery. The second doctrine of (contemporary 'neoclassical') inductivism is *the doctrine of the possibility of inductive logic*: it postulates that it is possible to assign – with the certainty of logic – to any pair of propositions a 'degree of confirmation', which characterizes the evidential support that the second proposition lends to the first. The third doctrine is that *this 'confirmation function' obeys the probability calculus.*[2]

Popper rejects inductive method and replaces it by his theory-dominated logic of discovery. He rejects the possibility of inductive logic, for it would have to rest on some synthetic *a priori* principle. Finally he proves that the statement that the confirmation function is probabilistic is not only unprovable but false.[3]

On this interpretation of Popper's position, a theory of acceptability$_3$ would only be inductivist if it claimed to be *a priori* true and/or if it was probabilistic; and there would be nothing wrong with a conjectural, non-probabilistic estimate of acceptability$_3$ of theories or with the non-inductive metaphysical speculations which may underlie such an estimate. But Popper's insistence, that his degree of corroboration – unlike Reichenbach's or Carnap's degree of confirmation – is analytic and must not be interpreted as being synthetic,[4] amounts to an opposition to *any* acceptability$_3$. This implies a sharp separation of scientific and practical rationality, which, indeed, both Popper and Watkins seem to advocate. Such a separation may indeed be 'fishy and hypocritical',[5] and leads to misinterpretations of what actually happens in technology.[6]

An escalation of the anti-inductivist crusade which makes a target of *any* concept of acceptability$_3$ can only vitiate its effectiveness. It

[1] Cf. Popper [1959], pp. 414–15 and 418.

[2] One may ask: was the three-headed inductivist dragon empiricist or rationalist? The high methodological regard for facts in the first head would suggest an empiricist; the synthetic *a priori* inductive principles in the second head suggest a rationalist. But in the light of Popper's philosophy one can explain the previously paradoxical fact that extreme apriorists are frequently extreme empiricists and vice versa (e.g. Descartes, Russell, Carnap): most kinds of empiricism and rationalism are only different variants (or components) of justificationism.

[3] These three points are summed up already in the first section of his [1934].

[4] *Ibid.*, §82.

[5] Cf. Watkins [1968*a*], p. 65.

[6] In most cases where Popper and Watkins allege that false theories are applied, one can show that the best-corroborated theories are in fact applied. Their argument looks plausible only because in those special examples there happens to be no difference between *applying* the best and the superseded next best theory.

should be explicitly abandoned, and it should be admitted that science is at least *a* guide of life.[1]

(The disarray of Carnap's camp is, in this respect, much worse. Carnap's apriorist metaphysics is hidden under a cloak of analyticity.[2] Carnapians have either to stick bravely to the 'analyticity' of their inductive logic, but then their probability cannot be a guide of life, as Salmon and Kneale keep reminding them;[3] or they can decide that inductive logic *is* a guide of life, but then admit that their inductive logic is an elaborate system of speculative metaphysics.[4])

This concludes the discussion of our three appraisals of theories. The first two appraise the growth achieved by a theory relative to some touchstone theory. Given the two theories, the first appraisal, based on excess empirical content, is a matter of logic, and can be said to be tautologous. The second appraisal has two interpretations: interpreted as a 'tautologous' appraisal it states that the new theory *survived* a test which the touchstone theory did not survive: this alone may serve as an appraisal of growth. Interpreted as a synthetic appraisal (with the fallible metaphysical lemma that excess corroboration means excess truth-content), the second appraisal hopefully guesses that the growth is real, that the new theory, at least in the 'field of application'[5] of the test, is nearer to the truth than its touchstone theory.

The third appraisal compares the total evidential support of theories. If we interpret it as a 'tautologous' appraisal, it merely draws up a balance-sheet of the victories and defeats of the research programmes that led to the compared theories. But then it may be misleading to call this an appraisal of 'evidential support', for why should even the greatest past victories give, without any additional metaphysical assumption, any real 'support' to the theory? They only give 'support' to the theory on the tentative metaphysical assumption that increasing corroboration is a sign of increasing verisimilitude. Thus we have two concepts of 'evidential support': one, 'tautologously', appraises the tests which the theory in its prehistory (or the 'research programme' leading to it) *has survived*; the other, with the help of metaphysical lemmas, synthetically appraises its *fitness to survive* (in the sense that having more verisimilitude, more of it will be able to 'survive').[6]

[1] One has to remember that since it is impossible to compare the degrees of corroboration of our most advanced theories, in many technological decisions pure epistemic considerations play very little part. The fact that theories frequently cannot be compared for reliability makes practical rationality *more independent* of scientific rationality than over-optimistic inductive logic suggests.

[2] Cf. *above*, pp. 160 and 165.

[3] Cf. Salmon [1968*a*], especially pp. 40ff.; and Kneale [1968], pp. 59–61.

[4] Bar-Hillel, in his [1968*a*] (pp. 66ff.) does not, I fear, make it clear where he stands.

[5] For 'field of application' cf. Popper [1934], *passim* (see the index of the English edition).

[6] In Popper's view, of course, survival does not imply *fitness for survival*. But, misleadingly, he uses the two terms as synonyms throughout his [1934] (cf. e.g. pp. 108 and 251 of the English edition).

7 THEORETICAL SUPPORT FOR PREDICTIONS VERSUS (TEST)-EVIDENTIAL SUPPORT FOR THEORIES[1]

The practical rule suggested by our considerations is this: '*Act upon the unrefuted theories which are "contained" in the extant body of science, rather than on theories which are not*'.

However this rule offers us – limited – guidance concerning only the choice of the most 'reliable' theory. But what about particular predictions? There we opt for the relevant prediction of the chosen theory.

The 'reliability' of particular propositions may then be characterized in two distinct steps: *first* we decide, if we can, which is the most 'reliable' theory among the relevant ones and *secondly* we decide, on the basis of this chosen theory, which is the most 'reliable' prediction for the given practical problem. Thus while theories may be said to be supported by evidence, 'predictions' are supported by theories.

Let us recall Carnap's important distinction between three possible approaches to defining degrees of confirmation or reliability. The first starts by defining the reliability of theories: the reliability of 'predictions' is derivative. This may be called the '*theoretical approach*'. The second proceeds in the opposite direction: it starts by defining the reliability of predictions: the reliability of theories is derivative. We may call it the '*non-theoretical approach*'. The third, finally, defines the reliability both of theories and predictions in one single formula.[2] This is the '*mixed approach*'. Carnap hesitated between the second and the third.[3] I propose the first, the theoretical approach. This has been consistently ignored by inductive logicians: despite the fact that it is generally used in actual practice. Instead of trusting a prediction according to *c*-values obtained by a complicated atheoretical method based on some formal language, the engineer will, as a matter of fact, prefer the predictions of the most advanced theory of the day.[4]

If the chosen theory is statistical, one can calculate the 'degree of reliability', or 'rational betting quotient' of any particular hypothesis within its range with the help of probability theory: our rational bet on h will be $p(h, T)$, p being logical probability, h being the prediction,

[1] 'Prediction' is here just shorthand for 'particular hypothesis'.

[2] Cf. *above*, p. 143. Carnap, of course, did not then foresee a fourth possibility: where one defines reliability *exclusively* for predictions.

[3] Carnap's 'qualified instance confirmation of theories' belongs to the second approach. But qualified instance confirmation can be defined only for theories of very simple logical form. So Carnap's second approach could not be carried out.

[4] Thus, I think, Carnap is wrong in emphasizing that 'inductive logic does not propose *new* ways of thinking, but merely to explicate *old* ways. It tries to make explicit certain forms of reasoning which implicitly or instinctively have always been applied both in everyday life and science' ([1953*a*], p. 189). But already with his 'non-theoretical approach' Carnap has departed from the reasonings which 'have always been applied...in science'. (Of course, there has been no form of reasoning from which somebody at some time has not deviated.)

and T being the chosen theory, usually of the form $P(h, s) = q$ (h denoting the predicted event, s the stochastic set-up, P physical probability).[1]

If the chosen theory is deterministic, one may bet all-or-nothing on the event predicted by the theory – with some reasonable safety margin.

This way of determining rational betting quotients, of course, is not open when there is no scientific theory at our disposal. In such cases one may use Carnap's Bayesian approach; but because of the arbitrariness in the choice of language, the arbitrariness of the prior distribution, and the dubious weight of the sparse accidental evidence, it will only yield an exact but irrational ritual. In cases where scientific theories are at hand, the theoretical approach yields intuitively rational betting quotients where Carnap's non-theoretical approach does not: this can be easily seen if one checks, one by one, our counterarguments to Carnap's values for rational betting quotients in section 5.

Of course, my 'theoretical betting quotients' are relative to the theory on which they are based. The *absolute* rational betting quotient on any proposition, whether universal or particular, is zero.[2] 'Theoretical betting quotients' are 'rational' but fallible: they depend on – and fall with – our extant theories. (One may, of course, calculate rational betting quotients for theories and for predictions given a language; but these betting quotients will depend on – and fall with – the language.[3] Moreover, one would need a theory of support – evidential or theoretical – for languages).[4]

APPENDIX. ON POPPER'S THREE NOTES ON DEGREE OF CORROBORATION

One of the main points of what I called the Popperian approach was that precise, numerical estimates of degrees of 'reliability' are so unreliable as to make any such estimates utopian; moreover, even non-numerical formal expressions are misleading if they suggest that they may lead to *general* comparisons of any real value.

Some students of Popper's work on his 'degree of corroboration', published between 1954 and 1959, may, however, wonder whether Popper himself is, on these terms, a 'Popperian'. Does he not offer *formulae* for his degree of corroboration? Does he not propose a precise, even infallible, *logical metric* in the important case of statistical theories? In view of these results inductive logicians do not know whether to count Popper as one who competes with them in devising

[1] In 'closed games' (cf. *above*, p. 169) T is fixed as a 'rule' of the game.

[2] It should be mentioned here that *all* empirical propositions are universal because of the universal names inevitably occurring in them (cf. Popper [1934], § 13, and [1963], p. 277). Only within the context of a given theory can one distinguish universal and particular propositions.

[3] For the 'fall' of a language cf. *above*, p. 162, n. 1.

[4] Cf. *above*, p. 163.

a priori metrics, or as one who would be against *any* such formulae. Many Carnapians regard Popper's formulae merely as new additions to the fast-growing literature on inductive logic. For instance, Kyburg, in his survey, 'Recent Work in Inductive Logic', lists Popper's formulae along with Carnap's, Kemeny's, and others' in a table and argues that the table 'shows the [large] extent to which the intuitions of these writers coincide'.[1] In one passage he lists Popper as one of the inductive logicians: 'Neither Barker, Popper, Jeffreys, nor any other inductive logician...'[2] In another passage however he states: 'There are writers (Popper, for example) for whom there is no such thing as inductive logic.'[3]

In order to clarify this problem, let me first outline Popper's aims in devising his formula for degree of corroboration.

Popper's main aim was to establish with a formal and conclusive argument that even *if* he grants that there can be a quantitative confirmation function defined over all propositions of a language, even *if* he grants that such a function can be expressed in terms of logical probability, then it cannot possibly be probabilistic, that is, it cannot obey the calculus of probability. It was a conversation with Janina Hosiasson in Prague in 1934 (when his book was already in print) which convinced him that such an argument was important.[4] This was the origin of Popper's intermittent work, in the late 1930s, on the axiomatization of probability theory.[5]

Popper, in his first note,[6] puts this argument in three stages: (*a*) he proposes ten adequacy requirements, or *desiderata*, each one supported by strong arguments, (*b*) he shows that they are consistent by displaying a formula which is expressed in terms of, but not identical with, logical probability and which does satisfy them, and (*c*) he shows that $p(h, e)$ does not satisfy some of them (or even some of Carnap's own *desiderata*). As to (*b*), since Popper has claimed that Carnap's own adequacy requirements were inconsistent, and since he attributed great importance to this inconsistency, he was obliged to produce a formula if for no other purpose than to show that his own *desiderata* were consistent.

But Popper's note makes it clear that *his desiderata are not comprehensive*: 'some intuitive *desiderata*...cannot be satisfied by any formal definition...one cannot completely formalize the idea of a sincere and ingenious attempt [at refutation]'.[7] Incidentally, he could have added an eleventh requirement: namely that there should be at least two

[1] Kyburg [1964], p. 258.

[2] *Ibid.*, p. 269.

[3] *Ibid.*, p. 249.

[4] Popper [1934], p. 263, n. 1.

[5] Kolmogorov's axiomatic system was unsuitable for his purpose: he needed a probability theory based on relative probabilities in order to be able to define $p(x, y)$ even if $p(y) = 0$: this enabled him to cope with universal propositions as second arguments (cf. his [1958], appendix *iv).

[6] Popper [1954–5]; reprinted in his [1959], pp. 395–402.

[7] Popper [1959], pp. 401–2.

theories with different degrees of corroborability and that there should be some possible evidence on which the more corroborable theory could obtain a higher degree of corroboration than the less corroborable one. This, of course, is a vital *desideratum* in the spirit of Popper's philosophy. But if he measures corroborability by logical improbability the *measure* of corroborability of *any* universal proposition must be 1, and then *differences in corroborability* of theories are not reflected in *differences in the measures of corroborability*, and, in this sense, this extra *desideratum* is not satisfied by his numerical formula. Because of this, his formula satisfies some of his most important *desiderata* only trivially for genuine theories, because the 'degree of corroboration' of genuine theories degenerates into their 'explanatory power.'[1] *This fact does not diminish the value of the formula as evidence for consistency, but destroys its value for providing an actual numerical metric.*

But Popper, in 1954, did not intend to use his formula for constructing a metric for corroboration. There is not the slightest indication in his first note that Popper might have changed his 1934 position, according to which 'the degree of corroboration of two statements may not be comparable in all cases' and 'we cannot define a numerically calculable degree of corroboration'.[2] Moreover, three years later, in 1957, in his second note,[3] he warns that '*there cannot be a satisfactory metric of p; that is to say, there cannot be a metric of logical probability which is based on purely logical considerations*'.[4]

Thus, throughout his first two notes on degree of corroboration, Popper regarded his formulae only in a polemical context, as mock-rival formulae, as it were, to combat inductive logic.

However, his third note, published in 1958, represents an interesting

[1] Popper's formula for degree of corroboration is

$$C(h, e) = \frac{p(e, h) - p(e)}{p(e, h) + p(e)}(1 + p(h)\,p(h, e)).$$

But if h is universal, $p(h)\,p(h, e) = 0$, and

$$C(h, e) = \frac{p(e, h) - p(e)}{p(e, h) + p(e)},$$

which he interprets as 'explanatory power': $E(h, e)$.

As Popper himself points out,

$$\frac{p(e, h) - p(e)}{p(e, h) + p(e)}$$

has 'defects' as degree of corroboration: it 'satisfies the most important of our *desiderata* but not all' (*ibid.*, p. 400).

Of course, if we do not interpret p as 'ordinary' logical probability but as a non-numerical function expressing what Popper calls the 'fine structure of probability', then $p(h)\,p(h, e)$ need not disappear. Indeed, our 'eleventh requirement' may be used to show that one cannot define degree of confirmation in terms of logical probability with real numbers as values. (For the 'fine-structure of probability' see Popper [1959], pp. 375–7.)

[2] Popper [1934], §82.

[3] Popper [1956–7]; reprinted in his [1959], pp. 402–6.

[4] Popper [1959], p. 404, Popper's italics.

change.[1] In this note Popper *did* elaborate a metric for degrees of corroboration of statistical theories given statistically interpreted evidence, a 'logical or absolute metric'[2] based upon purely logical considerations, which he found 'entirely adequate'.[3]

This result, of course, was more than Popper had originally planned. It was an unintended side result which seemed to turn his negative, critical, mock-rival programme into a positive rival programme. It seemed to Popper that Carnap would never be able to have degrees of rationality of beliefs 'measured, like degrees of temperature, on a one-dimensional scale'.[4] But he thought that in *his* new positive programme – at least in the special but important case where the theories are statistical theories and the evidence interpreted statistical reports – 'all these difficulties disappear',[5] and *his* method 'allows us to obtain *numerical results* – that is numerical degrees of corroboration – in all cases envisaged either by Laplace or by those modern logicians who introduce artificial language systems, in the vain hope of obtaining in this way an *a priori* metric of their predicates'.[6]

This left Popper with something of a problem. As he himself put it in the postscript to his three notes in 1959:

> It might well be asked at the end of all this whether I have not, inadvertently, changed my creed. For it may seem that there is nothing to prevent us from calling $C(h, e)$ 'the inductive probability of h, given e' or – if this is felt to be misleading, in view of the fact that C does not obey the laws of the probability calculus – 'the degree of the rationality of our belief in h, given e'.[7]

The answer, of course, depends on Popper's interpretation of his degree of corroboration. If he had interpreted it as a tautologous measure of growth and if he had condemned any synthetic interpretation as 'inductivist', then at least he would have given a clear answer. But *Popper seems to regard this an open problem.* In one sentence he says that if a theory has a high degree of corroboration, then 'we tentatively "accept" this theory – but only in the sense that we select it as worthy to be subjected to further criticism, and to the severest tests we can design'.[8] This remark suggests that he has the tautologous interpretation in mind with only methodological implications: high corroboration means high testworthiness but not high trustworthiness. But the next, concluding sentence adds a rider: 'On the positive side, we may be entitled to add that the surviving theory is the best theory – and the best tested theory – of which we know.'[9] But what is the 'best theory' *apart* from being 'best tested'? The one which is most 'trustworthy'? There is no answer.[10]

[1] Popper [1957–8]; reprinted in his [1959], pp. 406–15.
[2] Popper [1959], p. 417.
[3] *Ibid.*
[4] *Ibid.*, p. 408.
[5] *Ibid.*
[6] *Ibid.*, p. 412, n. *3.
[7] *Ibid.*, p. 418.
[8] *Ibid.*, p. 419.
[9] *Ibid.*
[10] Incidentally, it is exactly in the third note on degree of corroboration and in this postscript to it that Popper agrees that his degree of corroboration may be interpreted

Of course, even if he had finally decided to interpret his 'degree of corroboration' as an estimate of verisimilitude, he could still maintain that a *fallible* estimate of 'reliability' does not make him an inductivist. But he seems to be undecided and he stresses only the – no doubt, important – difference between the inductivist and Popperian *interpretation* of e in such formulae: in the Popperian interpretation '$C(h, e)$ can be interpreted as degree of corroboration only if *e is a report on the severest tests we have been able to design*'.[1] But this leaves wide open the problem of the philosophical significance of his $C(h, e)$, especially, when it offers a *metric* for statistical h's.

But whether Popper regarded his $C(h, e)$ – in his new, positive non-polemical interpretation – as a measure of evidential support in the 'tautologous' or in the 'synthetic' sense, it seems to conflict with my thesis that 'degrees of corroboration' can be compared *only* where the one theory supersedes the other.[2] Popper's metric seems to assign precise numerical values to *all* statistical hypotheses, given statistically interpreted test-evidence.

The simple solution of the seeming conflict is that Popper's metric measures only one narrow aspect of corroboration of statistical theories.

(1) First, one has only to remember that in Popper's formulae the measures of corroborability of any genuinely universal theory are the same, and therefore a theory which has much less empirical content than another, may, on this account, still achieve the same numerical degree of corroboration.[3] But then Popper's numerical appraisal of the corroboration of statistical theories pays no 'due regard to the degree of testability of the theory'[4] – and thus it is unsatisfactory. If we do want to pay due regard to the degree of testability of the theory, we may possibly opt for a 'vectorial' appraisal, consisting of Popper's content *and* of his 'degree of corroboration'; but then the linear ordering, let alone the metric, disappears.

(2) Popper's *metric* fails also on a second, independent ground. To recapitulate, the formula he offers is

$$C(h, e) = \frac{p(e, h) - p(e)}{p(e, h) + p(e)}.$$

as 'a measure of the rationality of our beliefs'. But this statement is certainly a slip; it goes counter to the general spirit of his philosophy, according to which belief, whether inductive or non-inductive, whether irrational or 'rational', has no place in the theory of rationality. *The theory of rationality must be about rational action, not about 'rational belief'*. (There is a similar slip in Watkins' [1968*b*] where he says that at least some cases in which it would be 'perverse' to believe that the more corroborated theory has less verisimilitude (p. 281).) But why is it perverse? Cf. Boltzmann's position, quoted *above*, p. 186, n. 1.)

[1] *Ibid.*, p. 418. Kyburg, in his survey (see *above*, p. 194) missed this point: his claim that Popper's intuition largely coincides with Carnap's or Kemeny's is no less absurd than if somebody alleged that two scientists largely agree on a controversial matter because they produce similar formulae, despite the fact that the symbols in the two formulae, *when interpreted*, have completely different meanings.

[2] Cf. *above*, p. 184.

[3] Cf. *above*, p. 195.

[4] Popper [1934], §82.

This formula, he claims, yields a metric if h is to be a genuinely universal statistical theory of the form $P(a, b) = r$ where P is his propensity (or any sort of objective, physical probability), a ranges over events (outcomes of an experiment), and b denotes the experimental set-up, stochastic mechanism, or 'population'. His e stands for a 'statistically interpreted report of h', or, briefly a 'statistical abstract of h', that is, a statement of the following form: 'In a sample [of a's] which has the size n and which satisfies the condition b (or which is taken at random from the population b), a is satisfied in $n(r+\delta)$ instances.'[1] We must plan the experiment in such a way that it should be possible that the outcome will make $|p(e, h)-p(e)|$ large (this is Popper's requirement of severity).

But I argued earlier that there is no such thing as the *absolute* probability of a hypothesis (other than zero): we may only calculate *relative* probabilities of particular hypotheses given some theory.[2] If so, the 'absolute probability' of e is zero, and $C(h, e)$ is 1. In Popper's account, however, $p(e) = 2\delta$, not zero.[3] But what Popper calls 'absolute probability of e' is, in fact, the relative probability of e given the theory h^* that all statistical abstracts of h of the same width are equiprobable. Thus Popper's expression 'absolute probability' is misleading: his formula should read:

$$C(h, e) = \frac{p(e, h)-p(e, h^*)}{p(e, h)+p(e, h^*)}.$$

(Similarly, Popper misleadingly calls $1-2\delta$ a measure of the content or of the precision of e. For let e be: 'In 1000 random choices from a population b the outcome is a 500±30 times'. Statements like (1) 'the population was b_1 (not b)'; (2) 'the outcome was a_1 (not a)'; (3) 'the choice was not random'; (4) 'in 1000 random choices from b the outcome was a 328 times', are all potential falsifiers of e and their total measure is 1; but if one restricts one's attention to the set of 'strictly statistical potential falsifiers' of the kind (4), as Popper seems to do in *this* note, one may correctly assign to this smaller set the measure $1-2\delta$. But then one should call it rather a 'measure of the strictly statistical content' of e).

Popper's formula is, in fact, a special case of the general formula

$$C(h, e, h') = \frac{p(e, h)-p(e, h')}{p(e, h)+p(e, h')},$$

where h' is a touchstone theory and e must be some severe test-evidence of h relative to h', that is, it should be possible for $|p(e, h)-p(e, h')|$ to assume a value near 1. *But I shall argue that this generalized version of*

[1] Popper [1959], p. 410. For the purpose of this discussion I do not question this theory of interpretation of statistical evidence.

[2] Cf. *above*, p. 193.

[3] Popper [1959], pp. 410–11, and especially n. *4 on p. 413.

Popper's formula is of some interest only if h' is a genuine rival theory with scientific interest, and not merely a Laplacean reconstruction of a state of ignorance.

Moreover, all that $C(h, e, h')$ can tell us is that h explains e better than h', or *vice versa*. But in order to obtain this much information we do not need this formula, but only need consider $p(e, h)$ and $p(e, h')$: we shall prefer that one of h or h' which has more explanatory power for some given severe test-evidence e. This, essentially, is Fisher's likelihood ratio method combined with a Popperian design of experiments. This method is not intended to provide us with an absolute metric for all h's given e, but only to select the best hypothesis from a well-defined class of competing hypotheses.[1] We shall appreciate statistical theories more if they have defeated several *real* rivals of scientific interest: but the cumulative effect of these victories cannot yield a linear ordering of *all* statistical hypotheses, let alone a metric on some absolute scale.

It must be realized that any absolute, universal metric for corroboration hinges on the arbitrary selection of *one* distinguished touchstone theory for h. In Popper's third note an equidistribution over the set of the samples of h seems to play the role of this distinguished touchstone theory; in Kemeny's and Oppenheim's work, for instance, $\bar{h}$ plays a roughly similar role.[2]

As soon as we consider different touchstone theories (of genuine scientific interest), the absolute universal metric disappears and is replaced by a mere partial ordering, which establishes a comparative–qualitative appraisal of competing theories. And this is exactly the crucial difference between inductive logic and modern statistical techniques. *The programme of inductive logic or confirmation theory set out to construct a universal logical confirmation function with one absolute metric, which, in turn, is based on one distinguished touchstone theory. This one touchstone theory usually takes the form of a Laplacean proto-distribution over the sentences of a universal formal language. But this atheoretical (or, if you wish, monotheoretical) approach is useless, and the programme of an absolute, universal confirmation function is utopian. Modern statistical techniques try at best to compare the evidential support of scientifically rival theories.* It is unfortunate that Popper, whose 1934 ideas anticipated much of the development of modern statistics, in 1958–9 proposed a universal, absolute logical metric for statistical theories – an idea completely alien to the general spirit of his philosophy.

Example. Let us calculate the degree of corroboration of the hypothesis h that the propensity of the heights of children (of a given age) in Indian families sharply to decrease as the number of children in the family increases, is near to 1.

Let us take as touchstone theory the hypothesis h^* that there is no such

[1] Cf. e.g. Barnard's elucidations in Savage and others [1962], pp. 82 and 84.
[2] Cf. their [1952], Theorem 18.

correlation and that the heights are probabilistically constant whatever the size of the family. Let us further assume that in fact the height of children is inversely proportional to the size of (Indian) families in all existing cases. Then for any large sample $p(e, h)$ will be near 1 and $p(e, h^*)$ will be near zero: any large sample will then be a severe test of h relative to h^*. If we acknowledged h^* as an *absolute* touchstone theory, we would have to say that h was very highly corroborated: $C(h, e) \approx 1$.

But suppose that a rival theory h' is proposed according to which the height of the children is – with propensity 1 – directly proportional to the average daily calory consumption. How can we plan a severe test of h relative to h'? On our previous e we could have $p(e, h) = p(e, h')$. But if so, what was crucial evidence relative to the touchstone theory h^*, becomes irrelevant relative to the touchstone theory h'. We have to take this time as test-evidence a set of events that will make $|p(e', h) - p(e', h')|$ high. Such test-evidence will be provided by a set of well-fed large families, because if h is true and h' false, or *vice versa*, $|p(e' h) - p(e', h')|$ may be near 1. To perform this experiment may take a generation, since we shall have to bring up well-fed large Indian families, which, on our original assumption, do not at present exist. But after having performed the experiment we may get $C(h, e', h') \approx -1$, so that h is decisively undermined in the light of h'.[1]

Our example shows that severity of tests and degree of corroboration of hypotheses depend on the touchstone theory. The same test may be severe relative to one touchstone theory but irrelevant relative to another; the degree of corroboration of a hypothesis may be high when it defeats one touchstone theory and low when it is defeated by another. It also shows that large amounts of extant evidence may be irrelevant in the light of some rival theories; but a small amount of planned severe evidence may be crucial. Finally it shows how hopeless are efforts to produce *absolute* numerical values of degree of corroboration of h by e.

All this is a commonplace for the Popperian philosopher, a commonplace for the working statistician; but it must sound preposterous for the atheoretical inductive logician.

[1] This is how severe test-evidence may resolve 'lack of identifiability'. Cf. Kendall and Stuart [1967], volume 2, p. 42.

9
On Popperian historiography*

If a theory of scientific rationality is too narrow, that is, if its standards are too high, then it makes *too much* of the actual history of science appear to be irrational – a caricature of its rational reconstruction. Historians who hold that scientific growth is the paradigm of rationality, tend, if guided by a narrow theory of rationality, *either* to give an impoverished, truncated account of history, *or* to twist historical facts in order to make the actual growth of science conform more with their image of rationality.[1] [Popper has not been completely immune to this temptation.] In particular, he has refused to notice two [historical] facts: (1) 'Crucial experiments' are frequently listed first as harmless anomalies, rather than 'refutations' (they usually get acknowledged as 'crucial' only after having been backed up by some new research programme in a victorious struggle against the old programme); and (2) All important theories are born 'refuted'. Of course, in the light of Popper's logic of discovery, the first fact is irrational: the *first* corroborated refutation must already be methodologically conclusive. The second fact would also make the acceptance of theories, however provisional, irrational. [No wonder then that these two facts tend to fade into the background in Popper's rational reconstruction of the history of science.]

Popper turns anomalies into 'crucial experiments' and exaggerates their instant impact on the development of science. In his presentation, great scientists accept refutations readily and this is the primary source of their problems. For instance, he claims – ignoring Lorentz's work after 1905 – that the Michelson–Morley experiment decisively overthrew classical ether theory, and he also exaggerates the role of this experiment in the emergence of Einstein's relativity theory.[2]

* This paper was probably written in the middle 1960s, and seems originally to have formed part of a larger paper. We publish here the later of two different versions found among Lakatos's manuscripts. We have, however, added some material from the other version as an appendix. Lakatos regarded this paper as in need of extensive revision and elaboration and had no plans to publish it as it stands. (*Eds.*)

[1] A similar situation may arise with *ethics*. A historian with Victorian standards of morality will either despair of the role of morality in history or his reconstruction of it will be hypocritical.

[2] Cf. Popper [1934], section 30 and Popper [1945], volume II, pp. 220–1. He stressed that Einstein's problem was not how to account for experiments 'refuting' classical physics and he 'did not...set out to criticise our conceptions of space and time'. But Einstein certainly did. His Machian criticism of our concepts of space and time, and,

(Although Popper never distorted history as much as Beveridge, who wanted to persuade economists to adopt an empirical approach by setting them Einstein as an example. According to Beveridge's falsificationist reconstruction, Einstein 'started [in his work on gravitation] from facts [which refuted Newton's theory, that is,] from the movements of the planet Mercury, the unexplained aberrancies of the moon'.[1] But of course, Einstein's work on gravitation (the 'theory of general relativity') grew out of a 'creative shift' in the positive heuristic of his special relativity programme, and certainly not from pondering over Mercury's anomalous perihelion or the moon's aberrancies.) It takes a naive falsificationist's simplifying spectacles to claim, with Popper, that Lavoisier's classical experiments refuted (or 'tended to refute') the phlogiston theory or that the Bohr–Kramers–Slater theory was knocked out by Compton's experiments. Popper also oversimplifies the refutation of the parity principle.[2]

Furthermore, Popper ignores the historical fact that theories are born refuted and that some laws, with known counterexamples, are further explained rather than rejected. Therefore, he tends to turn a blind eye on all anomalies known before the one which later was enthroned as 'crucial counterevidence'. For instance, he thinks, mistakenly, that 'neither Galileo's nor Kepler's theories were refuted before Newton'.[3] The context is significant. Popper holds that a most important pattern of scientific progress is when a crucial experiment leaves one theory unrefuted while it refutes a rival one. But, as a matter of fact, in most, if not all, cases where there are two rival theories, both are known to be simultaneously infected by anomalies. Since Popper's methodology does not offer rational guidance in such situations, he submits to the temptation to simplify the situation into one to which his methodology is applicable.

A catastrophical consequence of a narrow methodology is that, as well as impoverishing actual problem situations, it invokes *external* – psychological, sociological – explanations because its *internal* framework of rational explanation fails too soon. Agassi, in a most interesting discussion, showed how inductivist historiography opens the door to the wild speculations of the vulgar-Marxists.[4] But the falsificationist historiography he advocates does not go far enough in improving the situation. For instance, the Popperian insistence on abandoning a

in particular, an operationalist criticism of the concept of simultaneity played an important role in his thinking. *But cf. Zahar [1973] and [1977] (*eds.*).

[1] Beveridge [1937]. Beveridge used this story in order to set an example for empirical economics. Lipsey, in his naive falsificationist period, selected this Beveridge quotation as the motto of his [1963]. (It is ironical that in the second, 1966, edition of his book, in which he announced that he had turned against falsificationism, he still retained the motto.)

[2] Popper [1963*a*], pp. 220, 239, 242–3.

[3] *Op. cit.*, p. 246.

[4] Agassi [1963], p. 23.

theory after the 'crucial experiment',[1] opens the door to those trendy 'sociologists of knowledge' who are trying to explain the further – and possibly unsuccessful – development of the rival programme as the irrational, wicked, reactionary obstinacy of established authority against enlightened revolutionary innovation. But as I have shown, such rearguard skirmishes are perfectly explicable *internally* from the point of view of my methodology of scientific research programmes.

It was Agassi, who, among Popper's followers, undertook the major enterprise of elaborating the historiographical implications of Popperian philosophy of science. The result was his well-known book: *Towards an Historiography of Science* (1963). Agassi offers a brilliant criticism of inductivist historiography, but his critical exposition of conventionalist historiography is unsatisfactory and, finally, the positive part of his book is, in fact, a devastating indictment of falsificationist historiography.

Popper never referred to Agassi's book, which is regarded by most historians as the standard Popperian text on historiography. I hope that he will take my present criticism as an opportunity either to defend or to disown it.

Agassi's main historiographical problem is, as he put it, 'how are facts discovered'?[2] He claims that the answer will depend on our view about the relation between extant theory and the observation involved in the discovery. Baconians hold that this relation is independence, and that discovery comes about when theory (i.e. bias) is eliminated. Discoveries are the imprints of nature on the *tabula rasa* of the scientific mind. Whewellians hold that the relation is one of *deducibility*, and that discovery comes about when a new theory enters which predicts a novel fact. Discoveries are verifications of new ideas. Popperians hold that the relation is one of *incompatibility*, and that discovery comes about when an old theory is tested and refuted. Discoveries cannot be made *before* the theories which they refute exist: 'all discoveries...are refutations of past theories...According to Popper, the very crux of the matter lies here: whether an observation is predicted on the basis of a new idea (Whewell) or not (Bacon), its novelty and surprise value depend on its contradicting a reasonable scientific theory'.[3]

Agassi claims that Bacon and Whewell are wrong and Popper is right. He points out that 'Popper's theory, if false, might be criticised by our...finding a case where an important discovery did not conflict with an important idea immediately preceding it'.[4] This, of course, is

[1] For Popper's occasional hesitation on this point cf. volume 1, chapter 1, p. 94, n. 5.

[2] This is the title of his [1959]; but what he seems to mean is rather how are *important* facts discovered. Agassi's whole treatment is somewhat impaired by a conflation of factual and normative. 'Discovery' is a normative, not simply a factual term. We may observe a fact, even state it, without making a 'discovery': 'discovery' means that the fact acquires importance.

[3] Agassi [1963], p. 64.

[4] *Ibid.*

the announcement of a historical research programme, which Agassi then starts off with real *élan*. He takes some factual discoveries and reconstructs the theories they tested and contradicted and which had been pushed into the background by Baconian tradition. But he does not notice that Popper's position, as he interprets it, might be criticized on a different ground: by the frequency of anomalies, that is, of observations which on his terms contradicted a theory and nevertheless did *not* add up to a '*genuine* discovery'.[1] In order to appreciate the nature of this criticism, one must make two points.

First, the normative element in the term '[factual] discovery' must be made explicit; one cannot leave it in the metaphysical darkness of qualifications like 'genuine'. A factual discovery is *genuine* or *important* – I propose – if it leads to a considerable change in the general problem situation, if it alters the rational choice of problems, if it shifts the balance of two competing research programmes. For instance, neither the 1831 discovery of Mercury's anomalous perihelion nor the 1887 Michelson–Morley experiment would then qualify as a *genuine* discovery. *Secondly*, one must realize that our appraisal of a *genuine* discovery is a rational appraisal; if a factual discovery creates a mad rush to change bandwagons, that does not lend to it *rational* importance. We have to wait and see whether the rush was rational – and this can only be seen with long hindsight.[2]

After having indicated that Agassi missed the most serious pattern of criticism of his historiographical position,[3] let us look at his historical case studies and see how he shapes the historical material to make it fit his theory.

(1) Agassi's first example is 'Hertz's [1887] error in undervaluing his discovery of the photo-electric effect'.[4] Agassi's problem is why it is that Hertz observed and described the effect but that nobody cared about it until Einstein's 1905 paper. His answer is that Hertz 'made a logical error: he thought that the effect is explicable by Maxwell's theory as a resonance effect'.[5] It was only Einstein who 'showed that [the effect was in] conflict with Maxwell's theory'.[6]

It is not Hertz, but Agassi who made a logical error. To think that an effect is explicable within a programme is a matter of methodological judgment and not of hard logical relations. Moreover, if we understand by photo-electric effect the effect of electrons being knocked out by photons, the discovery of this 'fact' could not have been made before Millikan and Einstein. All that Hertz accidentally observed was an inexplicable current. An inexplicable current cannot

[1] Agassi [1959], p. 2.

[2] Cf. volume 1, chapter 1, pp. 86–7.

[3] Of course, this pattern becomes obvious only in the light of my methodology of scientific research programmes.

[4] Agassi [1963], p. 64.

[5] Agassi's claim that Hertz thought the effect was a resonance effect is false. The resonance theory came only after Millikan: cf. Richtmyer [1955], p. 98.

[6] Agassi [1963], p. 64.

be logically inconsistent with Maxwell's programme; only Einstein's programme, with its theory of photons and of their interactions with electrons, is inconsistent with it. The inconsistency was simply not there in Hertz's time: he observed a mere anomaly.[1] The *discovery* came in a Whewellian way: as a *confirmation* of a new research programme which superseded – and contradicted – Maxwell. Agassi's explanation of the neglect of Hertz's observation in terms of Hertz's lack of logical acumen is simply false. My explanation is that an insignificant anomaly (about lightwaves and electric currents) was turned into a major discovery when reinterpreted in the light of a new theoretical framework as a fact about photons and electrons.

(2) According to Agassi, Michelson's experiment was immediately hailed as a major discovery. But it only *refuted* the ether theory, it did not *verify* anything at the time. '*With this*' – claims Agassi – '*the whole philosophy of verification collapsed.*'[2] One wonders if Agassi has ever read Michelson. Michelson stated that he had proved Stokes's theory. Moreover, he was for years in despair that his experiments and conclusions were ignored: Agassi's claim that it had, in itself, an 'immense impact'[3] as a falsifying experiment is nonsense.

(3) Agassi elsewhere also mentions the Hahn–*Meitner* discovery of nuclear fission as a discovery which refuted a theory without verifying another: a 'spectacular case of counterexpectation'. I suppose he is referring to the Hahn–*Strassman* experiment. But the real story is completely different. What Hahn and Strassman discovered was not 'nuclear fission' but that, upon bombarding uranium barium seems, inexplicably, to appear. It was only Meitner and Frisch who interpreted the Hahn–Strassman anomaly as nuclear fission: this interpretation was elaborated into an independently testable hypothesis by Bohr and Kalckar. This hypothesis, in turn, was corroborated by Frisch and many others.[4]

Agassi is fascinated by the problem of why factual discoveries *surprise* the discoverer: he thinks that this is a refutation of the Whewellian idea that great factual discoveries are verifications. But – apart from the limited relevance of the psychological reactions of scientists to considerations of rationality – Agassi's surprise examples describe accidental discoveries, to my mind, of anomalies of the type which occur each day; only subsequent reinterpretations enthrone a few of them – including those which Agassi selected – as crucial experiments. But

[1] For a definition of 'anomaly' cf. volume 1, chapter 1, p. 72, n. 3.

[2] Agassi [1959], p. 3. Also cf. his [1963], p. 64.

[3] Agassi [1959], p. 4.

[4] The story is correctly told in Richtmyer *et al.* [1955], pp. 539–41 and especially Wehr and Richards [1960], p. 305. *Although he is not very explicit here, Lakatos's point is perhaps that the importance of a discovery may be realized only some time after the first performance of the experiment which is subsequently taken as establishing the discovery; and this, if true, seems to undermine the claim that the importance of discoveries rests on their refuting already existing theories. (See also the remarks on Oersted in n. *, p. 206.) (*Eds.*)

the term '*factual discovery*', with its normative implications, must be reserved for the re-enactment of the experiment in the light of a new, rival programme, unless we want to perpetuate in the history of science the thousands of minor anomalies which drove their discoverers into hysterical excitement – and were then totally forgotten.

(4) Agassi's favourite case study is Oersted's discovery of electromagnetism. This is understandable – Oersted's discovery used to be quoted as one of the great *accidental* discoveries;[1] and only recently was it argued that it was instead a great Whewellian discovery by a man who devoted a life's effort to producing evidence of the essential unity of electric and magnetic forces.[2] Agassi set out to show that both accounts are mistaken: Oersted's discovery was due to a sudden decision to test the Newtonian theory that all forces are central, a theory which until that moment he firmly believed to be true,[3] and his decision was rewarded on the spot by an immortal discovery – that this Newtonian theory was false. The discovery was to him 'so shocking that for a few months he did not publish it; he was perplexed and bewildered'.[4] Alas, Agassi's interpretation of history is, strictly speaking, untestable. We shall never find out what was in Oersted's mind when he made the famous last-minute rearrangement of his wires. But there are several considerations which undermine Agassi's interpretation.

First, if Oersted had been so clearly aware that he had refuted Newton's pancentrism, why did he never say so in his different meticulous accounts of the story? Why did he never criticize Ampère's Newtonian interpretation of the effect? Finally, Agassi's statement that Oersted was shocked and did not publish his result for months is a figment of Agassi's imagination. According to the clear available evidence, he was overjoyed, and rushed into a brief, superficial publication.* Also, Agassi's statement that Oersted was a Newtonian and

[1] E.g. Lenard [1933], p. 186. [2] E.g. Stauffer [1957].

[3] According to Agassi 'Oersted was a kind of Newtonian' (Agassi [1963], p. 72).

[4] Agassi [1959], p. 4.

* In fairness to Agassi, it should be pointed out that the claim which Lakatos attacks here is to be found only in his short (and 'popular') [1959] paper. In his [1963] monograph (see, e.g. p. 74), Agassi argues the entirely different claim that Oersted really made his discovery only in July 1820 (rather than some months earlier as is usually suggested). Since Oersted published shortly after this, the 'months' delay' disappears.

Oersted's own account (from his *Autobiography* quoted in Stauffer [1957], pp. 49–50) is quite unambiguous. He distinguishes between his discovery that the electric current has *some* effect on the magnetic needle and his discovery of the 'law governing the effect'. The former discovery is supposed to have been made in the famous lecture early in 1820. There was certainly a delay of some months between *this* discovery and publication in July 1820. But July 1820 was precisely when Oersted had, on his own account, become confident that he had discovered the 'law governing the effect'. (Oersted speaks explicitly of 'rushing' to publish once he had discovered the law.) Oersted explains the delay as caused by a delay in repeating the experiments which, early in 1820, were disclosing 'only a very weak effect'. This he explains in turn as due to his being 'burdened by daily routine for several months' and to his having

firmly believed that all forces are central forces – a vital ingredient for Agassi to explain Oersted's alleged surprise – seems to be mere fancy.[1]

(5) Finally, I mention Agassi's references to Galvani's and Röntgen's discoveries. In these cases he has no idea of how to reconstruct the refuted theory. In Galvani's case he is content with exhortations: 'It might be interesting to try to reconstruct Galvani's deep thoughts, to show that they led to some disappointed expectation concerning the frog's leg, and that the discovery was the refutation of these deep thoughts whose contents he did not mention.'[2] In Röntgen's case he self-confidently asserts that 'Röntgen *was* testing *some* hypotheses concerning the characteristics of the *various* emissions of cathode tubes'.[3] Historiography turns into falsificationist metaphysics!

In my view, of course, neither Galvani's, nor Röntgen's 'discovery' had *immediate* relevance. Röntgen actually thought that he 'discovered' longitudinal ether vibrations.[4] It is only with *hindsight* that a 'discovery' *may* become a real discovery: if it becomes embedded in a progressive research programme. If it does not it may remain, possibly forever, in the curiosity shop of the history of science. Indeed, if Agassi took his line completely seriously, he should have elevated from anonymity to excellence those astronomers who first observed deviations from Keplerian ellipses and later from Newtonian orbits; or those who published hundreds of accidental observations about radiation, fluorescence, ESP, etc. All of these contradicted 'reasonable' scientific theories. In Agassi's view, one of the main advantages of (naive) falsificationism is that one knows instantly that one has learned something. He has great contempt for those who are 'wise after the event'.[5] In my view this contempt is utopian.

All this, I hope, shows the poverty of Agassi's falsificationist interpretation of factual discoveries, of learning from experience. Agassi makes it crystal clear that his historiographical interest centres on factual discoveries because he holds that science is learning from experience and that one learns from experience by refuting, with the help of experience, past theories. This theory leads to a radical rewriting of history in the name of a wrong-headed rationality principle: that theories must be abandoned in face of the discovery of contrary factual evidence, and that the history of science is a history

'a certain tendency to procrastinate and to utilize his free moments to live in the world of thoughts' (*ibid.*). (*Eds.*)

[1] I find it ironical that Pearce Williams, a professional historian, should have said in his review, that '[Agassi's] analysis of Oersted's discovery of electromagnetism, while conjectural, sheds considerable light upon this epoch-making event.' To my mind, Agassi's analysis is a backward step compared with Stauffer's.

[2] Agassi [1963], p. 66.

[3] *Op. cit.*, p. 67, my italics.

[4] Röntgen [1895]. Indeed, Röntgen was a physicist of mediocre ability; his lucky achievement was vastly exaggerated by the wide technological use to which his X-rays were put. Since, in Agassi's historiography, accidental discoveries are reconstructed as strokes of critical genius, lucky hacks become heroes of discovery.

[5] Cf. e.g. his [1963], pp. 48–51.

of simple trial and error, of theory and refuting experiment. This is a rationality theory – and theory of learning – which is, no doubt, a great advance on some earlier theories and which I attributed in my earlier papers to 'Popper$_1$', the naive falsificationist. But Popper also contains elements of a still more advanced Whewellian 'Popper$_2$' whom Agassi ignores.[1]

APPENDIX ON 'ULTRA-FALSIFICATIONISM'

According to 'ultra-falsificationists' (and Popper has certainly never been one of them), the only rational motive for efforts to replace a theory is its experimental defeat; and a negative experimental result, while refuting one theory, ushers in, so to speak, *the* successor. This position constitutes a rearguard skirmish by anti-speculative conservatives. Ultra-falsificationists hold that it is irrational to indulge in a proliferation of theories before the dominant theory has been knocked out by a crucial experiment. And even then, they hold, one must not propose erratic quixotic phantasies: the true scientist's speculation is *informed* guessing and what else can inform a guess but the crucial experiment itself? The growth of science does not follow a simple Darwinian pattern of blind mutations and natural selection. The 'mutations' must not be blind, but designed to explain the truth content of refuted theories together with the counterevidence. Thus there must be a *constant and instantly effective* interaction between experiment and theory. For instance, according to the anti-speculative ultra-falsificationist, Rutherford's alpha-scattering experiments 'refuted' Thomson's 'full' atom-model and literally *showed* – almost by experimental proof as it were – that atoms were largely empty and even that they were minute planetary systems. The ultra-falsificationist might agree that Balmer's formulae may also have 'led', perhaps without Rutherford's experiments, to Bohr's programme; but he would insist that without either Rutherford's or Balmer's 'facts' the whole theoretical development would have been inconceivable.

It needs to be emphasized that interpreted facts have two very different functions in the growth of science. They may serve as tests of already proposed theories, and corroborate or undermine them; this function is part of the logic of discovery. They may also serve as stimuli to new theories; this function is part of the psychology of discovery. But visions and dreams may also act as stimuli. In the logic of discovery – the appraisal of theories – the pedigree of theories does not matter; in the psychology of discovery and in heuristics experiments play a much lesser role than most people believe.

Anti-speculative falsificationism has played a considerable role in the

[1] Agassi later realized that occasionally the learning process leads to the elimination of the 'observation' rather than of the theory. But he could never make head or tail of this phenomenon. (Cf. his [1966].)

historical misinterpretation of crucial experiments. To return to the origin of Einstein's special relativity theory, according to the folklore Michelson decisively refuted the ether theory and led Einstein, by the hand as it were, straight to relativity theory: 'Michelson's failure to detect the motion of the earth through a luminiferous ether *led* Einstein to the theory of relativity'.[1] According to Planck, Michelson's experiment 'compelled' or 'directed' modern physics to relativity theory.[2] But as a matter of fact, Einstein was unaware of the Michelson–Morley experiment or of its explanation by Lorentz. This fact, recently convincingly established,[3] deeply disturbs the anti-speculative falsificationists, and there has been a protracted controversy about the reliability of Einstein's autobiographical statements.

A most interesting document in this controversy is Adolf Grünbaum's [1961]. Grünbaum's case rests on a passage of Einstein's celebrated [1905], in which Einstein referred to 'unsuccessful attempts to discover any motion of the earth relatively to the "light medium" (aether)'. Grünbaum argues that

> it is surely incumbent upon all those historians of relativity theory who *deny* the inspirational role of the Michelson–Morley experiments to tell us *specifically* what *other* 'unsuccessful attempts to discover any motion of the earth relatively to the "light medium"' Einstein had in mind here. This obligation should also have been shouldered by the mature *reminiscing* Einstein himself when authorising the statement given by Polanyi [that he did not know, in 1905, of the Michelson–Morley experiment].

But this passage in Einstein's paper may refer to the long succession of experiments for testing the effects of the earth's orbital velocity on terrestrial optical phenomena, by Fizeau, Respighi, Hoek, Airy and Mascart, between 1850 and 1872.[4]

Grünbaum's interest is not simply in historical detail. He finds it absurd to think that 'actual experimental results played no role at all when [Einstein] groped his way to the principle of relativity'. 'If so' – argues Grünbaum – 'there would be the serious question whether the theoretical guesses of an Einstein can be regarded to have been genuinely more educated – as opposed to just more lucky – than the abortive phantasies of those quixotic scientific thinkers whose names have sunk into oblivion.'

This anti-speculative position has amusing effects. In 1960 Bernard Jaffe wrote a little book on Michelson, whom he greatly admired as the man whose 'ether drift experiment banished the notion of the ether'.[5] He wrote to Einstein asking him about his debt to Michelson. Einstein's answer was this:

> It is no doubt that Michelson's experiment was of considerable influence upon my work insofar as it strengthened my conviction concerning the *validity* of

[1] Gamow [1966], p. 37, my italics.
[2] Planck [1929].
[3] Cf. Holton [1960].
[4] Cf. Whittaker [1951].
[5] Jaffe [1960], p. 1.

the principle of the special theory of relativity. On the other side I was pretty much convinced of the validity of the principle *before* I did know this experiment and its result. In any case, Michelson's experiment *removed practically any doubt* about the validity of the principle in optics, and showed that a profound change of the basic concepts of physics was inevitable.[1]

Jaffe also found the text of a little banquet speech by Einstein in 1931 in Pasadena, when he addressed Michelson – who was then 80 years old – with these words:

It was you who led the physicists into new paths, and through your marvellous experimental work paved the way for the development of the Theory of Relativity. You uncovered an insidious defect in the ether theory of light, as it then existed, and stimulated the ideas of H. A. Lorentz and FitzGerald, out of which the Special Theory of Relativity developed. Without your work this theory would today be scarcely more than an interesting speculation; it was your *verifications* which first set the theory on a real basis.[2]

Jaffe's conclusion is that 'Einstein publicly attributed his theory to the experiment of Michelson'.[3] But Jaffe misread the texts he quotes. Einstein makes it crystal clear that he regarded Michelson's work as a corroboration of his programme, and therefore as a major encouragement for his *post*-1905 work, but not as having anything to do with his *pre*-1905 work.

Thus it is untrue that Einstein was 'led' by Michelson's experiments to his theory of special relativity. Also his work on gravitation (the 'theory of general relativity') grew out from the positive heuristic of his special relativity programme, not from the refutation of Newton's gravitational theory by Mercury's anomalous perihelion!

The reason why I criticize conservative 'ultra-falsificationism' in this paper is not that I think that Popper is an 'ultra-falsificationist' but that his position does not provide a sufficient platform to combat it, for he too overestimates, makes too direct, the role of empirical refutations in the rational growth of science.

[1] *Ibid.*, pp. 100–1, my italics.
[2] *Ibid.*, pp. 167–8, my italics.
[3] Jaffe [1960], p. 101. Grünbaum quotes this statement with approval ([1963], p. 381) and (p. 380) similarly misinterprets Einstein's [1915].

10

Anomalies versus 'crucial experiments'

(A Rejoinder to Professor Grünbaum)*

INTRODUCTION

I am grateful for Professor Grünbaum's criticism of the 'anti-falsificationist' features of my methodology of scientific research programmes, and I am glad to have the opportunity to reply. I have to start by trying to clarify a basic misunderstanding. My paper opened with the question: 'Exactly how and what do we learn about scientific theories from experiment?'[1] Later I made what Grünbaum called my 'provocative claim' that 'we cannot learn from experience the falsehood of any [scientific] theory'.[2] Now if one interprets 'theory' as 'proposition (fallibly) mirroring a fact', then because of the (epistemologically unbridgeable) gap between fact and proposition my claim is far from being provocative: it is an orthodox common-place. It says that if crucial experiments are to provide experimental *disproof*, there can be no crucial experiments. If I have a provocative claim, that claim is a stronger one; namely that no experimental result, in isolation, can ever *defeat* a 'theory', whether in my sense (that further work on it is irrational) or in Grünbaum's sense (that the experiment should change our rational belief into rational disbelief). That is, there are no 'crucial experiments' even in either of these two weaker senses.

1 THERE HAVE BEEN NO CRUCIAL EXPERIMENTS IN SCIENCE

I explained and elaborated my negative position concerning 'crucial experiments' in several papers between 1968 and 1971, and I tried to put it in a nutshell (admittedly with the aid of dozens of back

* This paper is a contribution to a debate between Lakatos and Professor Grünbaum concerning the status of crucial experiments. In 1973 Lakatos read a paper at Pennsylvania State University (published as Lakatos [1974*d*]), to which Grünbaum replied. The present paper is Lakatos's rejoinder to that reply. Grünbaum's reply, part of a larger paper entitled 'Falsifiability and Rationality', remains unpublished, but he has kindly agreed to the printing here of Lakatos's quotations from that paper (references are to the page numbers of the typescript (Grünbaum [1973])). It should not, however, be assumed that these quotations express Grünbaum's current views. Lakatos regarded the paper published here as a rough draft. His introductory footnote reads: 'I should like to acknowledge the constructive criticisms of *previous* versions I received from Peter Clark, Colin Howson, John Watkins, John Worrall, and also from Adolf Grünbaum.' (*Eds.*)

[1] Lakatos [1974*d*], p. 309.

[2] *Op. cit.*, p. 310.

references) at our Penn symposium.[1] In my view, in science we do not learn simply from conjectures and refutations. *Mature science is not a trial-and-error procedure, consisting of isolated hypotheses, plus their confirmations or their refutations.*[2] *The great achievements, the great 'theories', are not isolated hypotheses or discoveries of facts, but research programmes. The history of great science is a history of research programmes, and not of trial-and-error, nor of 'naive guessing'.*[3] No single experiment can play a decisive, let alone 'crucial', role in tilting the balance between rival research programmes. Of course, I do not deny that scientists occasionally confer, generally with hindsight, the honorific title 'crucial experiment' on some experiments which were successfully explained in one programme and not so successfully (i.e. only in an *ad hoc* way[4]) in another. Neither do I deny that some experiments have a decisive psychological effect in the war of attrition between two programmes, and that they may *cause* a collapse of the one and the victory of the other.[5] An anomaly may well have a paralysing effect on the imagination and determination of the scientists working in the research programme which is affected by it;[6] but I claimed that none of these anomalies, whether called 'crucial experiments' or not, are objectively crucial. Where the falsificationist sees a crucial negative experiment, I 'predict' that there was none. I predict that behind any alleged single fatal duel between theory and experiment one will find, as a matter of historical fact, a complex war of attrition between two research programmes[7] during which one may establish what the relative strengths (i.e. imaginative resources and empirical luck) of the two armies were at any given time. I also proposed (and initiated) a historiographical research programme to test all this.[8]

[1] I am referring to my [1968*c*], [1971*c*], and volume 1, chapters 1, 2 and 3. Professor Smart, I am afraid correctly, rebuked me in his [1972] for my predilection for cross-, forwards- and back- self-references, which tend to make my papers difficult to follow. But while being apologetic for this style of exposition, I am unrepentant concerning the content.

[2] If isolated hypotheses did constitute scientific *achievements*, Hegel, for instance, would have to be regarded as a great scientist and a forerunner of Einstein, since he muttered something about the interrelation of time and space.

[3] Cf. my [1976*c*], esp. pp. 70–82. This discussion of informal mathematics has clear implications for scientific explanation.

[4] For a discussion of three different types of adhocness cf. my [1968*c*], pp. 375–90, esp. p. 389, n. 1; and also volume 1, chapter 1, p. 88, nn. 1, 2 and 4.

[5] Cf. the distinction – and its implied division of labour – between internal and external history proposed in volume 1, chapter 2, and *above*, p. 114.

[6] Cf. my discussion of Mercury's perihelion, of the Michelson–Morley experiment, of the Lummer–Pringsheim experiment and of the alleged crucial experiments in favour of some theories of beta-decay in volume 1, chapter 1. Holton's interesting [1969], which was published when my [1970*a*] was being printed, also supports my conclusions (although, I am afraid, not *his*): cf. Zahar [1973].

[7] Cf. e.g. volume 1, chapter 1, p. 18.

[8] For a general discussion of this historiographical research programme cf. volume 1, chapter 2. * Some contributions to this historiographical research programme are to be found in Howson (*ed.*) [1976]. – (*Eds.*)

My position has clear implications for a theory of *scientific* learning.[1] The old problem – 'how and what do we learn scientifically from experience?' is solved in a novel way: '*In science* we learn from experience not about the truth (or probability) nor about the falsity (or improbability) of "theories", but about the relative empirical progress and degeneration of scientific research programmes.'

This solution involves a methodological and epistemological problemshift, in the course of which the problem of appraisal and of learning itself is reinterpreted, 'shifted', and the term 'scientific theory' is reinterpreted ('explicated') as 'scientific research programme'.[2]

Professor Grünbaum, in his paper, barely offers any opposition to my thesis. If he had wanted to challenge it at all seriously, he would have had to take one or more *concrete historical examples* of so-called crucial experiments and show that their role is the one described in his scheme. But he does not even attempt this. Indeed, in the last section of his paper, he goes out of his way to emphasize that he has never regarded the Michelson–Morley experiment as a 'crucial' one. In the first part of his paper, he also mentions that those who thought that Pasteur's 1862 experiments were 'crucial' and heralded final defeat for the idea of spontaneous generation of life, were mistaken. My earlier papers contain many more such examples. But then, on what point does Grünbaum *disagree* with my position?

This transpires only in the second half of his paper, notably in the section entitled '*Critique of Universal Falsificationist Agnosticism*'.[3] He could have given it the title: '*A Defence of Occasional Falsificationism*', since he argues that *at least in some exceptional cases, anomalies can constitute crucial negative experiments* and knock out theories.[4] I shall

[1] Cf. e.g. volume 1, chapter 1, p. 38, n. 2.

[2] For the relation between methodological and epistemological aspects of the problem cf. especially volume 1, chapter 3.

[3] I have to remind the reader again and again that, apart from my negative thesis about 'crucial' experiments, I also offer a new theory of scientific appraisal and criticize Popper's overkill of inductivism. I therefore regard as misleading the labelling of *my* position as 'universal agnosticism'.

[4] As *universal* falsificationism crumbles, falsificationists generally tend to withdraw from universal to occasional falsificationism: they try to demarcate insignificant anomalies from crucial negative experiments. For instance, Noretta Koertge recently tried to define a special class of 'striking anomalies'. (Cf. Koertge [1971]; but see my [1971*c*], pp. 177–8.) Popper now starts to distinguish between 'real discrepancies' and ordinary discrepancies. 'The first real discrepancy can refute [a theory]'. But, in his view, while a black swan refutes 'all swans are white', Mercury's perihelion constitutes an 'extremely small discrepancy' to Newton's theory and does *not* refute it. But what *general criterion* does he offer between 'real' discrepancies which refute and 'extremely small' ones which do not? (Cf. Popper [1971*a*], p. 9.) In the same interview he says that 'a theory belongs to empirical science if we say what kind of event we should accept as a refutation'. But then he has *either* to demarcate by a *general criterion* real discrepancies from apparent ones, *or* to specify 'real discrepancies' for each individual theory in an *ad hoc* way. But the latter approach can hardly avoid Polanyiism, for what would provide this piecemeal demarcation if not the authority of the expert scientist?

first examine his *specific* alleged counterexamples to my no-crucial-experiment thesis and then his *general* characterization of these exceptional examples.

Grünbaum's specific counterexample is as follows:

Suppose that one or more previously successful theoreticians develop a theory which comprises an aerodynamics and which makes a rich variety of daring, as yet untested predictions. Let this theory T be such as to entail that the existence of flying craft of any kind in the earth's atmosphere is *physically impossible.* In particular T entails that flying craft operated by humans cannot exist. It would then seem to follow that, UNLESS WE ARE ALL CONSTANTLY HALLUCINATING, those of us who are not hallucinating at certain times should neither observe airplanes, airships, blimps, helicopters and the like in the air from the ground at these times, nor should we perceive ourselves flying through the air in them at those times. Our central hypothesis H here is the assertion of the aerodynamic impotence of flying craft, while the relevant piece of background knowledge or auxiliary A is that at least some of the time, some of us are *not* hallucinating and can then be identified as such. Finally, the pertinent recalcitrant observational statement or so-called basic factual proposition is that some non-hallucinating observers do see flying craft in operation, or, if you will, that there are flying machines. Note that in thus observationally asserting the existence of flying craft, this basic factual proposition does *not* take on the complicated theoretical onus of specifying whether one or another of these machines is heavier than air or lighter than air.

I submit first of all that this example meets both Lakatos' challenges to me. Its basic proposition or observation statement is reliable at least in the sense of being *very* much more likely to be true than false and yet asserts a recalcitrant fact. Lakatos himself came to the congress at which our papers were presented in at least one aircraft from London, just as surely – at least qualitatively speaking – as that I am *not* Napoleon. And the auxiliary that *some* of us are *not* hallucinating at least some of the time does seem to satisfy the requirement of being so highly probable in at least a qualitative sense as to be beyond reasonable doubt. Incidentally, this requirement is assumed capable of being met in courts of law. But when combined with the very highly probable auxiliary, the central hypothesis H of aerodynamic impotence does entail that no actual non-hallucinating observer should ever see a flying craft in operation, a proposition which is contradicted by our reliable basic observation statement.

In the second place, I submit that the far-fetched character of this example is especially well suited to be a counterexample to Lakatos' very strong claims. For note that the theory T containing the denial of the existence of aircraft can be rationally indicted with at least qualitatively overwhelming probability as false but *without waiting* until the research program to which T belongs becomes degenerative or regressive as demanded by Lakatos.[1]

(Cf. volume 1, chapter 2, p. 137). Musgrave recognizes the problem (Musgrave [1973]) and redefines falsificationism so that it means that anomalies constitute *one* of the many sources of problems. I do not know a single philosopher of science (even including Polanyi) who would have denied this. If this is what remains of naive falsificationism, then we can just as well forget about it. (In his [1972] (p. 38, n. 5), Popper replies to my criticism from which it transpires that *he has now given up his universal demarcation criterion* and all that he wants is that the scientist should always specify for his theory, in an *ad hoc* way, at least one potential falsifier of his own choice. He seems to claim that for psychoanalytic theories this *cannot* be done. Why not?)

[1] Cf. Grünbaum [1973], pp. 62–3.

The first interesting feature of Grünbaum's 'counterexample' to my thesis is that *it describes an entirely imaginary case*. Does this imply that in past history of science he has found no crucial experiment? Indeed, the history of science is so rich, that his failure to produce an *actual* example should already make the history-orientated philosopher suspicious.[1]

Nevertheless, imaginary examples *can* be important. But Grünbaum unfortunately says very little about his theory *T*. All that we know about *T* is that it is a part of a *theoretically progressive* research programme *R*; it has a 'rich variety of daring predictions'; but that it also has one 'absurd' consequence: that there are no such things as flying crafts.[2] But if this is sufficient reason to dismiss *T*, then Copernican theory should have been dismissed because it implied the equally absurd consequence that our peaceful, stable Earth *was* a flying, spinning craft circling wildly around the Sun; Newton's theory should have been dismissed once it was shown that it implies the collapse of the planetary system into the Sun in one's lifetime. (When Grünbaum claims that theories like *T* may be declared as falsified 'in courts of law',[3] he should remember that Copernican theory was declared as falsified in the court of law of the Holy Inquisition exactly on his criterion.)

Grünbaum's 'counterexample' thus carries no weight whatsoever against my no-crucial-experiment stand. The greatest research programmes are characterized by the fact that at the time of their birth their hard cores were inconsistent with some 'factual' statements and auxiliary hypotheses accepted at the time. That is, *all* great research programmes were similar to Grünbaum's example: at their inception they 'raped the senses', they ran foul of the 'factual' and corroborated theoretical knowledge of the time. Yet, they were not ruled out. Grünbaum's *T* need not be ruled out either. Had Grünbaum's quasi-falsificationism[4] been followed, there would have been no scientific progress. If Grünbaum's claim is that although crucial experiments have never occurred, they will in the future, then his position is certainly more provocative than mine.

[1] As I have always claimed, paraphrasing Kant, (1) '*history of science without philosophy of science is blind*' and (2) '*philosophy of science without history of science is empty*'. (Cf. Crombie [1961], p. 458, where Hanson quoted it from me, then cf. my [1963–4], p. 3, and volume 1, chapter 2, p. 102.) Grünbaum seems to be an ally on the first score but not on the second.

[2] That *R* is progressive *and* that it is anomaly-laden can be judged – contrary to Grünbaum's claim – instantly ('without waiting'). What one cannot know instantly is when, *if ever*, scientists will start calling the anomaly a 'crucial experiment'. But this is surely a matter for external history which has no relevance for the purely normative discussion between Grünbaum and me.

[3] Cf. Grünbaum [1973], p. 63.

[4] I am puzzled why Grünbaum insists on replacing the apt Kuhnian term 'naive falsificationism' by 'quasi-falsificationism'. But whatever it is called, it remains naive, utopian.

2 THE IMPOSSIBILITY OF GRÜNBAUMIAN CRUCIAL EXPERIMENTS AND THE POSSIBILITY OF APPRAISING SCIENTIFIC GROWTH WITHOUT THEM

Professor Grünbaum challenges me to produce a *general proof* that crucial experiments are in principle impossible. He writes:

It seems to me that Lakatos' universal falsificationist agnosticism is rendered *at least gratuitous* by the fact that he has given no general proof to *rule out* the existence of *any and all* collectively incompatible trios of statements H, A and O' as follows: (i) In at least a qualitative, comparative sense of more or less, which does *not* necessarily require quantitative ascriptions of an exact numerical degree of corroboration to *each* member of the trio, the *prior corroboration* of H is very much lower than the prior corroboration of A and than the resulting corroboration of O' and (ii) A and O' are each corroborated at least to the extent that each one *is so very much more likely to be true than false as to be beyond reasonable doubt.* I maintain that if any such collectively incompatible trios exist, then it is at least generally *rational* to presume strongly that the pertinent H is false, whereas the denial of A in order to uphold H would be *irrational.*[1]

That is, according to Grünbaum (1) there exist trios H, A and O' which satisfy his two requirements; and (2) *if* they exist, *then* 'it is at least generally *rational* to presume strongly that the pertinent H is false, whereas the denial of A in order to uphold H would be irrational'.

But I deny Grünbaum's premise (1); and also his inference (2) *in any interesting interpretation.*

(1) I have argued at length elsewhere that corroboration comparisons are only possible in those (very exceptional) cases where one theory supersedes another – i.e. where the theories involved are *rivals.*[2] If my argument is correct, then, since Grünbaum's H, A and O' do not compete with each other, their degrees of corroboration are incommensurable.[3] But then Grünbaum's trios do not exist.

(2) Let us now assume, for the sake of argument, that there is some acceptable inductive logic which assigns corroboration values to H, A and O' in the way Grünbaum suggests.[4]

Let us imagine, again for the sake of argument, that the degrees of corroboration of the three hypotheses are as follows: $c(H, e) \approx 0$,

[1] Cf. Grünbaum [1973], p. 59, partly my italics. (H is the hypothesis under test, A the set of relevant auxiliary theories and O' reports some *prima facie* refuting experiment.)

[2] Cf. this volume, chapter 8, esp. pp. 184–5. How can one reasonably claim that Mendelian genetics, say, is more or less corroborated than the theory of beta-decay?

[3] Of course I am aware that inductive logicians try to construct measure functions which enable us to make such comparisons of theories even in very different fields; but, by now, the degeneration of these programmes for constructing inductive logics should be obvious.

[4] Grünbaum claims that there can be an assignment of such values in the light of which 'A and O' are each corroborated at least to the extent that each one is so very much more likely to be true than false as to be beyond reasonable doubt'. Even inductive logicians have grave doubts about whether inductive logic can do anything like this towards solving Hume's problem. (Cf. e.g. Salmon [1966], p. 132.)

$c(A, e) \approx 1$ and $c(O', e) \approx 1$, where e is the total evidence. (If H is the hard core of a programme and A the auxiliary belt, O' would be an 'anomaly' for the programme.) Let us now imagine that a scientist replaces A by an A' such that A is a limiting case of the more general A' (A and A' are inconsistent[1]) and H and A' predict successfully some novel facts. In my view *this replacement constitutes progress* ('a progressive problemshift'), even though H and A' and O' may be again inconsistent. Grünbaum surely admits that progress was then achieved by irrationally 'upholding' H and 'denying' A. But then 'rational belief' is irrelevant for the scientist!

My arguments do not amount to the logically watertight 'general proof' required by Grünbaum. Such a *proof*, of course, cannot ever be produced. First, as far as (1) is concerned, one can always assign, in a consistent fashion, confirmation values to any finite set of statements. My arguments, that all programmes of inductive logic which set out to perform this task have degenerated, do not constitute a strictly logical *proof* that no inductive logic will ever succeed. As to (2), Grünbaum might well accept my argument, but reject my interpretation of his terms 'rational presumption of falsehood' and 'irrational upholding'. *He can claim that while 'strongly presuming the falsity of H', the scientist should not be too disturbed by it and should develop his programme (based on H) regardless. He can say that he would never ask the scientist not to work on a discredited H.* And indeed, later in his paper he writes: 'As Laurens Laudan and Philip Quinn have independently pointed out to me, we must be mindful here of the distinction between the rationality and irrationality of *belief* in a hypothesis on the one hand, and the rationality or irrationality of *pursuing* some kind of provisional research work on it, on the other.'[2] Quite. But the 'distinction' is then between two redundant irrelevancies. For (1) *belief and disbelief, whether rational or non-rational, play no role whatsoever in the rational appraisal of science, and Grünbaum's (and others') vast work in defining degrees of rational belief (or rather, rational degrees of belief) in scientific theories serves no purpose whatever*; and (2) *nobody has yet offered a theory of rationality for the individual scientist telling him which, among competing research programmes, he should choose to work on, or whether and when he should try to start one on his own.* All that Grünbaum says of the latter is that it 'would be unwise to put all one's research eggs into one basket';[3] but, of course, that the monopoly of one paradigm is undesirable was the main Popperian criticism of Kuhn's 1962 approach;[4] and this trivial

[1] That is, the limiting case usually is an 'ideal', *counterfactual* one, like an 'ideal' gas.

[2] Cf. Grünbaum [1973], p. 87.

[3] For the individual, it is usually *wise* to put all one's research eggs into *one* basket, in the sense that it is unwise to give up a programme too easily; to acquire competence in the techniques of a serious research programme takes many of one's best years.

[4] Cf. primarily Watkins [1970], p. 34ff. The desirability of theoretical pluralism is also an obvious corollary both of Feyerabend's 'epistemological anarchism' and of my 'methodology of scientific research programmes'.

statement does not help the individual scientist in his decisions *in the least.* Once Grünbaum agrees – as he now does – to separate rational belief and rational appraisal, the *second* part of his argument, i.e. (2), is valid, but in an uninteresting sense. And logic alone cannot *prove* that a (consistent) philosophical endeavour is uninteresting. But one can *argue* – if not by inexorable deductive logic – that the endeavour is irrelevant and, indeed, possibly harmful. Before I do this in some detail,[1] I should like to clarify briefly my views on *practical advice.*

Note. One further comment: I do accept the existence of trios H, A and O', such that A and O' are better corroborated than any of their respective rivals; neither has been superseded, but, of course, neither is 'beyond reasonable doubt'. I accept that the inconsistency of H, A and O' constitutes a problem, since I respect deductive logic and accept consistency as a regulative principle. The question then arises *which* of the three is to be replaced to restore consistency. Since I claim that the comparison of the relative degrees of corroboration of H, A and O' is impossible, for me *none of the three is a privileged candidate for 'progressive' (non-ad hoc) replacement.*

3 ON PRACTICAL ADVICE

Professor Grünbaum, with the aid of another imaginary example, puts to me a challenge concerning practical action. He devises an imaginary example of a budding research programme designed to establish a new, more reliable, way of distinguishing between acute leukaemia and mononucleosis.[2] There is an old, moderately successful programme and a new, promising one which, however, contains no 'significant corroboration of [its] speculation': that is, on my standards, it is degenerating.[3] Grünbaum asks, on the assumption that the two rival research programmes suggest conflicting advice, which is to be followed in actual medical practice? He is 'curious how Lakatos' view of scientific rationality adjudicates an example of this kind'.[4] Well, I have discussed this problem of *practical advice* at some length in several papers. My practical advice is: one should act in any field according to the most 'trustworthy' or most 'reliable' theories in the given field.[5] We construct the body of 'most reliable' knowledge *from the body of scientific knowledge.* However, the latter is always inconsistent because of the presence of anomalies: any scientist accepts ('accepts$_1$' and 'accepts$_2$') an *inconsistent* set of propositions on which he works: he

[1] Cf. *below*, section 4.

[2] Cf. Grünbaum [1973], p. 64.

[3] It may be confusing that on my definitions, 'budding' programmes are 'degenerating'. But, alas, the phenomena of juvenile and senile behaviour bear considerable resemblance, as we have ample opportunity to learn from many contemporary youth movements.

[4] Cf. Grünbaum [1973], p. 65.

[5] For the concept of 'trustworthiness' or 'reliability' (or 'acceptability$_3$') cf. this volume, chapter 8, section 3.

does not 'eliminate' either the hard core, the auxiliary theories, the falsifiable versions of the programme or the anomalies. The '*body of* [*most reliable, or*] *technological knowledge*' on the other hand, is *consistent* for it derives at any time from the 'body of scientific knowledge' by mutilating the research programme in an *ad hoc* way; any applied scientist in 1900 accepted ('accepted$_3$') only part of Newtonian astronomy, and not when it was applied to cases like that of Mercury's perihelion. Thus the applied scientist (e.g. the medical practitioner) 'works with' a consistent trunk of scientific knowledge.[1]

Professor Grünbaum seems to have missed my separation of scientific–methodological 'acceptance$_1$' and 'acceptance$_2$' on the one hand and practical 'acceptance$_3$' on the other.[2] At this point I have to underline strongly the fact that Grünbaum misrepresents my position in a most puzzling way. He ascribes to me an 'epistemological asceticism',[3] as if I did not assume that Einstein's programme was epistemologically superior to Newton's or as if I put Greek mythology and quantum physics epistemologically on a par. This, of course, is the *exact opposite* of my position. Having tried to show the weakness of Popper's pure game of science,[4] my methodology of scientific research programmes provides a novel positive solution to the 'Duhem–Quine thesis', a solution[5] which Grünbaum completely ignores while criticizing other solutions.

[1] For the concepts 'acceptance$_1$', 'acceptance$_2$', 'acceptance$_3$' and the contraposition of 'scientific' and 'technological' bodies of knowledge, cf. *ibid.* But let me remind the reader that the 'body of technological knowledge' is constructed from the 'body of scientific knowledge' with the help of *ad hoc* (content reducing) stratagems. No doubt the less one says the safer one is. But then the *practical* rationality involved in constructing the 'body of technological knowledge' is alien to the scientific rationality involved in constructing the 'body of scientific knowledge'. Also it is 'more likely than not' that all technological knowledge ever constructed and ever to be constructed by man is false even if it 'works'.

[2] The section of Professor Grünbaum's paper 'Action and Rational Presumptions of Knowledge' shows that he is not familiar with my defence of an 'inductive principle' as a basis for action. Cf. this volume, chapter 8, sections 1, 3 and also volume 1, chapter 3, section 2(*b*). In the latter I substitute 'research programme' for 'theory'.

[3] Cf. Grünbaum [1973], p. 68.

[4] Grünbaum describes my position as 'rejectionism'; as if I regarded science *merely* as a game without epistemological relevance. But I do superimpose a conjectural inductive principle on the scientific game and then strict agnosticism vanishes. I argued this both in this volume, chapter 8 and in volume 1, chapter 3. Grünbaum's challenge on practical advice indicates that he thought that I share *Popper's* 1934 epistemological agnosticism. But I do not.

[5] This, of course, to repeat a point made earlier, must be seen against the background of my criticism of Popper's overkill of inductivism, against the background of the 'plea for a whiff of inductivism' in volume 1, chapter 3.

4 THE CHARACTERISTIC OF SCIENCE IS NOT RATIONAL BELIEF BUT RATIONAL REPLACEMENT OF PROPOSITIONS

An important clue to Grünbaum's misunderstanding of my intentions may well be his equation of science with the set of 'rational beliefs'. Grünbaum describes 'believing or disbelieving' merely as a kind of 'locution'[1] and does not define it. But since he rejects my Popperian 'disbelief in beliefs' when it comes to appraising scientific knowledge, and since he devotes several pages to his dissent, there is obviously a philosophical rather than a mere semantic disagreement between us. But what makes his 'rational believing' different from my 'accepting' ('$accepting_1$', '$accepting_2$', '$accepting_3$') or 'preferring'?

The Popperian position that scientific rationality has nothing to do with 'rational belief' is well-known.[2] But I propose a further argument to support it. Science progresses through the competition of research programmes, not simply through conjectures and refutations. But a programme is a complex entity, a special case of a problemshift (that is, a *series* of propositions), plus mathematical and observational theories *and* heuristic techniques which provide the tools for forging ahead. A research programme as a whole cannot be either true or false. How can one 'rationally believe' that a *programme* is 'likely to be true'? Grünbaum may retort that the scientist *might* rationally believe or disbelieve the programme's *hard core*. But there is no need to believe (rationally or non-rationally) in the hard core of the programme on which one is working. Newton disbelieved his own action-at-a-distance programme in its realist interpretation; Maxwell elaborated kinetic theory and Planck quantum theory with actual disbelief.[3]

But could not one rationally believe the synchronic cross-sections of a programme, 'the body of scientific knowledge'? Alas, this body has always been *inconsistent*.[4] How can one rationally believe an inconsistent set of propositions?

The advocate of science as rational belief might still retort: 'What about your own admittedly consistent "body of technological knowledge"? Do you not describe it yourself as "reliable" and "trustworthy"? Why is *that* not worthy of "rational belief"?' In response I can only point out that science grew by progressive replacements from Greek myths. It might well have grown from medieval or, say, Zande, beliefs. The propositions in the 'body of technological knowledge' are, at best, the latest products of such progressive problemshifts. But what should make us rationally believe the latest link in a progressive problemshift whose initial proposition, after all, was backed by mere

[1] Cf. Grünbaum [1973], p. 86.

[2] Popper, like the early Carnap, wanted to exorcize the term 'belief' and even 'rational belief' because these terms were originally associated with psychologism. Cf. e.g. Carnap [1950], pp. 37–51; and Popper [1972], *passim*.

[3] Cf. volume 1, chapter 1, p. 43.

[4] Cf. this volume, chapter 8, p. 176.

animal belief? *At what point in a problemshift comes the sudden change from animal belief to rational belief?* At no point. We may claim that progressive problemshifts do move us 'more likely than not' towards Truth rather than away from it. But this inductive principle which confers an epistemological status on our *convention* as to how to appraise problemshifts, is, in turn, backed by mere animal belief. Therefore problemshifts receive their epistemological rationality from animal belief (or, if you wish, from a bare postulate – an intellectual theft, as Russell used to characterize such 'posits'). (*Note.* The inductive principle was never progressively replaced.)

This concludes my – Humean – argument against 'rational belief' in science. I 'accept' the body of technological knowledge, but I do not 'rationally believe' in it.

If Grünbaum agrees with this argument, and still wishes to find some use for the term 'rational belief', I shall cease to object: words, after all, are mere conventions.[1]

Having said this, let me make a few more points about scientific rationality.

Rationality (and here I follow Popper) is not concerned with propositions in isolation (whether the proposition is 'basic', 'scientific' or 'metaphysical') but with the way we modify them and hence, their relation to other propositions.[2] This is one of Popper's greatest innovations. But in Popper's work, as in that of many great thinkers, the new is slightly marred by the remains of the old. His demarcation criterion is formulated in terms of falsifiability or unfalsifiability of *propositions* rather than in terms of the progressive or degenerating character of *problemshifts* (i.e. series of propositions which are the result of progressive or *ad hoc* modifications). But a careful reading of his text enables one to extricate the emerging powerful new idea in spite of the old-fashioned terminology and in spite of the frequent slips into the conceptual framework of the past.[3]

This shifting of the *locus* of scientific appraisal from propositions to problemshifts introduced a historical dimension into scientific appraisal. (The problem of when a hypothesis-*replacement* is '*ad hoc*', i.e.

[1] But I would object to the usage 'rational belief in a hard core' as *too* misleading – I would restrict it, even in the qualified sense, to our body of technological knowledge'.

[2] Cf. Popper [1934], section 20: 'my criterion of demarcation cannot be applied immediately to a *system of statements*. . . *only with reference to the methods applied* to a theoretical system is it at all possible to ask whether we are dealing with [science or pseudoscience]'. As Bennett put it succinctly: '[According to Popper] whether someone's intellectual behaviour counts as "scientific" or not depends not upon where he gets his hypotheses from but rather upon what he does with them when he has them.' Cf. Bennett [1964], p. 35 and also Latsis [1972], p. 240.

[3] Cf. this volume, chapter 8, pp. 178ff, and my [1968*c*] and volume 1, chapter 1, pp. 88–9. In order to separate the novel element in Popper from the old, I separated $Popper_2$ from $Popper_1$ in my [1968*c*]. (The idea of 'problemshifts' is implicit in $Popper_2$. Research programmes, on the other hand, are special problemshifts, which go beyond Popper's own ideas.)

irrational, degenerating, bad, has never been discussed with more attention and detail than by Popper and myself.)

The rational man, of course, cannot help but hold animal beliefs in propositions (like 'Napoleon is dead' or 'All men are mortal') which still stand waiting for a progressive replacement; he differs from the irrational man only in not taking his beliefs too seriously, in not attributing to them rationality. In particular – let me repeat – when the scientist works on a research programme, he has no need whatsoever to believe in its 'hard core'. Furthermore, even if we decide to confer the honorific title 'worthy of rational belief' on a proposition which stands as the latest in a progressive problemshift, we still should not describe either the proposition or *even the progress* as being 'beyond reasonable doubt'.

We should take note of another typical Popperian warning and be more modest and cautious both about the extent and about the safety of scientific progress. Let us consider, for instance, the average Elizabethan Englishman. He had no science whatsoever: science was non-existent before the seventeenth century.[1] Now while medieval man had no science, he certainly held beliefs with considerable truth content. Some of these (like 'God exists', or even 'All swans are white'[2]) might even have been nearer to the Truth than contemporary quantum theory. He had many such beliefs about the weather, the soil, about the power of barons and bishops, and he had beliefs about astrology and witchcraft. Of course, some of the Elizabethan beliefs have been replaced by new beliefs, such as the animal beliefs held by, say, the average American liberal. Some of these new beliefs, like beliefs about mental illness instead of belief in witchcraft, or beliefs about Capitalism and Socialism instead of beliefs about King and Church, are unscientific replacements of animal beliefs.

But not all belief changes since Elizabethan times were mere changes in fashion. There was the Scientific Revolution. *But the Scientific Revolution was not marked by a sudden emergence of true or highly verisimilar beliefs replacing false or improbable beliefs. Newtonian science and contemporary relativity theory may well have lower verisimilitude than some of the 'wisdom' of Elizabethan times. The Scientific Revolution was marked – and I wonder whether Professor Grünbaum will agree – by the emergence of scientific research programmes and their scientific appraisal. The characteristic of science is not a special set of propositions – whether provenly true, highly probable, simple, falsifiable, or worthy of rational belief – but a special*

[1] 'Medieval science' was invented by Duhem in order to rehabilitate the Catholic Church, and found a following among vulgar-marxists. The latter held that socialist science is better than bourgeois science, bourgeois science better than medieval feudal science and feudal science better than the science of slavery. If so, they had to invent some feudal science in order to build a bridge between Archimedes and Galileo, which they did by conferring the title 'scientific' on artisanship.

[2] Note that none of these propositions belong to a progressive research programme; therefore, in my sense, they are not scientific.

way in which one set of propositions – or one research programme – is replaced by another.

There is no ultimate proof that, even where Elizabethan beliefs were replaced in the course of progressive problemshifts (like beliefs about heat, magnetism), we have been heading towards the Truth. We can only (non-rationally) believe, or, rather hope, that we have been. Unless hope is a 'solution', there is no solution to Hume's problem.[1]

[1] I should like to mention that Popper's [1971*b*], so devastatingly criticized by Professor Grünbaum, was, as a matter of fact, an attempt to reply to my [1968*a*] (this volume, chapter 8) and [1971*a*] (volume 1, chapter 3); cf. the *Note* at the end of volume 1, chapter 3.

11

Understanding Toulmin*

INTRODUCTION

Human Understanding is Professor Toulmin's fifth book in the tradition of Wittgenstein's later philosophy. He applied this philosophy first to ethics in 1950,[1] then to the philosophy of science and logic in 1953[2] and 1958.[3] The main theme of his present three-volume *magnum opus*, of which this is the first volume, was foreshadowed by his earlier short book *Foresight and Understanding*, published in 1961.[4]

Frankly, I liked his earlier books much more than the new one. J. O. Wisdom once wrote about Wittgenstein's philosophy: 'One has the sense of wandering about the corridors of a maze; and the maze has no definite centre. This mode of presentation, leading one through a maze, whose "centre" is the discovery that there is no centre, in itself conveys some philosophical message.'[5] One senses the same philosophical message when reading Toulmin, and, in the case of *Human Understanding*, the maze is, of course, bigger and more complex than in his earlier books. I am afraid that this message cannot be conveyed in the form of a short 'digest' and then criticized sharply. Indeed to give a short digest, or sharp criticism, of a work written in the Wittgensteinian tradition is *bound* to lead to failure. What I shall do instead is to specify one single central problem with which philosophy of science traditionally has been concerned, and then attempt to see where Toulmin stands with regard to this problem.

This central problem is that of the (normative) appraisal of those theories which lay claim to 'scientific' status. This, it seems to me, is

* At the time of his death Lakatos was engaged in writing a review of Stephen Toulmin's book *Human Understanding*. He had written, and discarded, three progressively more detailed versions and in the summer of 1973 he began work on a fourth. This last, and longest, version was never quite completed, and Lakatos remained unhappy with some aspects of it. He wanted to approach Toulmin's work by placing it in the context of some general epistemological problems and traditions. Consideration of these general problems in fact takes up most of the fourth of Lakatos's manuscripts, and he felt that this made it inappropriate as a review of *Human Understanding*. We have deleted most of the general material and worked this into a separate paper which appears *above* as chapter 6. There is however some unavoidable overlap between the two papers. What appears here is based on the third of Lakatos's manuscripts, but with many points modified and extended using material from the other versions particularly the fourth. It was first published as a review of Toulmin's book in *Minerva*, **14**, pp. 126–43. (*Eds.*)

[1] Toulmin [1950].
[2] Toulmin [1953*a*] and [1953*b*].
[3] Toulmin [1958].
[4] Toulmin [1961].
[5] Wisdom [1959], p. 338.

the primary problem of the philosophy of science. To neglect it, or even to assign it a merely secondary role, implies philosophical surrender to a strictly descriptive sociology and history of science.

I shall first sketch the three major philosophical traditions in the approach to this one problem. I shall argue that Toulmin's basic position, for all its question-marks, ambiguities and contradictions, is clearly derived from one of these three traditions, that of 'élitism'. But his élitism is burdened with Wittgenstien's brand of pragmatism. I shall show that Toulmin's return in *Human Understanding* to the more conventional Darwinian brand of élitism is a natural escape from one of the most unpleasant ideas in Wittgenstein's philosophy: the idea that philosophers ought to constitute a 'thought-police'. But, I shall suggest, the only function of the Darwinian metaphors is to dress the Hegelian Cunning of Reason in trendy scientific garb. Toulmin's metaphors remain metaphors: they have no explanatory power. Throughout I shall try to expose those features of Toulmin's position which make it, in my opinion, untenable.

1 THREE SCHOOLS OF THOUGHT ON THE NORMATIVE PROBLEM OF APPRAISING SCIENTIFIC THEORIES

Scepticism: One school of thought on the problem of appraisal can be traced back to the Greek tradition of Pyrrhonian scepticism, and is now known as 'cultural relativism'. Scepticism regards scientific theories as just one family of beliefs which rank equal, epistemologically, with the thousands of other families of beliefs. One belief-system is no more 'right' than any other belief-system, although some may have more *might* than others. There may be changes in belief-systems but no progress. This school of thought, temporarily muted by the stunning success of Newtonian science, is today regaining momentum particularly in the anti-scientific circles of the New Left; its most influential version is Feyerabend's 'epistemological anarchism'. According to Feyerabend, philosophy of science is a perfectly legitimate activity; it may even influence science. Note that this view is different from Mao's 'let a hundred flowers bloom'. Feyerabend does not wish to impose a 'subjective' distinction between flowers and weeds on anybody. *Any* belief-system – including those of his opponents – is free to grow and influence any other belief-system; but none has epistemological superiority.[1]

Demarcationism: The second school of thought, which is the main rival to scepticism, is preoccupied with *positive* solutions of the *demarcation problem*.[2] This school originated in Greek 'dogmatism' (a

[1] I am here referring to the Feyerabend of the 1970 vintage, as best seen in his [1970], [1972] and [1975].

[2] I do not use the term 'demarcation problem' in Popper's strict sense, as referring to the problem of a black-and-white demarcation between science and pseudoscience.

nickname given to Stoic philosophy by Pyrrhonians: I use 'dogmatism' to denote the position that objective *knowledge* – infallible or fallible – is possible). It is an apriorist tradition. Leibniz, Bolzano and Frege belonged to this tradition, and, in this century, Russell and Popper. Carnap's early work also belongs to this 'demarcationist' line, as does my methodology of research programmes. In the demarcationist tradition, philosophy of science is a watchdog of scientific standards. Demarcationists reconstruct *universal* criteria which explain the appraisals which great scientists have made of *particular* theories or research programmes. But medieval 'science', contemporary elementary particle physics, and environmentalist theories of intelligence might turn out not to meet these criteria. In such cases philosophy of science attempts to overrule the apologetic efforts of degenerating programmes.[1]

Demarcationists differ over what precisely the universal criteria of scientific progress are, but they share several important characteristics. First, they all believe in the third of Frege's and Popper's three worlds. The 'first world' is the physical world; the 'second world' is the world of consciousness, of mental states and, in particular, beliefs; the 'third world' is the Platonic world of objective spirit, the world of ideas.[2] Demarcationists appraise the *products* of knowledge: propositions, theories, problems, research programmes, all of which live and grow in the 'third world'[3] (whereas the producers of knowledge live in the first and second worlds). In line with this, demarcationists also share a critical respect for the articulated. They readily agree that articulated knowledge is only the tip of an iceberg: but it is exactly this small tip of the human enterprise wherein rationality resides. Finally, demarcationists share a democratic respect for the layman. They lay down *statute law* of rational appraisal which can direct a lay jury in passing judgment. Of course, no statute law is either infallible or unequivocally interpretable. Both a particular ruling and the law itself can be contested. But a statute book – written by the

I use it in a general sense, as referring to the problem of appraisal of rival theories. (Of course, Popper proposed a *continuous* range of appraisal for theories according to their different degrees of empirical content, corroboration and verisimilitude; but his *main* interest was to argue for his identification of [white] science with 'falsifiability' and of [black] pseudoscience – or non-science – with 'unfalsifiability'.)

[1] For such militant demarcationism see Popper on psychoanalysis in his [1963*a*], pp. 37–8; or Urbach [1974]. For an attempt to show up a non-empirical branch of knowledge as degenerating scholasticism, see my own treatment of inductive logic in this volume, chapter 8.

[2] An exposition of this vital distinction may be found in Popper [1972], pp. 106–90; and especially in Musgrave's important unpublished doctoral thesis [1969].

[3] Most demarcationists agree that propositions are true if they correspond with the facts, and thus subscribe to the correspondence theory of truth. Most of them carefully distinguish truth and its fallible signs: whether or not a statement corresponds to the facts is an entirely separate question from whether or not we have reason to believe it. (See Popper [1934], section 84; and Carnap [1950], pp. 37–51.) One of Toulmin's fundamental mistakes is to miss this vital distinction.

'demarcationist' philosopher of science – is there to guide the outsider's judgment.[1]

Elitism: Toulmin belongs to neither of these schools, but rather to a third school of thought which is at present perhaps more influential than either of the previous two. This school – élitism – is, like demarcationism, a version of 'dogmatism'; but an undemocratic, authoritarian one. Unlike the sceptics – and like most demarcationists – its proponents acknowledge the vast superiority of Newton's, Maxwell's, Einstein's and Dirac's achievements over astrology, Velikovsky's theories, and pseudoscience of all sorts. But they claim, contrary to the demarcationists, that there is, and there can be, no statute law to serve as an explicit, universal criterion which separates progress from degeneration, science from pseudoscience. In their view, science can only be judged by case law, and the only judges are the scientists themselves. If these authoritarians are right, academic autonomy is sacrosanct and the layman, the outsider, must not dare to judge the scientific élite. If they are right, the subject of (normative) philosophy of science should be abolished as *hubris*. Polanyi advocated such views – and so did Kuhn.[2] Oakeshott's conservative conception of politics also falls in this third category. According to Oakeshott, one can *do* politics, but there is no point in philosophizing about it.[3] According to Polanyi, one can *do* science but there is no point in philosophizing about it. Only a privileged élite has the craft of science, just as – according to Oakeshott – only a privileged élite has the craft of politics. All élitists lay great stress on the inarticulable, on the 'tacit dimension' of science. But if the 'tacit dimension' plays a role in normative appraisal, the layman obviously cannot be a judge. For the tacit dimension is shared and understood (*verstanden*) only by the élite.[4] Only they can judge their own work. Thus in this tradition we have a combination of élitism and of a cult of the unarticulated and, indeed, of the inarticulable.

But if one theory is better than another if the scientific élite prefers it, it is vital to know who belongs to the scientific élite. While élitists claim that there are no acceptable universal criteria for appraising the third-world *products* of scientific activity, they may (and do) offer universal criteria for appraising the *producers* of science (primarily their 'second-world' mental states) – rules for deciding if certain persons or

[1] For Toulmin's sharp condemnation of demarcation criteria, see the volume under review, pp. 254–60, and also his [1972].

[2] Polanyi's original problem was to provide arguments for protecting academic freedom from the communists of the 1930s, 1940s and 1950s; see Polanyi [1964], pp. 7–9. Kuhn's problem was very different: the breakdown of traditional inductivist and falsificationist accounts of scientific growth. See Kuhn [1962], introduction.

[3] For a critical discussion of Oakeshott's philosophy see Watkins [1952], pp. 323–7.

[4] Elitism is closely related to the doctrine of *Verstehen*. For this see Martin [1969], pp. 53–67. This doctrine is, of course, alien to the 'positivistic' criteria for a satisfactory explanation, like the one I offered in volume 1, chapter 1, p. 33. ('*Positivismus*', by the way, seems to be the German swear-word for what I call 'demarcationism'.)

communities belong to the élite. As a consequence, while for the demarcationists the philosophy of science is the watchdog of scientific standards, for élitists this role is to be performed by the psychology, social psychology or sociology of science.

The first two modern élitists were Bacon and Descartes. Bacon thought that the scientific mind was one purged of 'prejudices', Descartes that it was one which had been through the torment of sceptical doubt. The nazis held that Aryan science was superior to Jewish science. Other élitists appraise communities for 'scientificness' rather than individuals. For some pseudo-Marxists, the quality of science depends on the structure of the society which produced it: feudal science is better than ancient slave science, bourgeois science is better than feudal science, and proletarian science is true.

This stress on the inarticulable shifts the problem of 'knowing that' to the problem of 'knowing how'; from knowledge expressed in propositions to knowledge expressed in skills and activities. This in turn leads from the classical conception of truth – a proposition is true if it corresponds with the facts – to *pragmatism* – a belief is 'true' if it gives rise to useful or effective action. (Demarcationists like Russell regarded this theory as part of the intellectual ancestry of fascism.[1]) Propositions, and thus the 'third world', are redundant.

All those who belong to this third, élitist, tradition run into a major problem over scientific progress. They believe that science can make real progress, but since they claim that there is no universal criterion of progress, they are bound to claim that any change in science means, by a Hegelian Cunning of Reason, progress in science. Might is right – at least among genuine scientists or within genuine scientific communities; selective survival is the criterion of progress.

[We shall see that it is to this third tradition that Toulmin belongs. He is an élitist. He appraises communities rather than theories; and he resorts to a form of historicism.] But one is bound to misinterpret Toulmin's book as a whole, unless one is aware of an all-pervasive but not sufficiently explicit, feature of it. This feature is Toulmin's devotion to one of the most obscurantist traditions in contemporary philosophy: the philosophy of the later Wittgenstein. It is to this philosophy, and the question of how it coheres with élitism, that I now turn.

2 TOULMIN AND THE WITTGENSTEINIAN 'THOUGHT-POLICE'

Gilbert Ryle condemns all generalizations as 'unclarifications'.[2] The Wittgensteinian Cavell, who deeply influenced Kuhn,[3] writes that

[1] See especially Russell [1935].
[2] See Naess [1968], p. 165.
[3] See Kuhn [1962], p. xiii: '[Cavell] is...the only person with whom I have ever been able to explore my ideas in *incomplete sentences*. That mode of communication attests

all that is available of Wittgenstein's philosophy are 'hints of echoes of shadows of Wittgenstein's intentions'.[1] This implies that understanding – even at the philosophical meta-level – is available only to a select é lite which grasps its tacit dimension. According to Anthony Kenny, '[Wittgenstein's] *Investigations* contains 784 questions, only 110 of these are answered and 70 of the answers are *meant* to be wrong'.[2]

The early Wittgenstein of the *Tractatus* rediscovered the fact that we look at the world through the spectacles of our conceptual frameworks, which are expressed in our language. This timeworn idea – which Wittgenstein no doubt learned from Bühler – is trivially true. Now there are apriorists who aim at manufacturing perfect spectacles. Others, like Popper, aim at appraising the relative merits of different spectacles. But the later Wittgenstein denied that one can distinguish the quality of one pair of spectacles from another: all that one can do is to clean the spectacles one has got. But this is also one's *duty*. Note the difference between Wittgenstein and Feyerabend: Feyerabend does not mind if some people go about with dirty glasses.[3]

According to the later Wittgenstein the spectacles through which we look at the world are 'language games'. To learn a language game requires more than the learning of a 'language' in the ordinary syntactical and semantic sense. For these games are not just semantic structures but social institutions: 'To obey a rule [of a language game], to make a report, to give an order, to play a game of chess are *customs* (uses, institutions)'.[4] Toulmin following Wittgenstein, defines 'concepts' not as 'Platonic' objects, i.e. designations of words, but as 'skills or traditions, the activities, procedure, or instruments of Man's intellectual life and imagination' (p. 11). 'Concepts are micro-institutions' (p. 352). The Wittgensteinian 'concepts' so much referred to by Toulmin, receive their meaning from their complex social use, from the game as a whole, which, in turn, constitutes a 'form of life'.[5] When Toulmin says that 'questions about concepts underlie questions about propositions', he means that deep questions about real-life scientific activity should precede the superficial, shallow questions about the truth or falsehood of propositions.

an understanding that has enabled him to point me the way through or around several major barriers.' (My italics.)

1 Cavell [1962], pp. 67–93, esp. p. 73.

2 Fann [1969], p. 109. I wonder whether Kenny's statement itself was meant to be wrong?

3 The metaphor of the spectacles is from Popper; see his discussion with Strawson and Warnock in Magee, B. (*ed.*) [1971].

4 For this and similar passages see Feyerabend [1955], section *ix*. This paper is from Feyerabend's near-Wittgensteinian period, which falls between his near-Dinglerian and near-Popperian periods; it is a useful, albeit over-sympathetic account of Wittgenstein's *Philosophical Investigations*. For a less sympathetic exposition see Gellner [1959].

5 Feyerabend [1955], section *xi*.

It is important to realize that for Wittgenstein and Toulmin *facts* play no clear-cut role in the acceptance of a scientific theory. For Wittgenstein a 'correct explanation' may be 'accepted' without 'agreeing with experience'. 'You have to give the explanation that is accepted. This is the whole point of the explanation.' Or 'The correct analogy is the accepted one'.[1] On Toulmin's account, the role of facts can be quickly described: science's 'explanatory techniques must be..."consistent with the numerical records"'. But, more importantly 'they must be also acceptable for the time being at any rate – as "absolute" and "pleasing to the mind"'.[2]

Let us now go on with our little dictionary of Wittgensteinian jargon. Another technical term of Wittgensteinian philosophy, besides 'language games' and their constituent 'concepts' is 'understanding'. In Wittgensteinian philosophy, 'understanding' means the learning of the social conventions and commitments of the language game. This learning includes learning to feel certain and to lose one's doubts about 'foundations'.[3] One can learn a game by being indoctrinated by one's parents, teachers,[4] surroundings and by experiences shared with language users who are sufficiently mature to act as judges or preferably to constitute a court martial.[5] Some doubts are permissible, but others – regarding the 'foundations' – show that the doubter has not understood the game, that he is possibly incapable of learning it, or that he is mentally deranged.[6] A language game 'is not reasonable (or unreasonable). It is there – like our life'.[7] 'Truth' for Wittgenstein, as for all pragmatists,[8] equals practical – i.e. social – acceptability, and the test of one's 'understanding' is whether one plays the game right. Thus one cannot agree or disagree with a language game, but only *understand* or *misunderstand* its 'concepts'. One can only be an *insider* or, alternatively, an *outsider*. 'Understanding *the* [*objective, outside*] *world*' is a pipe dream, a 'castle in the air'. Wittgensteinian–Toulminian 'understanding' is '*human* understanding': understanding the human

[1] Wittgenstein [1966], pp. 18, 25.

[2] Toulmin [1961], p. 115. It is, by the way, amusing that this first apparently minimal requirement of consistency with the 'numerical records', taken by Toulmin as a trivial concession to empiricism, is one that no major scientific theory seems ever to have satisfied. Newton's theory, for example, was never consistent with all the known facts. See volume 1, chapter 1, pp. 49–52.

[3] See Wittgenstein [1969], paras 446, 449.

[4] For a description of Wittgensteinian teacher–pupil relations, see the rather frightening account by Wittgenstein [1951], pp. 310–22, also p. 106. For the Wittgensteinian practice of teacher–pupil relations, see the hair-raising account in Pascal [1973]. Strawson puts the Wittgensteinian philosophy of education clearly: 'Of course, in the instructor–pupil situation, explanations are in place; but the purpose of the explanation is to get the pupil to do as we do, and find it equally natural': Strawson [1954], esp. p. 81.

[5] Wittgenstein [1969], paras 453 and 557.

[6] *Ibid.*, paras 155–6.

[7] *Ibid.*, para. 559.

[8] The later Wittgenstein certainly was a pragmatist. The meaning of a proposition (or rather a 'speech act'), and hence its truth value, are given by the social context of the 'game'.

conceptual world or sub-world in which we live and manage to conform and thus survive. 'Understanding' is shorthand for 'understanding how to play the language game correctly'. Such understanding can hardly be learned from books. One has to live in the language user's society, sit at a master's feet, steal his *Zettels*, watch his gestures. Then 'understanding' may come, but only through total involvement. Any slight wavering in commitment and understanding is lost.

In the subject index of Toulmin's *Human Understanding* the word 'understanding' does not appear. Wittgensteinians, of course, never define terms. Their meaning lies in their polymorphous, undefinable use. It is part of Wittgenstein's and Toulmin's brain-teasing techniques however, to keep asking 'abstract' questions[1] such as 'What is explanation?'[2] or 'What is science?' and the question then trickles out in an incoherent monologue ending with 'etc.'. Indeed, full stops result in systematically misleading expressions. In Toulmin's words: 'A nutshell definition of science [as any nutshell definition] inevitably floats around on the surface. An investigation of any depth forces us to recognise that the truth [about anything] is much more complex...'[3] Gellner called this technique 'Wittgensteinian polymorphism': 'There is a very great variety in the kinds of use that words have... [therefore] general assertions about the use [and meaning] of words are impossible'.[4] This 'cult of caution'[5] gives Wittgenstein and Toulmin the right *not* to define their technical terms, and to be elusive even after they have absent-mindedly defined them. Of course, if a philosopher – or scientist – emphasizes that his 'activities' cannot be fully expressed in any finite sequence of propositions, and that therefore he bears no responsibility for his own summaries or crisp aphorisms, criticizing his view is not exactly easy. Neither is criticism made easier by the claim that certain questions are 'limiting questions'. This is yet another technical term and denotes questions to which there is no answer inside the game. Asking a 'limiting question' shows that you have not yet learned the rules; although asking them and being rebuffed may help in learning the rules.

I hope I have made it clear that Toulminian 'understanding', like a Toulminian 'concept', is a Wittgensteinian technical term.[6] Realizing

[1] 'Abstract' is a swear-word in Wittgensteinese. See Toulmin [1974], 1.41.

[2] See, for example, Toulmin [1961], p. 14.

[3] *Ibid.*, p. 15.

[4] See Gellner, *op. cit.*, p. 30.

[5] *Ibid.*, p. 209.

[6] One might think that Wittgensteinians should not use technical terms; but 'consistency' – like relevance – is for them irrelevant. For instance, Toulmin considers 'abstraction' to be an unforgivable crime when he criticizes Hempel, Carnap, Popper and me: see, for example, Toulmin [1966], pp. 129–33. But when he starts to forge abstract models (like '*compact* intellectual discipline') he says: 'In itself, of course, the abstract character of our account is no basis for an objection': *Human Understanding* (p. 361). In fact he sets out to defend the necessity and fruitfulness of abstractions. (*Ibid.*, p. 362.)

this, one can be on guard and catch him out when he introduces a new technical term without proper warning. I have already discussed 'language game' – now, incidentally, re-christened 'discipline' – 'concept' and 'understanding'. Another crucial esoteric term is 'rationality'. Since his so-called 'concepts' are skills, abilities, know-how and, indeed, action, it is not surprising to find that his so-called 'rationality' means conformity with the tribal habit in one's (conceptual) world by trying to achieve a (tribally meaningful) action. One has to be wary of Toulmin when he utters sentences which may sound Popperian. For instance, one may think he had plagiarized from Popperians his frequently occurring phrase: 'A man demonstrates his rationality, not by a commitment to fixed ideas, stereotyped procedures, or immutable concepts, but by the manner in which, and the occasions on which, he changes those ideas, procedures and concepts'.[1] But, properly interpreted, this is an anti-Popperian doctrine. Although Toulminian 'rationality' is shown in 'respond[ing] to new situations with open minds', the response must be in conformity with the language game. Thus man's 'rationality' is to be judged by putting him into unexpected, unusual situations, where it requires personal ingenuity to find the proper tribal response. His task is then to apply the tribal closed mind by using an individual open mind. This is the Wittgensteinian–Kuhnian idea of puzzle-solving in normal science. The 'rational' man uses his open, ingenious mind to find the paradigmatic solution acceptable within the tribe. A 'rational' man is quick at 'understanding'.

But such 'rationality' changes from tribe to tribe, from language game to language game, from society to society. And there are different societies. Western society differs from, say, Soviet or Azandan tribal society. For Wittgensteinians they are different, indeed, mutually untranslatable language games with different realities defined by them. For them, in the Zande language game there *is* witchcraft. For them, in Western society there *is* God. But in the Soviet Union there exists, I repeat, *there exists*, neither witchcraft nor God.[2]

But even in one single society there can be different language games; games which display a 'conceptual diversity', as Toulmin puts it. In Western society there is moral language, scientific language, religious language, commercial language, etc. But where are the borders to be drawn? No Wittgensteinian tries to individuate language.[3] Toulmin's 'concept', for instance, seems to be a smaller unit than the original Wittgensteinian 'language game'. But exactly how much smaller? And how many 'concepts' make up that new-fangled Toulminian unit, a 'conceptual population'? There is also, I suppose, bound to be a

[1] This is the motto of his book. See also p. 486.

[2] This is St Anselm's method of ontological proof, revived in Wittgensteinian scholasticism as the famous 'paradigm case argument'. See Watkins [1957], pp. 25–33.

[3] See Kenny [1973], pp. 164–5.

minimal size for a 'conceptual population' to make up a language game, for a language game must be a powerful, established social institution; it must stand up to strains, tensions, must become a 'way of life' before it deserves to be called a 'language'. Perhaps all these are 'limiting questions'.

One thing at least is clear about language games – they ought to be completely autonomous. Each game brings with it its own standards. 'The kind of certainty is the kind of language game.'[1] Wittgenstein admits that one cannot always prevent a struggle among different language games, but he, for one, wants to stop it wherever possible. He is fanatically against war, and on this point his cultural relativism is modified by a strong normative, 'therapeutic' element. A private individual must not start a language game. If he does, and stops following the unwritten and inarticulable rules of the established game to which he belongs, he will see problems where 'in fact' there are none – problems like the problem of induction, the body–mind problem, the free will–determinism problem, etc. The heretic must be brainwashed, restored to mental health by that therapeutic activity which is called philosophy, and which is the 'legitimate heir' to old-fashioned philosophy.[2]

While the 'party of new professional revolutionary philosophers' supplies 'red guards' and 'grand inquisitors', to ensure conceptual stability in each separate language game and can order deviants to mental asylums, established societies must live in 'peaceful coexistence', each behind its solid 'iron curtain'. Missionaries must not be sent to try to convert members of alien cultures; 'cold war' and Radio Free Europe are banned. The *status quo* is sacrosanct and any effort to set up a new language game is *anathema*. This, incidentally, lands Wittgensteinians with the standard problem of all defenders of the *status quo*. If the established order is sacrosanct, in 1917 one had to defend Tsarism and in 1937 one had to defend Bolshevism. *At which point* should one turn one's coat? When does East Germany or Israel become sufficiently established to qualify for membership in the United Nations with 'immutable borders'?

Some Wittgensteinian apologists claim that Wittgenstein, in his more relaxed moments, was willing to allow for mild evolutionary changes to allow a language game some adaptability to a changing or expanding environment. This is certainly Toulmin's position, but, as for Wittgenstein, we have to wait for further studies of the 'master's thought' to clarify this issue. But Wittgenstein's conservatism, whatever its subtleties, remains an Orwellian idea. There can be no doubt that Wittgenstein wants professional counter-revolutionaries to guard the given closed social structure and to determine its permissible degree of flexibility. As Toulmin put it: 'If I may stand a famous remark of

[1] Wittgenstein [1951], p. 224.
[2] Wittgenstein on Freud [1966].

Karl Marx on its head: the point is not to change the world but to understand [i.e. to accept] it.'[1]

There is no need in this context to draw a more detailed 'map' – 'map' is another Wittgensteinian technical term – of this Orwellian world. Let me now focus on one country: science and philosophy of science.

Science is one of the legitimate language games. Philosophy of science cannot be one. The main crime of old-fashioned philosophers of science – and of mathematics and logic – was to try to set themselves up as a new language game, autonomous from science. Moreover, traditional philosophers wanted to set up an improper language game, with explicit – Wittgensteinians say 'mechanical' – rules to tell science and pseudoscience apart and explicit criteria for progress or degeneration within science. These interlopers even tried to separate language from its social context and devised their disembodied 'third world of ideas'.[2] Frege's anti-psychologistic logic polluted common sense; Popper's hypothetico-deductive model and Carnap's inductive logic tried to pervert even science itself and landed it with pseudo-problems. These demarcationists' external standards, like consistency, falsifiability, weight of evidence, etc., constituted serious dangers for the very life of science. Wittgenstein's charge against 'demarcationists' is that they are foreign missionaries in the land of science. Philosophy of science must leave science as it is. He thought that the mathematical language game was perverted by those incompetent users of mathematical language called 'mathematical logicians' and he scribbled a whole volume – let us say, 'performed a book-act' – in order to exorcize them.[3] Professor Toulmin, one of the Wittgensteinian 'grand inquisitors', has led two celebrated crusades, one against deductive and one against inductive logic and has butchered Carnap, Tarski, Hempel and Nagel, in turn.[4]

In a healthy, closed scientific community there is no need for Wittgensteinian philosophers of science. Healthy science simply works. Trouble only arises when old-style philosophers, outsiders, manage to corrupt the community.[5] New-style philosophers, who may well be

[1] Toulmin [1957], p. 347. For Toulmin's equation of understanding and acceptance see above, pp. 230–1.

[2] For Wittgensteinians, Frege's anti-psychologism was a capital crime. The young Wittgenstein's attempts to construct a non-psychologistic, correspondence theory of truth was another capital crime. Wittgenstein, and most sects of the Oxford movement, turn the clock back to pragmatism (see, e.g., Wittgenstein [1969], p. 422; and Naess [1968], p. 156). Incidentally, in this sense, Kuhn too was very much influenced by Wittgenstein's philosophy (Kuhn [1962]).

[3] 'Mathematical logic has completely deformed the thinking of mathematicians and philosophers'. (Wittgenstein [1956], p. 48.)

[4] See especially Toulmin [1953], *loc. cit.*; and [1966].

[5] This idea is strongly expressed by W. H. Watson: 'The soul of physics is given by physicists who think about it, who do experiments, discuss it, write about it and teach it. This is the only soul worth having. The rest is a sort of pathological morbidity that keeps a man from learning about nature, and discourages real participation in that

normally members of the scientific community, then have to come out from their watchtowers, nip revolutions in the bud, restore conceptual stability and then retire again, just keeping quietly on the alert. It is *not* possible to introduce *good* change, but only corruption and degeneration, from the outside. Indeed, according to Wittgenstein, the history of mathematics and of science *is* full of such degeneration caused by interlopers; presumably only the establishment of the 'new revolutionary thought-police' can maintain its health. The 'struggle for peace' never ceases; degeneration and 'irrationality' are possible and – contrary to Hegel – might is not always right.

3 TOULMIN'S DARWINIAN SYNTHESIS OF HEGEL AND WITTGENSTEIN

Let us, at last, turn our full attention on Toulmin himself. My claim will be, remember, that Toulmin firmly belongs in the 'élitist' tradition. But, because of Wittgenstein's influence on him and his attempt to avoid some Wittgensteinian problems, Toulmin's élitism is of a special kind.

Toulmin inherits pragmatism from Wittgenstein. Indeed for Toulmin the major mistake of most philosophers of science has been to concentrate on ('third-world') questions of 'logicality' about propositions and their provability, confirmability, probability or falsifiability, rather than on questions of 'rationality' concerning skills and social activities – 'conceptual populations' and 'disciplines' – and their 'cash value' – the practical profits and losses they incur.[1]

For Toulmin, fruitless scholastic questions about whether a conclusion is a logical consequence of a set of premises – questions concerning relations between propositions – should be replaced by questions about whether one's *actions* are appropriate in view of the information at one's disposal. A valid inference is not one in which the conclusion stands in a certain 'third-world' relation to the premises, nor even is it one in which the rational man cannot but believe the conclusion if he believes the premises. It is rather one in which the *action* based on the premises is appropriate, i.e. successful. According to Toulmin, 'logic is not... *la science de la pensée*, but *l'art de penser*'.[2]

creative process. In good health it is not natural. Philosophy, as Wittgenstein once remarked, ought to liberate us from the idea that there is a kind of academic doctor who can do things for physicists and other scientists that they are incapable of doing for themselves' ([1967], p. xi).

[1] This basic idea is repeated in each of Toulmin's papers, in each chapter of his book. Just one sample: 'The important thing about drawing the proper conclusion is to be ready to *do* the things appropriate in view of the information at one's disposal: an actuary's respect for logic is to be measured less by the number of well placed yes-feelings he has than by the state of his profit and loss accounts' (Toulmin [1953], p. 95).

[2] *Ibid.* Toulmin has for many years been trying to bring about a counter-revolution in logic. He is recommending us to give up one of the most marvellous progressive

For Toulmin, questions about the truth and falsity, confirmation, corroboration, falsification, etc., of propositions should be replaced by questions about the 'adequacy', 'the practical effects', the 'power', the 'survival value' of 'concepts', that is, of skills.[1] All this is pragmatism, pure and simple.[2]

Conflict and change pose difficulties for the pragmatist. If different people find that different 'explanatory procedures' yield them 'understanding', straightforward pragmatism seems to land us in an extreme form of subjectivism or cultural relativism. Wittgenstein 'solved' this problem by establishing a 'thought-police' to eliminate dissenters and heretics from each community. But for Toulmin 'conceptual change' – so long as it is not too violent – is not only possible, it is sometimes even desirable. This is Toulmin's major divergence from Wittgenstein. Toulmin disbands Wittgenstein's cruel 'thought-police', but only at the expense of introducing the – admittedly gentler but scarcely more acceptable – Cunning of Reason.

Toulmin's Cunning of Reason sees to it that in the Darwinian struggle for existence, at least within a properly constituted scientific 'discipline', those 'conceptual-variants' survive which are right. Even 'master scientists' cannot accept any old concept, for the Cunning of Reason imposes an 'external objective restraint' on them. If scientists make a wrong move, there is a Hegelian self-correcting mechanism which will expose their lack of judgment, so that, in the 'long run' – in fact, only in the 'full run' – reason prevails.

Thus for Toulmin – unlike for the sceptic Wittgenstein! – *might is right*; selective survival is the criterion of progress. The last chapter in the first volume of Toulmin's *Human Understanding* is entitled 'The Cunning of Reason'. The last sentence could have been written by Hegel himself: 'One thing at least can now be said. As those "rational transactions" to which we have committed ourselves continue to work themselves out in the course of subsequent history, the same verdict of historical experience which earlier thinkers called the Cunning of Reason will, in the long run, penalize all those who – whether knowingly, or through negligence – continue playing according to out-dated strategies.' He applies social Darwinism to science: the fittest are those

research programmes in the history of human knowledge – mathematical logic – which provides the most efficient weapons of objective criticism mankind has ever produced. And he recommends us to replace it by woolly, 'élitist', Wittgensteinian 'inference tickets'.

[1] Thus Toulmin ([1974], 1.22) recommends us to 'redirect our attention away from the accumulation of true propositions and propositional systems to the development of progressively more powerful concepts and explanatory procedures'.

[2] I specified above (p. 228) that one of the most characteristic features of pragmatism is its denial of the existence of the 'third world'. Toulmin's denials, since he does not have a clear idea of the 'third world' are rather tortuous. Perhaps his clearest statement is in his review of Carnap's *Logical Foundations of Probability* where he criticizes Carnap for putting 'logical relations on one footing with minerals', i.e. for ascribing to 'third world' objects as genuine an existence as first world ones. (Toulmin [1953], pp. 86–99.)

which survive. 'The question "What gives scientific ideas merit and how do they score over their rivals?" can be stated briefly in the Darwinian formula: "What gives them survival-value"?'[1] Toulmin's problemshift makes 'demarcationist' philosophers like myself redundant: 'It is not for the philosopher to impose [his judgment] on science'.[2] 'The philosopher' – he goes on – 'must [only] analyse the standards by which scientific variants are judged and found worthy or wanting.'[3] Or: 'What is sound in science is what has proved sound, what is "justifiable" is what is found justifiable.' (p. 259). Thus the philosopher must not lay down standards of his own; he is allowed only to analyse the scientist's own standards. But this surely turns him from a *philosopher* into a *descriptive historian* – whereupon he will find his humble services rewarded by the Royal Society.[4] One wonders why Toulmin still keeps talking of *philosophy of science* when the philosopher is only allowed to record, describe, or at best, 'analyse' the scientist's standards?[5] This is surely the social historian's business. For Toulmin's worship of history the following sentence is perhaps most characteristic: 'A historian [cannot] justly criticise earlier scientists for not jumping straight to the views of 1960.'[6] Does this mean that we *needed* the Dark Ages in order to get from Archimedes to Galileo? (This is, of course, the Catholic–Hegelian view.) Toulmin is indeed committed to just this since in his view *all change – within the scientific community – is progress, and the speed of actual progress is its necessary speed.*

Toulmin proposes to reveal the true principles of objective normative reason in his third volume[7] while discussing the purely descriptive ecology of concepts in the first two volumes.[8] But if Toulmin really believes in his Hegelian Cunning of Reason, the third volume of his *magnum opus* need not be written. If progress is guaranteed by the Cunning of Reason, the description of change is the description of progress.

But what if there is conflict within the scientific community over some proposed change? What about, for instance, the long disagreement between Newtonians and Cartesians? Or the disagreement between Einstein and Bohr? Only one of the two factions in such debates can be right. Toulmin's answer is that in such cases where two

[1] Toulmin [1961], p. 111. 'In the long run' deals with all fundamental controversies in what we used to call 'history of science'; these controversies are deplorable but necessary lapses of science into non-science.

[2] *Ibid.*, p. 110.

[3] *Ibid.*, p. 110.

[4] In Great Britain the Royal Society supports historians of science: it refuses to finance philosophy of science.

[5] But even the descriptive accuracy of Toulmin's Darwinian metaphors can be challenged. It is challenged in a review to which I should like to refer the interested reader: Cohen [1972], pp. 41–61.

[6] Toulmin [1961], p. 110.

[7] *The Rational Adequacy and Appraisal of Concepts.*

[8] *The Collective Use and Evolution of Concepts*, and *The Individual Grasp and Development of Concepts.*

or more proposed 'strategic redirections' are in conflict, only history will decide. Here he brings in the time-worn *ad hoc* stratagem of historicism: the Long Run. In 1687 it became clear to everybody that Copernicus was right and his opponents wrong. In the nineteenth century it became clear to everybody that Newtonians were right about dynamics and Cartesians definitely wrong; it also then became clear to everybody that Newton's optics was wrong. Today, but only today, it is clear that while 'Newton's dynamical theories retained a legitimate intellectual authority of their own until the year 1880 or later, the influence of the *Opticks* was already having a narrowing effect before the end of the eighteenth century. By 1800, in fact the continued authority of the *Opticks* represented little more than the magisterial dominance of a great mind over lesser ones.... If we cite both the *Principia* and the *Opticks* as illustrating a single theory of scientific change, we must recognise that they serve as paradigms in significantly different senses of the term' (p. 111).[1] But does Toulmin's introduction of hindsight really solve the problem? An apparently defeated research programme may, at some stage in the future, stage a comeback. At that point, the 'verdict of history' would seem to be reversed. How do we know that the hindsight we have the advantage of is *long enough* hindsight? Toulmin, it seems, ought to hold that 'true rationality' is revealed only in the 'closed down long run,' on the day of Final Judgment, when we are all dead.

But if so, one's rational reconstruction of history keeps changing. Indeed, Toulmin's third volume containing the 'absolute' evaluation cannot be written before the extinction of mankind, and certainly not by 1976, as Toulmin announced. If Toulmin means that in 'the light of the final unfolding of reason' one can *explain* which parts of the serpentine lead steeply upwards, and which ones were mere roundabouts, then for this insight one has to wait till the end of human history. Only when history has reached its final consummation – Hegel's Prussian State – can the layman finally understand what purpose certain seemingly monstrous deviations served in the 'march of history'. The Cunning of Reason, as Georg Lukacs, in his more optimistic moments used to say, arrives at the mountain top via a twisting, winding road and not via the steep and direct route. One can reach true understanding of history only on arrival at the summit. As far as I can see, Toulmin agrees with this: 'If we take the trouble to understand exactly and in detail... [a completed human enterprise] ...we shall then – but only then – be in a position [first] to understand what *counted* for them [i.e. for those involved in the enterprise] as an intellectual "achievement" or theoretical "improvement" [progress] and how far – in that particular problem-situation – they were *justified* in applying the principles of judgment and criteria of choice

[1] The idea that the development of optics was retarded by Newton's authority is shown to be untenable in Worrall [1976].

they did' (p. 318). While the Darwinian struggle between conceptual populations is going on we may feel lost in the maze of the Kafkaesque twists and turns, we may not be able to see 'either the general "method" or a definite end'. 'Yet science is none the less *rational* for all that.'[1] This will be realized when looking back from the summit, from which vantage point everything will turn out to be justified and rational, but only with hindsight. Articulable, final understanding, like Minerva's owl, flies only after dusk.

Thus at the *end* of history, it will be clear which scientific changes constituted scientific progress. But then it seems Toulmin cannot write his third volume (or any *normative* history of science) before the end of history.

He tries to get out of this difficulty by introducing his particular brand of 'élitism'. According to Toulmin, a privileged élite has a hot line to the Cunning of Reason. The line is not perfect and they cannot foresee the future infallibly, but it is reasonably good. The 'supreme court justices' can make 'rational bets'.[2]

Toulmin's 'élitism' exactly conforms with my definition. According to him 'conceptual judgments' are a matter of case law, not of code law, of precedents, not of principles. Thus there is an élite, which has tacit partial knowledge of which paths lead towards the top.[3]

The authority of the élite is needed not only in 'cloudy' cases[4] when 'strategic redirections' are called for but also in small, tactical problems where changes are proposed within the context of the same 'explanatory ideals' (I suppose these correspond to my creative changes within the same research programme). Even in such 'clear cases'[5] choices between proposed 'conceptual variants' demand 'a balance of gains against losses', and so call for 'judgment':[6] 'judgment' which only belongs to those 'scientists whose authoritative standing in the profession is based on the range of their *experience*...in the business of "making sense...of the relevant aspects of Nature"'.[7]

Given both the Cunning of Reason and an élite which has privileged access to its workings, a member of this élite can give rational advice even without the benefit of hindsight. Galileo knew that Copernicus was right even though there was no explicit evidence for it at the time.[8]

[1] Toulmin [1974], 5.43.
[2] *Ibid.*, 3.41.
[3] I have already mentioned Toulmin's emphatic opposition to demarcationism, to the idea that progress can be judged according to a universal demarcation criterion. (See *above*, p. 127, n. 1). 'There is no universal recipe' for scientific appraisal (Toulmin [1961], pp. 14–15). Or: 'As to this question [of which "conceptual variants" can be progressively incorporated into science] no universal formula or decision procedure can be given.' ([1971], p. 552.) Toulmin's derision of 'inductivism...verification, falsification, confirmation, corroboration, refutation', and of my methodology of scientific research programmes follows from his élitism. (*Ibid.*, [1972], p. 480.)
[4] Toulmin [1974], 3.32.
[5] *Ibid.*, 2.4.
[6] *Ibid.*, 2.41.
[7] *Ibid.*, 3.11. 'Relevant' is not time-dependent here.
[8] As Polanyi put it: 'He [that is, the great scientist] can have a tacit foreknowledge of yet undiscovered things. This is indeed the kind of foreknowledge the Copernicans

And if a historian of science is also a member of this same élite, he may well write Toulminian rational history. The élite's judgment is not subjective, since it is under the external constraint of the Cunning of Reason,[1] or, in old-fashioned Cartesian terms, he is guided by a benevolent God's helping hand.[2]

If only the élite can sniff out progress, it is important to know who the prophets are – we must not be misled by false prophets. Thus, like all élitists Toulmin demarcates persons and communities rather than achievements. Also, since Toulmin the pragmatist regards science as an activity, it is important to know who is acting scientifically and who is not. Thus the logic of his élitism forces Toulmin to adopt psychologism and sociologism.[3] Despite the fact that these doctrines were long ago discredited by Frege, Husserl and the Vienna Circle, Toulmin embraces both with gusto. The clearest statement of Toulmin's commitment to 'psychologism' is his appraisal of Wittgenstein: 'Wittgenstein's very personality was the expression of a highly articulated, though largely unverbalizable, personal point of view.'[4] The trouble with Wittgenstein's 'London opponents' like Popper and Gellner, was that they judged Wittgenstein's intellectual products by looking merely at his writings, instead of by looking at their producer: 'The real man, the real *philosopher*, [and therefore his philosophy] had escaped them.'[5]

But for Toulmin science is a communal activity. His main concern, therefore, is to demarcate scientific communities rather than scientific individuals. In doing so he follows the tradition of Wittgenstein, Polanyi and Kuhn by characterizing the scientific community as a closed society. Toulmin gives five 'connected' criteria for when, in his new terminology, a 'rational enterprise' constitutes a 'compact discipline':

> (1) The activities involved are organised around and directed towards a specific and realistic set of agreed collective ideals. (2) These *collective ideals* impose corresponding demands on all who commit themselves to the professional pursuit of the activities concerned. (3) The resulting discussions provide disciplinary loci for the production of 'reasons', in the context of

must have meant to affirm when they passionately maintained, against heavy pressure, during one hundred and forty years before Newton proved the point, that the heliocentric theory was not merely a convenient way of computing the paths of planets, but was really true.' ([1967], p. 23.) According to Toulmin, Kepler exemplified this Polanyian foreknowledge. (Toulmin [1974], 4.32.)

[1] *Ibid.*, 3.4.

[2] Descartes needed *God's* guiding hand in order to recognize the validity of an inference. Today a Turing machine will do. Toulmin does not want to hear about Turing machines. He wants to restore privileged access to logic: 'In logic as in morals, the real problem of rational assessment – telling sound arguments from untrustworthy ones, rather than consistent from inconsistent ones – requires experience, insight and judgment' (Toulmin [1958], p. 188).

[3] See *above*, p. 228.

[4] Toulmin [1969], p. 59. See also Toulmin [1953], pp. 94–7, where he wholeheartedly embraces psychologism, at least in its 'sophisticated', i.e. pragmatist, form.

[5] Toulmin [1969], p. 59.

justificatory arguments whose function is to show how far procedural innovations measure up to these collective demands, and so improve the current repertory of concepts or techniques. (4) For this purpose, professional forums are developed, within which recognised 'reason-producing' procedures are employed to justify the collective acceptance of novel procedures. (5) Finally, the same collective ideals determine the criteria of adequacy by appeal to which the arguments produced in support of those innovations are judged (p. 379).[1]

The picture here described is that of a society without radical alternatives, where one can only 'improve' but not replace 'the current repertory of concepts', a society whose membership depends on oaths of loyalty to specific doctrines ('commitment to collective ideals') and where only 'professional forums' can judge the implications of these doctrines for specific cases. In this closed society critical reappraisal and modification are allowed only if done by 'qualified judges'. The layman is powerless, the élite self-perpetuating.

4 CONCLUSION

[In choosing élitism, burdened with pragmatism and historicism, Toulmin has, in my opinion, chosen just about the worst of all possible philosophical worlds. But I should like to conclude with one or two specific criticisms, which may strike home even with someone who is inclined to go along with Toulmin's choice.]

First, I find it interesting that Toulmin gives, in this book, very few examples of actual changes in the history of science which are allegedly unrecognizable as progress on any universal criterion. In some of the few examples he does give, he has already been refuted. For instance, in his view, no acceptable statute law can be given on which Copernicus's theory is progress over Ptolemy's. And no statute law can be given to show the superiority of the 'relativistic concept of momentum' over 'Newton's *quantitas motus*'. But recently one statute law has been shown to account for both cases.[2]

Secondly, checking his five conditions for a genuinely scientific community – or 'compact discipline' as Toulmin calls it – whose collective aim is 'explanation',[3] it turns out that Catholic theology, Soviet Marxism and scientology are all better paradigm cases than, say, quantum mechanics. If, within an apparently unified discipline, 'explanatory ideals' conflict, then for Toulmin the discipline lacks consensus and has to be demoted into a 'non-genuine' discipline (pp. 382–3). On these terms, Newtonian physics became scientific only after the Cartesians admitted defeat, and accepted Newtonian explanatory

[1] One wonders how Toulmin could betray the Wittgensteinian tradition with such an *explicit*, although admittedly woolly, *general* characterization.

[2] See volume 1, chapter 4, and Zahar [1973].

[3] See *above*, p. 240. Toulmin's best imaginary example of a 'compact discipline' is a 'Royal College of Prostitution' ([1972], p. 405). But this college narrowly escapes being 'scientific' – its ideal is not an explanatory one.

ideals (p. 381). Toulmin's notions of 'explanatory ideal' and of a 'sufficient' degree of consensus are never spelt out and I found them rather elusive. But it seems rather easy to differ over explanatory ideals; in fact Newton and Leibniz, Bohr and Einstein, and Delbrück and Luria did. But given that, within a scientific community, the deviation from consensus must be 'no more than marginally significant', such communities are bound to be very small religious 'microcommunities'. The Newtonian community will, it seems, have to be different from the Cartesian one and the Bohrian community different from the Einsteinian one.[1]

But the major specific difficulty with which I should like to confront Toulmin is this. How does he know *which* 'ecology of "conceptual populations"' to study? One cannot study the history of science without some point of view, which will amount, willy nilly, to some – provisional – definition of science. The plan of Toulmin's *magnum opus* – 'true rationality' as I have said, is to come in the third volume – seems to suggest that he thinks that true rationality will emerge from the descriptive history analysed in the first two volumes. But would the same principle of 'true rationality' emerge if Toulmin had devoted his attention to astrology, witchcraft or the Mafia rather than to physics and to chemistry?

In excluding things like the history of witchcraft from the history of science, Toulmin has already used a general demarcation criterion, a statute law for scientific progress, the whole idea of which he holds in such complete contempt. It seems that if Toulmin's volume on impartial rationality is to be written at all, it should have come first. If scientific rationality consists in more than mere survival for a given time in some given place – say Berlin, 1933, or Moscow, 1949 – any principle of rationality is bound to specify some norm which demarcates science from pseudoscience. But then, rational appraisal must *precede*, not *follow*, full-scale empirical history. As I have put it, 'internal (normative) history is primary and external (descriptive-empirical) history is secondary'.[2]

I agree with Toulmin that no demarcation criterion is absolute. I am a fallibilist with regard to demarcation criteria, just as I am a fallibilist with regard to scientific theories. They are both subject to criticism and I have specified criteria not only by which one research programme can be judged better than another, but also by which one demarcation criterion can be judged better than another.[3] But I do not draw the Wittgenstein inference from fallibility of propositions to their dismissal. I do not panic: I do not switch from articulated propositions to inarticulable skills of doing and judging science. For to do so is to reintroduce through the back door a pragmatist version

[1] Kuhn preceded Toulmin in retreating to this curious position; for a discussion of Kuhn's retreat, see Musgrave [1971], esp. p. 289.

[2] Volume 1, chapter 2.

[3] See volume 1, chapter 3.

of justificationism with the help of a Hegelian Cunning of Reason. I want clear theses in both science and in philosophy of science where logic can assist criticism and help to appraise the growth of knowledge. Toulmin's improvement of human understanding has no use for logic, since logic is part of the 'Platonic-propositional' approach which he utterly condemns. It is largely because of my conviction that without deductive logic there can be no genuine criticism, no appraisal of progress, that I stick to old-fashioned Popperian-type *Criticism and the Growth of Knowledge* and that I cannot be persuaded to replace it by Toulminian – to my mind uncritical, woolly and confused – *Human Understanding*.

Part 3

Science and Education

12

A letter to the Director of the London School of Economics*

Dear Director,

The Majority Report of the Machinery of Government Committee . . . contains the principle that students, as well as staff, should determine the general academic policy of the School.[1] This principle is clearly inconsistent with the principle of *academic autonomy*, according to which the determination of academic policy is exclusively the business of academics of some seniority. The implementation of this latter principle has been achieved – and sustained – in a long historical process. I came from a part of the world where this principle has never been completely implemented and where during the last 30–40 years it has been tragically eroded, first under Nazi and then under Stalinist pressure. As an undergraduate I witnessed the demands of Nazi students at my University to suppress 'Jewish–liberal–marxist influence' expressed in the syllabuses. I saw how they, in concord with outside political forces, tried for many years – not without some success – to influence appointments and have teachers sacked who resisted their bandwagon. Later I was a graduate student at Moscow University when resolutions of the Central Committee of the Communist Party determined syllabuses in genetics and sent the dissenters to death. I also remember when students demanded that Einstein's 'bourgeois relativism' (i.e. his relativity theory) should not be taught, and that those who taught such courses should confess their crimes in public. There can be little doubt that it was little more than coincidence that the Central Committee stopped this particular campaign against relativity and diverted the students' attention to mathematical logic and mathematical economics where, as we know, they succeeded in thwarting the development of these subjects for many years. (I am fortunate that I did not have to witness the humiliation of University Professors by the students of Peking University during their 'cultural revolution'.)

Invoking these ghastly memories may seem out of place in this

* This letter, first published in C. B. Cox and A. E. Dyson (*eds*): *Fight for Education, A Black Paper*, was written during the student troubles at the LSE in 1968 (*eds.*).

[1] The Committee on the Machinery of Government of the LSE included Governors, academics and students. In February 1968 they published two reports: a Majority Report and a Minority Report, written by two students, David Adelstein and Dick Atkinson.

country. It will be said that here there is no political force or motivation behind students' demands. Unlike the demands of Hitler's, Stalin's and Mao's youth, their aim is to improve rather than erode the university tradition of informed research and competent teaching.

But is this so? The 'Minority Report' which the LSE Students' Union adopted, has an underlying philosophy which may have been taken directly from the posters of Mao's 'Cultural Revolution'. As Adelstein, one of its authors, put it:

> Student representation on governing bodies is only the *beginning*, and representation can be good or bad – it can give a false sense of unity. The *next thing* is for students to begin to run their own courses, initially through their own societies, and then to demand that they should run a particular part of a course: its content, how it is taught, and who teaches it.
>
> The *next step* is for students to appoint their own teachers and to do some teaching themselves. *Ultimately*, students should work for a certain amount of the time. Academic and intellectual problems become meaningful if they are associated with practical life...
>
> I accept the word militancy, but it means for me that one is prepared to consider any action that will achieve one's end, which is in accordance with one's ends. One would not rule out any mode of action because it has not been accepted in the past...
>
> We do initiate unconstitutional action.
>
> We do not accept constitutional limits because they are undemocratic. When democracy fails, this is the only way of doing it...[1]

Should one leave such an extremist manifesto of a member of the Machinery of Government Committee of the London School of Economics without comment? Can one accept the 'beginning' stage of this programme without argument, without having to fear that this is only the thin end of the wedge? According to the Majority Report, we can. I shall argue that we cannot.

1 The crucial shortcoming of the Majority Report is that it does not demarcate between two completely different sets of student demands.

The *first set of demands* are for free expression of student complaints and criticism and for guarantees that these complaints and criticism will get a proper hearing; also for participation in decision making in matters in which they are, nearly, equally or even more competent than the academic staff. These demands were originally opposed – and in many places still are – by the champions of the paternalistic *in loco parentis* conception of University authority, but they no longer meet opposition at the LSE, in my opinion rightly so.

The *second set of demands* are completely unjustified demands for *student power* – as opposed to demands for *student rights of criticism* – concerning appointments, establishing new chairs, positions, designing syllabuses and, in general, concerning the content of teaching and research. The policy of the 'revolutionaries' is to blur the distinction. This policy has achieved considerable success, mainly because of the

[1] *The Times*, 18.3.68, my italics.

widespread but unproven assumption that if it had not been for 'revolutionary' militancy, even the justified demands might not have been satisfied to the degree they are satisfied now. But whether this is true or not, it does not alter the plain and sad fact that these militants have no interest whatsoever in the apolitical and constructive student demands. They only advocate the demands for freedom for political expediency, in order to win the students' support for their demands for (their) power. They are turning surreptitiously the justified revolt against academic paternalism into a political revolt against academic autonomy. This is why it is so important to draw a sharp line between the two kinds of demands. The main fault of the Majority Report is that it fails to do so.

It is worth while mentioning that, for instance, the National Executive of the Association of University Teachers, in a recent decision, made the demarcation quite clear. They agree that

(1) At departmental level there should be students on joint staff–student committees of the departmental board or board of studies;

(2) In general, on any committees dealing with matters such as residential accommodation, refectory and catering, student welfare, there should be student representatives elected by the students themselves;

(3) There should be a Senate Student Affairs Committee with roughly equal staff and student numbers and this committee should advise directly the Senate and other sub-committees of Senate when matters of importance to students arise for discussion.

But they oppose students' participation on the Council and Senate: 'The undergraduate, who by definition is still learning what the content of his subject is, is not in a position to take decisions on matters like curricula...'

Of course, this does not mean that they are 'not in a position' to *criticize* such matters. But the students of our School do have the right to criticize, both in private (for instance to the Dean of Undergraduate Studies) and in public (for instance in *Beaver*[1] or in departmental staff–student committees) the content and method of teaching and of research, or even individual courses, classes, appointments etc. and request their discussion. The problem is rather that they have not yet made real use of this right. They should be encouraged – and even helped – to make the best of it.

But there is a world of difference between the right to criticism and consultation and the power to participate in decision making. No academic would deny the right of the Government or of students to *criticize* any aspect of University life, or to have access to relevant information. But no academic would agree that Parliament (or Party Politbureaus) should have a voice in *deciding* about appointments, syllabuses, etc.

There are no arguments for Student Power that would not be

[1] *Beaver* is the newspaper of the LSE Students' Union.

arguments also for Government Power. Students may be part of the academic community in an important sense in which the Government are not; but they receive their education at considerable expense to the taxpayers, whose representatives may therefore be said to have more right to interfere with University life than the students whose education they finance. Militant students frequently use the analogy that the consumer ought to be able to influence the production of the goods he buys: they do not notice that in this analogy the real consumer is the State – *they* are the goods to be delivered. It is the thin wall of academic autonomy and nothing else that protects students from political interference in the age of State-financed education.

This is perhaps the most important reason why we should resist student power while accepting student freedom to criticize: because we resist government power while accepting government freedom to criticize. Of course there will be people both among students and politicians who believe that freedom of criticism is useless without power. But the history of universities is full of evidence to the contrary. Indeed, the main danger is that academics, fully aware that academic autonomy has no real power basis, are too quick rather than too slow in yielding to outside criticism and pressure: the latest evidence is exactly this Majority Report. I contend that, if students were now denied membership on Council and Senate, and if they were asked in three years' time which of their constructive criticisms and proposals have not been seriously considered, the answer would be 'none'.

One may ask whether the qualifying adjective 'constructive' was not a device for ignoring arbitrarily some of the criticisms. This objection brings me to another demarcation between student demands. *My first demarcation was between freedom of criticism and power in decision making.* Freedom of criticism and expression of demands must be, of course, unlimited and extend to both 'constructive' and 'destructive' criticisms and demands. But if we look at the *concrete content* of students' demands we find that they can be separated into two classes. Some students want better teaching facilities, more rational examination structure, better coordinated syllabuses, better lectures, classes, seminars, reading lists, better library facilities, etc. etc. These students want universities to serve better the old ideal of expanding and transmitting knowledge. Other students want to demolish universities as centres of learning and turn them into avant-garde centres of social and political conflict and commitment, whatever this may mean. As the Adelstein–Atkinson report puts it: 'Discovery itself lies in action'. *My second demarcation is then between the 'constructive' demands which seek to improve the University as we know it and the 'destructive' demands which seek to destroy it.* It is sad that these two have been conflated.

Now my two demarcation lines coincide: I claim that *those who concentrate on 'constructive' demands are content with student freedom while those who concentrate on 'destructive' demands want student power.*

2 As a matter of fact, there is not one single argument in the Majority Report as to *why* students should be admitted to Senate and Council; nor are obvious counterarguments refuted or even mentioned.

I should like therefore to set out some of these counterarguments in concrete form. The first argument is implicit in what I have already said. Academic autonomy is being assaulted with different degrees of vehemence and of success all the time, everywhere (the most recent examples being the purge of Greek Universities by a junta of colonels, and the dismissal of seven 'liberal–Zionist' professors of Warsaw University). Therefore the principle of academic autonomy must be clearly stated and argued for. I do not know, however, of a coherent and convincing defence of it published anywhere. The reason is simple: good academics prefer doing their research and teaching to writing manifestos in defence of academic autonomy as long as its erosion is bearable. However, when it becomes unbearable it is too late to try to defend it publicly because this has become politically impossible. This is why it is vital to stand up when the corrosion *starts* and let the argument penetrate to those countries where they cannot be published any more.

I have little doubt that a forceful defence of academic autonomy will be understood and appreciated by the majority of our students. The absence of any such effort is one of the most perplexing features of the whole affair.

But let us now consider the more immediate practical consequences of students' membership in Senate and Council.

The contribution of 'constructive' student members may be useful but very limited, and, indeed, it can scarcely be expected to exceed the contribution which they can make through the already available channels without being members of Senate or Council. Let us remember that it takes at least a year even for a senior academic member to become a competent member, and students will have to leave the Senate as soon as they become more acquainted with its problems and procedures. What will be the contribution of the 'destructive' members? Once on Council and Senate they will follow what used to be called by some Komintern leaders the 'salami tactic': piecemeal slicing of the academic tradition. They will fight first for increased student membership, then for the erosion of 'reserved subjects';[1] they will propose additional items on the agenda protesting against discontinuation of teaching contracts for their favourite academic misfits, demanding new Chairs in alienation, in cultural revolution, or in American (but not Communist) war crimes in Vietnam, etc. etc. They will fight for an *increased* role of Lay Governors in the

[1] The Majority Report recommended that students should normally be excluded from the discussion of a list of 'reserved subjects' which included academic appointments, promotions, etc., and also 'any other category or item of business declared by the Chairman [of the respective committee] to be a "reserved subject"'.

School's affairs, with the difference that they will want to have as Governors the representatives of the Trade Unions, of the 'avant-garde culture' etc., rather than the present 'City type'. They will devote full-time energy to further their aims which we shall be able to counter only by abandoning academic life for the full-time defence of academic life. They will ruthlessly and systematically use the pressure of Students' Union meetings on the Senate and there can be no doubt that they will exploit to the limit any mistake made by the Chairman; they will issue distorting statements, etc. etc. I do not think they will win their cases; but very soon the most sensitive agenda will be discussed and agreed upon in a Director's informal caucus before the Senate meetings and driven through without discussion to avoid Maoist filibuster. They will be sprung on the Senate under 'Any other business' in order to avoid previous build-up of student pressure. But then the students will feel – rightly – deceived and mass protests will follow. The militants will *not* be isolated.

I think this clearly shows two things. *First*, it shows the inconsistency of those who are pressing simultaneously for the elimination of Lay Governors, for increased student representation; *and* for more democratic Government of the School. But surely students are as incompetent in strictly academic matters as Lay Governors; and nothing is more dangerous to a democratic Senate than an (even tiny) minority openly dedicated to its destruction and thereby creating a siege psychology. *Secondly*, it shows that student membership, in the presence of a Maoist minority, does not lessen but increases the danger of student riots. This, in fact, already shows at the Free University of Berlin where the extremists, having fought their way into the Senate, now are pressing for the second 'stage' of their revolution: for one third representation for students, one third for junior and temporary staff.

One may, of course, hope that 'destructive' elements will not be elected to Senate and Council. Some signatories of the Majority Report privately pin their hopes on the Students' Union Reform which would turn it into a fairly representative body – which now it is certainly not – and thereby reduce the chances of a Maoist party appearing on the Senate. But such a Reform, I understand, is being successfully obstructed and even if it is passed, will not close the doors to the political extremists. For, let us face it, whatever electoral structure the Students' Union will adopt, there will be very few future academics among the students who will offer themselves as candidates on Senate and Council. Serious undergraduates who want to profit maximally from their short three years' course will not normally stand and canvass for election. At least a sizeable number of student members will belong to a group of 'activists', who openly seek to destroy the universities as places of learning and to turn them into centres of political commitment and who openly confess that they want to use membership of Senate and Council for furthering their political aims.

The adoption of the Majority Report would be a considerable encouragement to the numerically weak group of student extremists. The LSE riots seem to have blinded LSE academics to the national weakness of this group. The National Union of Students has not demanded student membership on Senates and Councils; but if the LSE academics accept it, how will they be able to resist their militant wing? But most academics in Britain, unlike the appeasers at LSE, *will* resist. Years of unrest will follow, both at the 'resistant' and the 'appeaser' universities. It is not at all impossible that a conservative reaction to the permissive mood of the 1960s could even bring a demand that Parliament should have watchdogs on Senates to ensure that the Universities give that sort of education which the taxpayers expect for their money.

It may be objected that I exaggerate the dangers. But I do not contend that the University tradition will *necessarily* be destroyed within a few years if we adopt the appeasers' position. I do claim, however, that it is a miracle that the University tradition was ever established and that it has survived until this day. There is nothing necessary about this survival: we have to fight against its erosion all the time in order to be in a better position when, with the periodical social and political crises, the onslaught on universities, as so frequently happens, becomes acute.

3 Of course, I do not think that academic autonomy is sufficient in itself to guarantee the growth of knowledge and to uphold and improve standards of university education. There are many dangerous ills which are *consistent* with academic autonomy. But seeking the cure for such ills through the erosion of academic autonomy is no better than curing the ills of parliamentary democracy through fascism, communism, or Maoism. I would propose a motion based on my demarcation between 'constructive' and 'destructive' student demands as follows:

This Board welcomes any proposal which would provide channels for an improvement of the staff–student dialogue on the content and method of teaching. It approves of the departmental staff–student committees and of student membership on School committees where students may have useful direct contribution. At the same time this Board firmly opposes any erosion of academic autonomy and upholds the principle that academic policy within the University should be determined solely by academics.[1]

Yours sincerely,
Imre Lakatos

28 March 1968

[1] The Academic Board of the LSE rejected both the Majority and the Minority reports. However, on 13 November 1968 the Board passed a motion that 'the responsibility on behalf of the School for the determination of matters involving general academic standards must rest, and be seen to rest, entirely with the academic staff of the School'.

13

The teaching of the history of science*

The article by Mr Hoskin in *The Times Educational Supplement* (7 July 1961) has made it clear that the history-cum-philosophy of science is now passing through a rather critical stage of its rapid growth because of the shortage of teachers. I should like to re-emphasize this point. Indeed I wonder whether the brakes should not be applied to diasporic provincial appointments and to the establishment of degree courses all over the country. Our subject is an extremely difficult one, being a boundary subject between logic, scientific method, philosophy, science, and history, and many may be tempted to think that a boundary subject has no area and that one can try rope-walking this boundary without being competent in the fields which it 'bounds'. We cannot tolerate a situation where historian-cum-philosophers of science learn their mathematics, science and history from popular expositions. This is why I think that we should pay due attention to building up *centres of research* to *educate* historians and philosophers of science besides *appointing* them, and to *building up* this new field of knowledge before, or at least simultaneously with, *spreading* the gospel.

My second point concerns the evaluation of the recent increase of interest in the subject. The undergraduate clamour for courses in history-cum-philosophy of science is not an argument for the granting of their wishes. They clamour not because they have suddenly become passionately interested in problems, such as that of the role of the Arabs in the preservation of antique tradition, but because they are unhappy about the way in which history on the one hand and science on the other are being taught. History teaching still turns a blind eye towards science, the most exciting and noble of human ventures, and science and mathematics teaching is disfigured by the customary authoritarian presentation. Thus presented, knowledge appears in the form of infallible systems hinging on conceptual frameworks not subject to discussion. The problem-situational background is never stated and is sometimes already difficult to trace. Scientific *education* – atomized according to separate techniques – has degenerated into scientific *training*. No wonder that it dismays critical minds.

Now history-cum-philosophy of science has to show up science in

* This short piece is reprinted from A. C. Crombie (*ed.*): *Scientific Change*, 1963. (*Eds.*).

history on the one hand and history in science on the other, and by doing this to exert an important *therapeutic influence* on both. If we fail to achieve this, we shall soon face a situation where an abundance of separate courses in history-cum-philosophy of science will turn the present two uncultures (to paraphrase Sir Charles Snow) into three, instead of helping to debarbarize both.

Now it seems to me that this therapeutic aspect has been narrowly interpreted. In the usual question, 'Can history of science give arts undergraduates a worthwhile insight into science?' the term 'arts' should be deleted. We cannot accept the present barbaric method of teaching science – not even for science students.

My third point is a minor reflection on the old plan of Science Greats. I think that to combine science and philosophy-cum-history of science at the undergraduate level would be very difficult. We certainly need intelligent people for many purposes who, while having all the advantages of the traditional arts education, will not be afraid of, or alien to, science. This could be achieved by developing a new honours school on the pattern of Literae Humaniores concentrating on the Seventeenth Century instead of on Classics. This is perhaps the last great epoch in the history of mankind of which a synoptic view may be achieved at undergraduate level.

14
The social responsibility of science*

Respect for Truth, Russell tells us, is of Judaeo-Christian origin. Science respects truth no more than Catholicism; they differ only about how to recognize truth. The authority for science is reason and experience, the authority for Catholicism is revelation. But Galileo and the Inquisition had one common ground: they both craved for Truth articulated in propositions whose truth and falsehood was independent of the person who uttered them or of the machine that printed them. In this sense, Catholics and scientists alike come under what Dr Ravetz calls the 'classical' tradition. They all pursued truth and differed only on how to evaluate propositions with respect to their truth or falsity, or with respect to their probability.

Science achieved its autonomy in a long struggle with the Church – leaving the Church and nonscientists to decide matters about God, morality, politics, but maintaining a rule of reason and experiment as far as the factual truth about the Universe is concerned. Scientists and philosophers disagreed about the relative proving power of intellect and experience and about the general criteria by which to judge scientific – and especially *rival* scientific – theories; but in specific cases they finally seemed to reach very good agreement: all scientists, for example, agree that Einsteinian mechanics is better than Newtonian.

The values and methods of appraisal of science have continually come under attack from the outside. Sceptics, like Hume, questioned its exaggerated claims to certain knowledge. Some, like Popper, questioned its claims even to (provably) probable knowledge. But science could live very well with the sceptics' criticisms as long as these sceptics did not present a rival set of aims and standards. As long as we agree that Newton, Faraday, Maxwell, Einstein represent peaks of human achievement, and their search for the Blueprint of the Universe should be allowed to go on without outside interference, we may disagree in our particular theories as to what makes their achievements great and objective, and as to what exactly the intellectual attitudes which we should learn from these achievements are.

The Romantics and pragmatists, however, *did* present a rival set of

* Lakatos delivered this paper as part of a debate with Dr J. R. Ravetz at a meeting of the British Society for Social Responsibility in Science on February 18, 1970. Lakatos did not intend to publish the paper, at least in its present form. (*Eds.*)

aims and standards for science. Not only did they claim, like the sceptics, that the intellect is powerless, they claimed that it must be replaced by feeling, sentiment, will. They praised the inarticulate, the inexpressible, and preached contempt for the articulated. But of course what is not articulated, what is not expressible in propositions, cannot be criticized, cannot be appraised impersonally. For these Romantics and pragmatists, the personal became paramount. Instead of putting forward clearly expressed ideas about nature and subjecting them to the severe criticism of facts, they praised the mystic communion with nature which they called *understanding*. They detested the *abstract and verbal* and praised the *particular and instinctive*. A romantic 'would be moved to tears by the sight of a simple destitute peasant family but would be cold to well-thought-out schemes for ameliorating the lot of peasants as a class'. He would cry on the sight of a televised group of maimed Vietnamese but will be completely unable to appraise untelevised evidence of 20 million Russians killed in concentration camps which he could never see.

The Romantics, from Rousseau through Fichte, Coleridge, and Hegel to Hitler, Stalin, Sartre, Heidegger and Marcuse, viewed science through eyes different from those of the scientist. Their question was not which theory was nearer to the truth. Hegel thought the English Newton perverted the deep and unutterable vision of his hero Kepler, the mystic German, and forced it into the Procrustean bed of vacuous mathematical formulas. Hitler distinguished German from Jewish science; it did not even occur to him to ask which was nearer to the truth. Stalin thought that proletarian, socialist science was superior to bourgeois science: he thought that bourgeois science serves the bourgeoisie, socialist science the proletariat, and he sent bourgeois geneticists to die in concentration camps. Professor Bernal thought at one time that one can find out which theory is more advanced by looking at the social class which created it. Slave science was worse than feudal science and feudal science worse than bourgeois science, etc. One consequence of this view has been an alliance of socialists with Catholics both with a vested interest in showing that (alleged) medieval science was not as bad as is usually thought.

So Romantics (and pragmatists) apply external standards to science and try to force it to conform to these standards. Marcuse claims that the idea that the aim of pure science is truth irrespective of social consequences, is a dangerous one. According to the New Left, certain types of research, like nuclear physics or genetics, must be stopped. The autonomy of the scientific community must be destroyed. It is the society which should completely determine the scientist's choice of problems, forbid some and lavishly finance others. The search for truth has no autonomous value.

So here comes my first question to Dr Ravetz. Where does he stand on this issue? Would he want to suppress certain branches of research and publication of results of scientific value? Would he want a totalitarian state to determine what scientists can, and must, do? Or should pure scientists like Newton, Maxwell, and Einstein be allowed to work free from fear?

In my view, science, as such, has no social responsibility. In my view it is society that has a responsibility – that of maintaining the apolitical, detached scientific tradition and allowing science to search for truth in the way determined purely by its inner life. Of course scientists, as citizens, have responsibility, like all other citizens, to see that science is *applied* to the right social and political ends. This is a different, independent question, and, in my opinion one which ought to be determined through Parliament. Of course, as a citizen I am all for *using* science to serve antipollution rather than to serve pollution, to serve the defence of liberty rather than to serve the subjugation of weaker people. And here I come to my second question to Dr Ravetz. In my view one of the most important social responsibilities of British people is to use science to defend the liberty of this country. In my view this can only be done by maintaining the high social prestige of applied nuclear scientists working for the army. Now what does Dr Ravetz want British engineers to manufacture: the nuclear umbrella for freedom or Chamberlain's umbrella for serfdom?

References

Abel, N. H. [1826*a*]: 'Untersuchungen über die Reihe

$$1+\frac{m}{1}x+\frac{m\cdot m-1}{1\cdot 2}x^2+\ldots$$'

Journal für die Reine und Angewandte Mathematik, **1**, pp. 311–39.

Abel, N. H. [1826*b*]: 'Letter to Hansteen', in S. Lie and L. Sylow (*eds.*): *Oeuvres Complètes*, vol. 2, pp. 263–5. Christiana: Grøndahl, 1881.

Adam, C. and Tannery, P. [1897–1913]: *Oeuvres de Descartes.* Twelve volumes. Paris: Leopold Cerf.

Agassi, J. [1959]: 'How are Facts Discovered?', *Impulse*, **3**, pp. 2–4.

Agassi, J. [1961]: 'The Role of Corroboration in Popper's Methodology', *Australasian Journal of Philosophy*, **39**, pp. 82–91.

Agassi, J. [1963]: *Towards an Historiography of Science.* Wesleyan University Press.

Agassi, J. [1966]: 'Sensationalism', *Mind.* N.S. **75**, pp. 1–24.

Alexander, H. G. (*ed.*) [1956]: *The Leibniz–Clarke Correspondence.* Manchester University Press.

Arnauld, A. and Nicole, P. [1724]: *La Logique, ou l'Art de Penser.* Translated into English by T. S. Baynes as *The Port-Royal Logic.* Edinburgh: Sutherland and Knox, 1861.

Ayer, A. J. [1936]: *Language, Truth and Logic.* London: Victor Gollancz.

Ayer, A. J. [1956]: *The Problem of Knowledge*, London: Macmillan.

Bar-Hillel, Y. [1955–6]: 'Comments on "Degree of Confirmation" by Professor K. R. Popper', *British Journal for the Philosophy of Science*, **6**, pp. 155–7.

Bar-Hillel, Y. [1956–7]: 'Remark on Popper's Note on Content and Degree of Confirmation', *British Journal for the Philosophy of Science*, **7**, pp. 245–8.

Bar-Hillel, Y. [1963]: 'Remarks on Carnap's Logical Syntax of Language', in P. A. Schilpp (*ed.*): [1963], pp. 519–44.

Bar-Hillel, Y. [1967]: 'Is Mathematical Empiricism Still Alive?', in I. Lakatos (*ed.*): [1967], pp. 197–9.

Bar-Hillel, Y. [1968*a*]: 'Inductive Logic as "the" Guide of Life', in I. Lakatos (*ed.*): [1968*a*], pp. 66–9.

Bar-Hillel, Y. [1968*b*]: 'The Acceptance Syndrome', in I. Lakatos (*ed.*): [1968*a*], pp. 150–61.

Bar-Hillel, Y. [1968*c*]: 'Bunge and Watkins on Inductive Logic', in I. Lakatos (*ed.*): [1968*a*], pp. 282–5.

Baumann, J. J. [1869]: *Die Lehren von Zeit, Raum und Mathematik*, volume 2. Berlin: G. Reiner.

Beck, L. J. [1952]: *The Method of Descartes: A Study of the Regulae.* Oxford: Clarendon Press.

Bell, E. T. [1939]: *Men of Mathematics.* London: Victor Gollancz.

Bell, E. T. [1940]: *The Development of Mathematics.* New York: McGraw-Hill.

Benacerraf, P. and Putnam, H. (*eds.*) [1964]: *Readings in the Philosophy of Mathematics.* Oxford: Basil Blackwell.
Bennett, J. [1964]: *Rationality.* London: Routledge and Kegan Paul.
Bernays, P. [1939]: 'Bemerkungen zur Grundlagenfrage', in F. Gonseth (*ed.*): *Philosophie Mathématique*, pp. 83–7. Paris: Hermann.
Bernays, P. [1965]: 'Some Empirical Aspects of Mathematics', in P. Bernays and S. Dockx (*eds*): *Information and Prediction in Science*, pp. 123–8. New York: Academic Press.
Bernays, P. [1967]: 'Mathematics and Mental Experience', in I. Lakatos (*ed.*): [1967], pp. 196–7.
Bernays, P. and Hilbert, D. [1939]: *Grundlagen der Mathematik*, volume 2. Berlin: Springer.
Beveridge, W. [1937]: 'The Place of the Social Sciences in Human Knowledge', *Politica*, **2**, pp. 459–79.
Boltzmann, L. [1896–8]: *Lectures on Gas Theory.* Translated by S. G. Brush. Berkeley and Los Angeles: University of California Press, 1964.
Born, M. [1949]: *Natural Philosophy of Cause and Chance.* Oxford: Clarendon Press.
Bourbaki, N. [1949*a*]: 'The Foundations of Mathematics for the Working Scientist', *Journal of Symbolic Logic*, **14**, pp. 1–8.
Bourbaki, N. [1949*b*]: *Topologie Générale.* Paris: Hermann.
Bourbaki, N. [1960]: *Eléments d'Histoire des Mathématiques.* Paris: Hermann.
Boyer, C. B. [1949]: *The Concepts of the Calculus.* New York: Columbia University Press.
Braithwaite, R. B. [1953]: *Scientific Explanation.* Cambridge University Press.
Braithwaite, R. B., Russell, B. A. W. and Waismann, F. [1938]: 'Symposium: The Relevance of Psychology to Logic', *Aristotelian Society Supplementary Volume*, **17**, pp. 19–68.
Broad, C. D. [1922]: Review of Keynes [1921], *Mind*, N.S. **31**, pp. 72–85.
Broad, C. D. [1952]: *Ethics and the History of Philosophy.* London: Routledge and Kegan Paul.
Broad, C. D. [1959]: 'A Reply to my Critics', in P. A. Schilpp (*ed.*): *The Philosophy of C. D. Broad*, pp. 711–830. New York: Tudor.
Brunschvicg, L. [1912]: *Les Etapes de la Philosophie Mathématique.* Paris: Librairie Félix Alcan.
Cajori, F. [1924]: *A History of Mathematics.* 2nd edition. New York and London: Macmillan, 1961.
Campbell, N. [1920]: *Foundations of Science.* New York: Dover, 1957.
Carnap, R. [1928]: *Scheinprobleme in der Philosophie.* Frankfurt am Main: Surkamp, 1966.
Carnap, R. [1930–1]: 'Die Alte und die Neue Logik', *Erkenntnis*, **1**. Translated into English in A. J. Ayer (*ed.*): *Logical Positivism*, pp. 133–45. London: George Allen and Unwin.
Carnap, R. [1931]: 'Die Logizistische Grundlegung der Mathematik', *Erkenntnis*, **2**, pp. 91–105. English translation in P. Benacerraf and H. Putnam (*eds.*): [1964], pp. 31–41.
Carnap, R. [1935]: Review of Popper's [1934], *Erkenntnis*, **5**, pp. 290–4.
Carnap, R. [1936]: 'Testability and Meaning', *Philosophy of Science*, **3**, pp. 419–71.
Carnap, R. [1937]: *The Logical Syntax of Language.* London: Kegan Paul. (Revised translation of *Logische Syntax der Sprache.* Vienna: Springer, 1934.)
Carnap, R. [1946]: 'Theory and Prediction in Science', *Science*, **104**, pp. 520–1.
Carnap, R. [1950]: *Logical Foundations of Probability.* Chicago University Press.

Carnap, R. [1952]: *The Continuum of Inductive Methods.* Chicago University Press.

Carnap, R. [1953*a*]: 'Inductive Logic and Science', *Proceedings of the American Academy of Arts and Sciences,* **80**, pp. 189–97.

Carnap, R. [1953*b*]: 'What is Probability?', *Scientific American,* **189**, pp. 128–30, 132, 134, 136.

Carnap, R. [1953*c*]: 'Remarks to Kemeny's Paper', *Philosophy and Phenomenological Research,* **13**, pp. 375–6.

Carnap, R. [1958]: 'Beobachtungssprache und Theoretische Sprache', *Dialectica,* **12**, pp. 236–47.

Carnap, R. [1960]: 'The Aim of Inductive Logic', in E. Nagel, P. Suppes and A. Tarski (*eds.*): *Logic, Methodology and Philosophy of Science,* pp. 303–18. Stanford University Press.

Carnap, R. [1963*a*]: 'Intellectual Autobiography', in P. A. Schilpp (*ed.*): [1963], pp. 1–84.

Carnap, R. [1963*b*]: 'Replies and Systematic Expositions', in P. A. Schilpp (*ed.*): [1963], pp. 859–1013.

Carnap, R. [1966]: 'Probability and Content Measure', in P. K. Feyerabend and G. Maxwell (*eds*).: *Mind, Matter and Method,* pp. 248–60. University of Minnesota Press.

Carnap, R. [1968*a*]: 'On Rules of Acceptance', in I. Lakatos (*ed.*): [1968*a*], pp. 146–50.

Carnap, R. [1968*b*]: 'Inductive Logic and Inductive Intuitions', in I. Lakatos (*ed.*): [1968*a*], pp. 258–67.

Carnap, R. [1968*c*]: 'Reply', in I. Lakatos (*ed.*): [1968*a*], pp. 307–14.

Carnap, R. and Stegmüller, W. [1964]: *Inductive Logic und Wahrscheinlichkeit.* Vienna: Springer.

Cauchy, A. L. [1813]: 'Recherches sur les Polyèdres', *Journal de l'Ecole Polytechnique,* **9**, pp. 68–86. (Read in 1811.)

Cauchy, A. L. [1821]: *Cours d'Analyse de l'Ecole Royale Polytechnique.* Paris: de Bure.

Cauchy, A. L. [1823]: *Résumé des leçons sur le calcul Infinitésimal.* Paris: de Bure. In *Oeuvres Complètes,* Séries 2, volume 4, pp. 5–261.

Cauchy, A. L. [1853]: 'Note sur les séries convergentes dont les Divers Terms sont des Functions Continues d'une Variable Réelle ou Imaginaire entre des Limites Données', *Comptes Rendus des Séances des l'Académie des Sciences,* **36**, pp. 454–9.

Cavell. S. [1962]: 'The Availability of Wittgenstein's Later Philosophy', *Philosophical Review,* **71**, pp. 67–93.

Cherniss, H. [1951]: 'Plato as Mathematician', *The Review of Metaphysics,* **4**, pp. 395–425.

Church, A. [1932]: 'A Set of Postulates for the Foundation of Logic', *Annals of Mathematics,* **33**, Second Series, pp. 346–66.

Church, A. [1939]: 'The Present Situation in the Foundations of Mathematics', in F. Gonseth (*ed.*): *Philosophie Mathématique,* pp. 67–72. Paris: Hermann.

Chwistek, L. [1948]: *The Limits of Science.* London: Kegan Paul.

Cleave, J. P. [1971]: 'Cauchy, Convergence and Continuity', *British Journal for the Philosophy of Science,* **22**, pp. 27–37.

Cohen, I. B. [1974]: 'Newton's Theory *vs* Kepler's Theory and Galileo's Theory', in Y. Elkana (*ed.*): *The Interaction Between Science and Philosophy,* pp. 299–338. New York: Humanities Press.

Cohen, L. J. [1968]: 'An Argument that Confirmation Functors for Consilience are Empirical Hypotheses', in I. Lakatos (*ed.*): [1968*a*], pp. 247–50.

Cohen, L. J. [1972]: 'Is the Progress of Science Evolutionary?', *British Journal for the Philosophy of Science*, **24**, pp. 41–61.

Cornford, F. M. [1932]: 'Mathematics and Dialectics in the Republic VI–VIII', *Mind*, **41**, pp. 37–52 and 173–90.

Courant, R. and Robbins, H. [1941]: *What is Mathematics?* Oxford University Press.

Couturat, L. [1905]: *Les Principes des Mathématiques*. Hildesheim: Georg Olms, 1965.

Crombie, A. C. (*ed.*) [1961]: *Scientific Change*. London: Heinemann.

Curry, H. B. [1958]: *Outline of a Formalist Philosophy of Mathematics*. Amsterdam: North Holland.

Curry, H. B. [1963]: *Foundations of Mathematical Logic*. New York: McGraw-Hill.

Curry, H. B. [1965]: 'The Relation of Logic to Science', in P. Bernays and S. Dockx (*eds.*): *Information and Prediction in Science*, pp. 79–98. New York and London: Academic Press.

Descartes, R. [1628]: *Rules for the Direction of the Mind*, in E. R. Haldane and G. R. T. Ross (*eds.*): [1911], pp. 1–77.

Descartes, R. [1637]: *Discourse on the Method of Rightly Conducting the Reason*, in E. R. Haldane and G. R. T. Ross (*eds.*): [1911], pp. 81–130.

Descartes, R. [1638]: 'Letter to Mersenne, 11 October 1638', in C. Adam and P. Tannery (*eds.*): [1897–1913], volume 2, pp. 379–405.

Descartes, R. [1644]: *Principles of Philosophy*, in E. R. Haldane and G. R. T. Ross (*eds.*): [1911], pp. 203–302.

Descartes, R. [1664]: *Description du Corps Humain*, in C. Adam and P. Tannery (*eds.*): [1897–1913], volume 2, pp. 223–90.

Dirichlet, P. L. [1829]: 'Sur la Convergence des Séries Trigonométriques que Servent à Représenter une Function Arbitraire entre des Limites Données', *Journal für die Reine und Angewandte Mathematik*, **4**, pp. 157–69.

Dorling, J. [1971]: 'Einstein's Introduction of Photons: Argument by Analogy or Deduction from Phenomena?', *British Journal for the Philosophy of Science*, **22**, pp. 1–8.

Drury, M. O'C. [1960]: 'Ludwig Wittgenstein', *The Listener*, January 28, pp. 163–5.

Du Bois Reymond, P. D. G. [1874]: 'Über die Sprungweisen Werthänderungen Analytischer Functionen', *Mathematische Annalen*, **7**, pp. 241–61.

Duhamel, J. M. C. [1865]: *Des Méthodes dans les Sciences de Raisonnement*, volume 1. Paris: Bachelier.

Duhem, P. [1906]: *La Théorie Physique; son Objet, sa Structure*. English translation of the second (1914) edition: *The Aim and Structure of Physical Theory*. Princeton University Press, 1954.

Eilenberg, S. and Steenrod, N. [1952]: *Foundations of Algebraic Topology*. Princeton University Press.

Einstein, A. [1905]: 'Zur Elektrodynamik der Bewegter Körper', *Annalen der Physik*, **17**, pp. 891–921.

Einstein, A. [1915]: 'Die Relativitätstheorie', in E. Warburg (*ed.*): *Physik*. Leipzig: Teubner.

Engels, F. [1894]: *Anti Dühring*. 3rd edition. London: Lawrence and Wishart, 1955.

Fann, K. T. [1969]: *Wittgenstein's Conception of Philosophy*. Oxford: Basil Blackwell.

Feferman, S. [1968]: 'Autonomous Transfinite Progressions and the Extent of Predicative Mathematics', in B. van Rootselaar and J. F. Staal (*eds.*): *Logic, Methodology and Philosophy of Science* III, pp. 121–35. Amsterdam: North Holland.

Feigl, H., Maxwell, G. and Scriven, M. (*eds.*) [1958]: *Minnesota Studies in the Philosophy of Science*, **2**, *Concepts, Theories and the Mind–Body Problem.* Minneapolis: University of Minnesota Press.

Feyerabend, P. K. [1955]: 'Wittgenstein's *Philosophical Investigations*', *Philosophical Review*, **64**, pp. 449–83.

Feyerabend, P. K. [1962]: 'Explanation, Reduction and Empiricism', in H. Feigl and G. Maxwell (*eds.*): *Minnesota Studies in the Philosophy of Science*, **3**, pp. 28–97. University of Minnesota Press.

Feyerabend, P. K. [1970]: 'Against Method', in M. Radner and W. Winokur (*eds.*): *Minnesota Studies for the Philosophy of Science*, **4**, *Analyses of Theories and Methods of Physics and Psychology*, pp. 17–130. University of Minnesota Press.

Feyerabend, P. K. [1972]: 'Von der Beschränkten Gültigkeit Methodologischer Regeln', in R. Bubner, K. Cramer and R. Wiehl (*eds.*): *Dialog als Methode*, pp. 124–71. Göttingen: Vanderhoeck and Ruprecht.

Feyerabend, P. K. [1975]: *Against Method.* (Expanded version of Feyerabend [1970].) London: New Left Books.

Fisher, R. A. [1922]: 'On the Mathematical Foundations of Theoretical Statistics', *Transactions of the Royal Society of London*, Series A, **222**, pp. 309–68.

Fourier, J. [1822]: *Théorie analytique de la Chaleur.* Translated into English as *The Analytical Theory of Heat.* New York: Dover.

Fraenkel, A. A. [1927]: *Zehn Vorlesungen über die Grundlegung der Mengenlehre.* Leipzig and Berlin: B. G. Teubner.

Fraenkel, A. A., Bar-Hillel, Y. and Levy, A. [1973]: *Foundations of Set Theory.* 2nd edition. Amsterdam: North Holland.

Frayne, T., Morel, A. C. and Scott, D. S. [1962–3]: 'Reduced Direct Products', *Fundamenta Mathematica*, **51**, pp. 195–228.

Frege, G. [1893]: *Grundgesetze der Arithmetik*, Volume 1. Jena.

Fries, J. F. [1831]: *Neue oder Anthropologische Kritik der Vernunft.* Heidelberg: Winter.

Galileo, G. [1630]: *Dialogue Concerning the Two Chief World Systems.* Translated by S. Drake. University of California Press, 1967.

Gamow, G. A. [1966]: *Thirty Years that Shook Physics.* Garden City, New York: Doubleday.

Giedymin, J. [1968]: 'Empiricism, Refutability, Rationality', in I. Lakatos and A. Musgrave (*eds.*): [1968], pp. 67–78.

Gellner, E. [1959]: *Words and Things.* London: Victor Golancz.

Gödel, K. [1931]: 'Discussion zur Grundlegung der Mathematik', *Erkenntnis*, **2**, pp. 147–8.

Gödel, K. [1938]: 'The Consistency of the Axiom of Choice and the Generalized Continuum Hypothesis', *Proceedings of the National Academy of Sciences*, **24**, pp. 556–7.

Gödel, K. [1944]: 'Russell's Mathematical Logic', in P. A. Schilpp (*ed.*): [1944], pp. 125–53. Reprinted in P. Benacerraf and H. Putnam (*eds.*): [1964], pp. 211–32.

Gödel, K. [1947]: 'What is Cantor's Continuum Hypothesis?', *American Mathematical Monthly*, **54**, pp. 515–25.

Gödel, K. [1964]: 'What is Cantor's Continuum Hypothesis?', in P.

Benacerraf and H. Putnam (*eds.*): [1964], pp. 258–73. Revised and expanded version of Gödel [1947].
Good, I. J. [1960]: Review of Popper [1959], *Mathematical Reviews*, **21** (2), pp. 1171–3.
Goodstein, R. L. [1951*a*]: *Constructive Formalism.* University College Leicester.
Goodstein, R. L. [1951*b*]: *The Foundations of Mathematics.* University College Leicester.
Goodstein, R. L. [1962]: 'The Axiomatic Method', *Aristotelian Society Supplementary Volume*, **36**, pp. 145–54.
Grattan-Guinness, I. and Ravetz, J. R. [1972]: *Joseph Fourier, 1768–1830.* Cambridge, Mass.: M.I.T. Press.
Grünbaum, A. [1961]: 'The Genesis of the Special Theory of Relativity', in H. Feigl and G. Maxwell (*eds.*): *Current Issues in the Philosophy of Science*, pp. 43–53. New York: Holt, Reinhart and Winston.
Grünbaum, A. [1963]: *Philosophical Problems of Space and Time.* Second edition, 1973. Dordrecht: Reidel.
Grünbaum, A. [1973]: 'Falsifiability and Rationality', *unpublished.*
Gulley, N. [1958]: 'Greek Geometrical Analysis', *Phronesis*, **33**, pp. 1–14.
Haldane, E. R. and Ross, G. R. T. (*eds.*): [1911]: *The Philosophical Works of Descartes*, volume 1. Cambridge University Press.
Hankel, H. [1874]: *Zur Geschichte der Mathematik in Altertum und Mittelalter.* Hildesheim: George Olms, 1965.
Hardy, G. H. [1918]: 'Sir George Stokes and the Concept of Uniform Convergence', *Proceedings of the Cambridge Philosophical Society*, **18/19**, pp. 148–56.
Heath, T. L. [1925]: *The Thirteen Books of Euclid's Elements.* Second edition. Reprinted, New York: Dover, 1956.
Hempel, C. G. [1945*a*]: 'On the Nature of Mathematical Truth', *American Mathematical Monthly*, **52**, pp. 543–56. Reprinted in P. Benacerraf and H. Putnam (*eds.*): [1964], pp. 366–81.
Hempel, C. G. [1945*b*]: 'Studies in the Logic of Confirmation', *Mind*, **54**, pp. 1–26, 97–121.
Hempel, C. G. [1965]: *Aspects of Scientific Explanation.* New York: The Free Press.
Hempel, C. G. and Oppenheim, P. [1945]: 'A Definition of "Degree of Confirmation"', *Philosophy of Science*, **12**, pp. 98–115.
Henkin, L. [1947]: *The Completeness of Formal Systems.* PhD Thesis. Princeton University.
Herbrand, J. [1930]: 'Les Bases de la Logique Hilbertienne', *Revue de la Métaphysique et de la Morale*, **37**, pp. 243–55.
Hesse, M. [1964]: 'Induction and Theory Structure', *Review of Metaphysics*, **18**, pp. 109–22.
Heyting, A. [1967]: 'Weyl on Experimental Testing of Mathematics', in I. Lakatos (*ed.*): [1967], p. 195.
Hilbert, D. [1923]: 'Die Logischen Grundlagen der Mathematik', *Mathematische Annalen*, **88**, pp. 151–65.
Hilbert, D. [1925]: 'Über das Unendliche', *Mathematische Annalen*, **95**, pp. 161–90. Translated into English in J. van Heijenoort (*ed.*): *From Frege to Gödel*, pp. 367–92. Harvard University Press, 1967.
Hintikka, K. J. J. [1957]: 'Necessity, Universality and Time in Aristotle', *Ajatus*, **20**, pp. 65–90.
Hintikka, K. J. J. [1968]: 'Induction by Enumeration and Induction by Elimination', in I. Lakatos (*ed.*): [1968*a*], pp. 191–216.
Hintikka, K. J. J. and Remes, U. [1974]: *The Method of Analysis.* Dordrecht: D. Reidel.

Holton, G. [1960]: 'On the Origins of the Special Theory of Relativity', *American Journal of Physics*, **28**, pp. 627–31 and 633–76.

Holton, G. [1969]: 'Einstein, Michelson, and the "Crucial" Experiment', *Isis*, **6**, pp. 133–97.

Houel, J. [1878]: *Calcul Infinitesimal*, volume 1. Paris.

Hume, D. [1739]: *A Treatise of Human Nature.* Oxford: Clarendon Press.

Huyghens, C. [1690]: *Treatise on Light.* University of Chicago Press, 1945.

Jaffe, B. [1960]: *Michelson and the Speed of Light.* London: Heinemann.

Jeffrey, H. and Wrinch, D. [1921]: 'On Certain Fundamental Principles of Scientific Enquiry', *Philosophical Magazine*, **42**, pp. 269–98.

Jeffrey, R. [1968]: 'Probable Knowledge', in I. Lakatos (*ed.*): [1968*a*], pp. 166–81.

Joachim, H. H. [1906]: *The Nature of Truth.* Oxford University Press.

Kalmár, L. [1959]: 'An Argument against the Plausibility of Church's Thesis', in A. Heyting (*ed.*): *Constructivity in Mathematics*, pp. 72–80. Amsterdam: North Holland.

Kalmár, L. [1967]: 'Foundations of Mathematics – Whither Now?', in I. Lakatos (*ed.*): [1967], pp. 187–94.

Kemeny, J. [1952]: 'A Contribution to Inductive Logic', *Philosophy and Phenomenological Research*, **13**, pp. 371–4.

Kemeny, J. [1955]: 'Fair Bets and Inductive Probabilities', *Journal of Symbolic Logic*, **20**, pp. 263–73.

Kemeny, J. [1958]: 'Undecidable Problems in Elementary Number Theory', *Mathematische Annalen*, **135**, pp. 160–9.

Kemeny, J. [1959]: *A Philosopher Looks at Science.* Princeton: Van Nostrand.

Kemeny, J. [1963]: 'Carnap's Theory of Probability and Induction', in P. A. Schilpp (*ed.*): [1963], pp. 711–37.

Kemeny, J. and Oppenheim, P. [1953]: 'Degree of Factual Support', *Philosophy of Science*, **20**, pp. 307–24.

Kendall, M. G. and Stuart, A. [1967]: *The Advanced Theory of Statistics*, volume 2. 2nd edition. London: Charles Griffin.

Kenny, A. [1973]: *Wittgenstein.* London: Allen Lane.

Keynes, J. M. [1921]: *A Treatise on Probability*, London: Macmillan.

Kleene, S. C. [1943]: 'Recursive Predicates and Quantifiers', *Transactions of the American Mathematical Society*, **53**, pp. 41–73.

Kleene, S. C. [1952]: *Introduction to Metamathematics.* Amsterdam: North Holland.

Kleene, S. C. [1967]: 'Empirical Mathematics?', in I. Lakatos (*ed.*): [1967], pp. 195–6.

Kleene, S. C. and Rosser, J. B. [1935]: 'The Inconsistency of Certain Formal Logics', *Annals of Mathematics*, **36**, pp. 630–6.

Klein, F. [1908]: *Elementary Mathematics from an Advanced Standpoint.* New York: Dover.

Kneale, W. C. [1949]: *Probability and Induction.* Oxford: Clarendon Press.

Kneale, W. C. [1950]: 'Natural Laws and Contrary to Fact Conditionals', *Analysis*, **10**, pp. 121–5.

Kneale, W. C. [1955]: 'The Necessity of Invention', *Proceedings of the British Academy*, **41**, pp. 85–108.

Kneale, W. C. [1961]: 'Universality and Necessity', *British Journal for the Philosophy of Science*, **12**, pp. 89–102.

Kneale, W. C. [1968]: 'Confirmation and Rationality', in I. Lakatos (*ed.*): [1968*a*], pp. 59–61.

Koertge, N. [1971]: 'For and Against Method', *British Journal for the Philosophy of Science*, **23**, pp. 274–90.

Kreisel, G. [1956–7]: 'Some Uses of Metamathematics', *British Journal for the Philosophy of Science*, **7**, pp. 161–73.
Kreisel, G. [1967*a*]: 'Informal Rigour and Completeness Proofs', in I. Lakatos (*ed.*): [1967], pp. 138–71.
Kreisel, G. [1967*b*]: 'Reply to Bar-Hillel', in I. Lakatos (*ed.*): [1967], pp. 175–8.
Kreisel, G. [1967*c*]: 'Comment on Mostowski', in I. Lakatos (*ed.*): [1967], pp. 97–103.
Kreisel, G. and Krivine, J. L. [1967]: *Elements of Mathematical Logic.* Amsterdam: North Holland.
Kuhn, T. S. [1962]: *The Structure of Scientific Revolutions.* Second edition. University of Chicago Press, 1970.
Kuhn, T. S. [1963]: 'The Function of Dogma in Scientific Research', in A. C. Crombie (*ed.*): [1961], pp. 347–69.
Kuhn, T. S. [1970*a*]: 'Reflections on my Critics', in I. Lakatos and A. Musgrave (*eds.*): [1970], pp. 231–78.
Kuhn, T. S. [1970*b*]: 'Postscript – 1969' to second edition of Kuhn [1962], pp. 174–210.
Kuhn, T. S. [1971]: 'Notes on Lakatos', in R. C. Buck and R. S. Cohen (*eds.*): *P.S.A., 1970, Boston Studies in the Philosophy of Science*, **8**, pp. 137–46. Dordrecht: D. Reidel.
Kyburg, H. [1964]: 'Recent Work in Inductive Logic', *American Philosophical Quarterly*, **1**, pp. 1–39.
Lakatos, I.: see Lakatos Bibliography, *below* pp. 274–6.
Latsis, S. [1972]: 'Situational Determinism in Economics', *The British Journal for the Philosophy of Science*, **23**, pp. 207–45.
Lehman, R. S. [1955]: 'On Confirmation and Rational Betting', *Journal of Symbolic Logic*, **20**, pp. 251–62.
Leibniz, G. W. F. [1678]: 'Letter to Conring, 19 March', in L. Loemker (*ed.*): *Leibniz's Philosophical Papers and Letters*, pp. 186–91. Dordrecht: Reidel, 1967.
Leibniz, G. W. F. [1687]: 'Letter to Bayle', in C. I. Gerhardt (*ed.*): *Philosophische Schriften*, **3**, p. 52. Hildesheim: George Olms, 1965.
Leibniz, G. W. F. [1704]: *Nouveaux Essais.* First published, 1765.
Lenard, P. [1933]: *Grosse Naturforscher.* Translated into English as *Great Men of Science.* London: G. Bell and Sons, 1933.
Lenin, V. I. [1908]: *Materialism and Empirio-Criticism*, in *Collected Works*, volume **13**. London: Lawrence and Wishart, 1938.
Levy, A. and Solovay, R. M. [1967]: 'Measurable Cardinals and the Continuum Hypothesis', *Israeli Journal of Mathematics*, **5**, pp. 234–48.
Lhuilier, S. A. J. [1787]: *Exposition Elémentaire des Principes des Calculs Supérieurs.* Berlin: G. J. Decker.
Lipsey, R. G. [1963]: *Positive Economics.* London: Weidenfeld and Nicolson. Second edition, 1966.
Lusin, N. [1935]: 'Sur les Ensembles Analytiques Nuls', *Fundamenta Mathematica*, **25**, pp. 109–31.
Mach, E. [1883]: *Die Mechanik in ihrer Entwicklung.* Translated into English as *The Science of Mechanics.* La Salle: Open Court, 1960.
Mackie, J. [1963]: 'The Paradox of Confirmation', *British Journal for the Philosophy of Science*, **13**, pp. 265–77.
Magee, B. (*ed.*) [1972]: *Modern British Philosophy.* London: Secker and Warburg.
Mahoney, M. S. [1968–9]: 'Another Look at Greek Geometrical Analysis', *Archive for the History of the Exact Sciences*, **5**, pp. 319–48.

Martin, D. A. and Solovay, R. M. [1970]: 'Internal Cohen Extensions', *Annals of Mathematical Logic*, **2**, pp. 143–78.

Martin, J. [1969]: 'Another look at the Doctrine of Verstehen', *British Journal for the Philosophy of Science*, **20**, pp. 53–67.

Masterman, M. [1970]: 'The Nature of a Paradigm', in I. Lakatos and A. Musgrave (*eds.*): [1970], pp. 59–89.

Mehlberg, M. [1962]: 'The Present Situation in the Philosophy of Mathematics', in B. M. Kazemier and D. Vuysje (*eds.*): *Logic and Language: Studies Dedicated to Professor Rudolf Carnap on the Occasion of his Seventieth Birthday*, pp. 69–103. Dordrecht: Reidel.

Merton, R. [1949]: 'Science and Democratic Social Structure', in *Social Theory and Social Structure*, pp. 604–15. New York: Macmillan. Enlarged edition, 1965.

Miller, D. W. [1974]: 'Popper's Qualitative Theory of Verisimilitude', *British Journal for the Philosophy of Science*, **25**, pp. 166–77.

Mises, L. von [1960]: *Epistemological Problems of Economics*. Princeton: Van Nostrand.

Mostowski, A. [1955]: 'The Present State of Investigations on the Foundations of Mathematics', *Rozprawy Matematyczne*, **9**. Compiled in collaboration with A. Grzegorczyk, S. Jaśkowski, J. Łoś, S. Mazur, H. Rasiowa, and R. Sikorski.

Musgrave, A. [1968]: 'On a Demarcation Dispute', in I. Lakatos and A. Musgrave (*eds.*): [1968], pp. 78–88.

Musgrave, A. [1969]: *Impersonal Knowledge*. Unpublished PhD thesis, University of London.

Musgrave, A. [1971]: 'Kuhn's Second Thoughts', *British Journal for the Philosophy of Science*, **22**, pp. 287–97.

Musgrave, A. [1973]: 'Falsification and its Critics', in P. Suppes (*ed.*): *Proceedings of the 1971 Bucharest International Congress for Logic, Philosophy and Methodology of Science*. Amsterdam: Elsevier.

Myhill, J. [1960]: 'Some Remarks on the Notion of Proof', *The Journal of Philosophy*, **57**, pp. 461–71.

Naess, A. [1968]: *Four Modern Philosophers*. University of Chicago Press.

Nagel, E. [1944]: 'Logic without Ontology', in Y. H. Krikorian (*ed.*): *Naturalism and the Human Spirit*. New York: Colombia University Press. Reprinted in P. Benacerraf and H. Putnam (*eds.*): [1964], pp. 302–21.

Nagel, E. [1963]: 'Carnap's Theory of Induction', in P. A. Schilpp (*ed.*): [1963], pp. 785–825.

Neumann, J. von [1927]: 'Zur Hilbertischen Beweistheorie', *Mathematische Zeitschrift*, **26**, pp. 1–46.

Neumann, J. von [1947]: 'The Mathematician', in R. B. Heywood (*ed.*): *The Works of the Mind*, pp. 180–96. Chicago: University of Chicago Press.

Newton, I. [1686]: 'Letter to Halley'. Quoted in D. Brewster [1855]: *Memoirs of the Life, Writings and Discoveries of Sir Isaac Newton*, volume **1**, p. 441. New York: Johnson Reprint Corporation, 1965.

Newton, I. [1713]: 'Letter to Roger Cotes, 28 March', in J. Edelston (*ed.*): *Correspondence of Sir Isaac Newton and Professor Cotes*, pp. 154–6. Cambridge University Press, 1850.

Newton, I. [1717]: *Optics*. New York: Dover, 1952.

Nidditch, P. H. [1954]: *Introductory Formal Logic of Mathematics*. London: University Tutorial Press.

Pascal, B. [1659]: *Les Réflexions sur la Géométrie en Général* (*De l'Esprit Géométrique et de l'Art et Persuader*). In J. Chevalier (*ed.*): *Oeuvres Complètes*, pp. 575–604. Paris: La Librairie Galliard, 1954.

Pascal, F. [1973]: 'Ludwig Wittgenstein; a Personal Memoir', *Encounter*, **41**, number 2, August, pp. 23–39.
Pearce Williams, L. [1963]: 'Review of Agassi' [1963], *Archives Internationales d'Histoire des Sciences*, **16**, pp. 437–9.
Planck, M. [1929]: 'Zwanzig Jahre Arbeit am Physikalischen Weltbilt', *Physica*, **9**, pp. 193–222.
Polanyi, M. [1964]: *Science, Faith and Society*. University of Chicago Press.
Polanyi, M. [1967]: *The Tacit Dimension*. London: Routledge and Kegan Paul.
Popper, K. R. [1934]: *Logik der Forschung*. Vienna: Springer. Expanded English edition: Popper [1959].
Popper, K. R. [1948]: 'Naturgesetze und Theoretische Systeme', in S. Moser (*ed.*): *Gesetz und Wirklichkeit*, pp. 43–60. Innsbruch and Vienna: Tyrolia Verlag.
Popper, K. R. [1949]: 'Note on Natural Laws and So-called "Contrary to Fact Conditionals"', *Mind*, **58**, pp. 62–6.
Popper, K. R. [1952]: 'The Nature of Philosophical Problems and their Roots in Science', *British Journal for the Philosophy of Science*, **3**, pp. 124–56. Reprinted in Popper [1963*a*], pp. 66–96.
Popper, K. R. [1955–6]: '"Content" and "Degree of Confirmation", a Reply to Dr. Bar-Hillel', *British Journal for the Philosophy of Science*, **6**, pp. 157–63.
Popper, K. R. [1956–7]: 'A Second Note on Degree of Confirmation', *British Journal for the Philosophy of Science*, **7**, pp. 350–3.
Popper, K. R. [1957]: 'The Aim of Science', *Ratio*, **1**, pp. 24–35. Reprinted in Popper [1972], pp. 191–205.
Popper, K. R. [1957–8]: 'A Third Note on Degree of Confirmation', *British Journal for the Philosophy of Science*, **8**, pp. 294–302.
Popper, K. R. [1959]: *The Logic of Scientific Discovery*. London: Hutchinson.
Popper, K. R. [1963*a*]: *Conjectures and Refutations*. London: Routledge and Kegan Paul.
Popper, K. R. [1963*b*]: 'The Demarcation between Science and Metaphysics', in P. A. Schilpp (*ed.*): [1963]. Reprinted in Popper [1963*a*], pp. 253–92.
Popper, K. R. [1968*a*]: 'On Rules of Detachment and so-called Inductive Logic', in I. Lakatos (*ed.*): [1968*a*], pp. 130–8.
Popper, K. R. [1968*b*]: 'Theories, Experience and Probabilistic Intuitions', in I. Lakatos (*ed.*): [1968*a*], pp. 285–303.
Popper, K. R. [1971*a*]: 'Interview with Bryan Magee', in B. Magee (*ed.*) [1972].
Popper, K. R. [1971*b*]: 'Conjectural Knowledge: My Solution of the Problem of Induction', *Revue Internationale de Philosophie*, **95–6**, pp. 167–97. Reprinted in Popper [1972].
Popper, K. R. [1972]: *Objective Knowledge*. Oxford: Clarendon Press.
Pringsheim, A. [1916]: 'Grundlagen der Allgemeinen Functionenlehre', in M. Burkhardt, W. Wutinger and R. Fricke (*eds.*): *Encyklopädie der Mathematischen Wissenschaften*, **2**, Erste Teil, Erste Halbband, pp. 1–53. Leipzig: Teubner.
Putnam, H. [1967]: 'Probability and Confirmation', in S. Morgenbesser (*ed.*): *Philosophy of Science Today*, pp. 100–14. New York: Basic Books.
Quine, W. V. O. [1941*a*]: 'Element and Number', *Journal of Symbolic Logic*, **6**, pp. 135–49. Reprinted in *Selected Logical Papers*, pp. 121–40. New York: Random House, 1966.
Quine, W. V. O. [1941*b*]: 'Review of Rosser: "The Independence of Quine's Axioms *200 and *201"', *Journal of Symbolic Logic*, **6**, p. 163.
Quine, W. V. O. [1953*a*]: 'On ω-inconsistency and a So-called Axiom of In-

finity', *Journal of Symbolic Logic*, **18**, pp. 119–24. Reprinted in *Selected Logical Papers*, pp. 114–20. New York: Random House, 1966.

Quine, W. V. O. [1953*b*]: 'Two Dogmas of Empiricism', in *From a Logical Point of View*, pp. 20–46. Harvard University Press.

Quine, W. V. O. [1958]: 'The Philosophical Bearing of Modern Logic', in R. Klibansky (*ed.*): *Philosophy in the Mid-Century*, volume 1, pp. 3–4. Firenze: La Nuova Italia.

Quine, W. V. O. [1963]: *Set Theory and its Logic*. Harvard University Press.

Quine, W. V. O. [1965]: *Elementary Logic*. Revised edition. New York: Harper Torchbooks.

Quine, W. V. O. [1972]: *Ontological Relativity and Other Essays*. New York: Columbia University Press.

Ramsey, F. P. [1925]: 'The Foundations of Mathematics', *Proceedings of the London Mathematical Society*, **25**, pp. 338–84. Reprinted in *The Foundations of Mathematics and other Essays*. Edited by R. B. Braithwaite. London: Kegan Paul, 1931.

Ramsey, F. P. [1926*a*]: 'Truth and Probability', in *Foundations of Mathematics*, pp. 156–98.

Ramsey, F. P. [1926*b*]: 'Mathematical Logic', *The Mathematical Gazette*, **13**, pp. 185–94. Reprinted in *The Foundations of Mathematics*.

Reichenbach, M. [1936]: 'Induction and Probability', *Philosophy of Science*, **3**, pp. 124–6.

Renyi, A. [1955]: 'On a New Axiomatic Theory of Probability', *Acta Mathematica Academiae Scientiarum Hungaricae*, **6**, pp. 285–337.

Rescher, N. [1958]: 'A Theory of Evidence', *Philosophy of Science*, **25**, pp. 83–94.

Richtmyer, F. K., Kennard, E. H. and Lauritsen, T. [1955]: *Introduction to Modern Physics*. Fifth edition. New York: McGraw-Hill.

Ritchie, A. D. [1926]: 'Induction and Probability', *Mind*, N.S. **35**, pp. 301–18.

Robbins, L. C. [1932]: *An Essay on the Nature and Significance of Economic Science*. Second edition, 1935. London: Macmillan.

Robert, A. [1937]: 'Descartes et l'analyse des Anciens', *Archives de Philosophie*, **13**, cahier 2, pp. 221–42.

Robinson, A. [1963]: *Introduction to Model Theory and to the Metamathematics of Algebra*. Amsterdam: North Holland.

Robinson, A. [1966]: *Non-Standard Analysis*. Amsterdam: North Holland.

Robinson, A. [1967]: 'The Metaphysics of the Calculus', in I. Lakatos (*ed.*): [1967], pp. 28–40.

Robinson, R. [1936]: 'Analysis in Greek Geometry', *Mind*, **45**, pp. 464–73. Reprinted in *Essays in Greek Philosophy*, pp. 1–15. Oxford: Clarendon Press, 1969.

Robinson, R. [1953]: *Plato's Earlier Dialectic*. Second edition. Oxford: Clarendon Press.

Röntgen, W. C. [1895]: 'Über eine Neue Art von Strahlen', *Sitzungsberichte der Würzburger Physikalische-Medicinischen Gesellschaft*, Jahrgan, 1895. Translation in *X-rays and the Electric Conductivity of Gases*, pp. 28–47. Edinburgh: Livingston, 1958.

Rosser, J. B. [1937]: 'Gödel's Theorems for Non-Constructive Logics', *Journal of Symbolic Logic*, **2**, pp. 129–37.

Rosser, J. B. [1941]: 'The Independence of Quine's Axioms *200 and *201', *Journal of Symbolic Logic*, **6**, pp. 96–7.

Rosser, J. B. [1953]: *Logic for Mathematicians*. New York: McGraw-Hill.

Rosser, J. B. and Wang, H. [1950]: 'Non-Standard Models for Formal Logics', *Journal of Symbolic Logic*, **15**, pp. 113–29.

Russell, B. A. W. [1895]: 'Review of G. Heyman's: *Die Gesetze und Elemente des Wissenschaftlichen Denkens*', *Mind*, **4**, pp. 245–9.

Russell, B. A. W. [1896]: 'The Logic of Geometry', *Mind*, **5**, pp. 1–23.

Russell, B. A. W. [1901*a*]: 'The Study of Mathematics' in *Philosophical Essays*. Page references are to reprint in *Mysticism and Logic*, pp. 48–58. London: George Allen and Unwin, 1917.

Russell, B. A. W. [1901*b*]: 'Recent Work in the Philosophy of Mathematics', *The International Monthly*, **3**. Reprinted as 'Mathematics and the Metaphysician', in *Mysticism and Logic*. London: George Allen and Unwin, 1917.

Russell, B. A. W. [1903]: *Principles of Mathematics*. London: George Allen and Unwin.

Russell, B. A. W. [1910]: *Philosophical Essays*. London: George Allen and Unwin.

Russell, B. A. W. [1912]: *Problems of Philosophy*. London: George Allen and Unwin.

Russell, B. A. W. [1919]: *Introduction to Mathematical Philosophy*. London: George Allen and Unwin.

Russell, B. A. W. [1924]: 'Logical Atomism', in J. H. Muirhead (*ed.*): *Contemporary British Philosophy: Personal Statements*, First Series, pp. 357–83. Reprinted in R. C. Marsh (*ed.*): *Logic and Knowledge*, pp. 323–43. London: George Allen and Unwin, 1956.

Russell, B. A. W. [1935]: 'The Revolt Against Reason', in *Philosophical Quarterly*, **6**, pp. 1–19. Reprinted as 'The Ancestry of Fascism', in *In Praise of Idleness*, pp. 53–68. London: George Allen and Unwin.

Russell, B. A. W. [1944]: 'Reply to Criticism', in P. A. Schilpp (*ed.*): [1944], pp. 679–741.

Russell, B. A. W. [1948]: *Human Knowledge: Its Scope and Limits*. London: George Allen and Unwin.

Russell, B. A. W. [1959]: *My Philosophical Development*. London: George Allen and Unwin.

Russell, B. A. W. and Whitehead, A. N. [1925]: *Principia Mathematica*, volume 1. Second edition. Cambridge: Cambridge University Press.

Rychlik, K. [1962]: *Theorie der Reellen Zahlen im Bolzano's Handschriftlichen Nachlasse*. Prague: Verlag der Tschechoslowakischen Akademie der Wissenschaften.

Ryle, G. [1954]: *Dilemmas*. Cambridge University Press.

Sacks, G. E. [1972]: 'Differential Closure of a Differential Field', *Bulletin of the American Mathematical Society*, **78**, pp. 629–34.

Salmon, W. [1966]: *The Foundations of Scientific Inference*. University of Pittsburg Press.

Salmon, W. [1968*a*]: 'The Justification of Inductive Rules of Inference', in I. Lakatos (*ed.*): [1968*a*], pp. 24–43.

Salmon, W. [1968*b*]: 'Reply', in I. Lakatos (*ed.*): [1968*a*], pp. 74–97.

Savage, L. J., *et al.* [1961]: *The Foundations of Statistical Inference*. London: Methuen.

Schilpp, P. A. (*ed.*) [1944]: *The Philosophy of Bertrand Russell*. Northwestern University Press.

Schilpp, P. A. [1959–60]: 'The Abdication of Philosophy', *Kant Studien*, **51**, pp. 480–95.

Schilpp, P. A. (*ed.*) [1963]: *The Philosophy of Rudolf Carnap*. La Salle: Open Court.

Schläfli, L. [1870]: 'Über die Partielle Differentialgleichung $\frac{dw}{dt} = \frac{d^2w}{dx^2}$,' *Journal für Reine und Angewandte Mathematik*, **72**, pp. 263–84.

Schlick, M. [1934]: 'Über das Fundament der Erkenntnis', *Erkenntnis*, **4**. Translated into English as 'The Foundation of Knowledge' in A. J. Ayer (*ed.*): *Logical Positivism*, pp. 209–27. London: George Allen and Unwin.
Seidel, P. L. [1847]: 'Note über eine Eigenschaft der Reihen, welche Discontinuirliche Functionen darstellen', *Abhandlungen der Mathematik–Physikalischen Klasse der Königlich Bayerischen Akademie der Wissenschaften*, **5**, pp. 381–94.
Shimony, A. [1955]: 'Coherence and the Axioms of Confirmation', *Journal of Symbolic Logic*, **20**, pp. 1–28.
Shoenfield, J. [1971]: 'Measurable Cardinals', in R. O. Gandy and C. E. M. Yates (*eds.*): *Logic Colloquium '69*, pp. 19–49. Amsterdam: North Holland.
Sidgwick, H. [1874]: *The Methods of Ethics*. London: Macmillan.
Sierpinski, W. [1935]: 'Sur une Hypothèse de M. Lusin', *Fundamenta Mathematica*, **25**, pp. 132–5.
Smart, J. J. C. [1972]: 'Science, History and Methodology', *British Journal for the Philosophy of Science*, **23**, pp. 266–74.
Smith, D. E. [1929]: *A Source Book in Mathematics*. New York: Dover, 1959.
Solovay, R. M. and Tennenbaum, S. [1967]: 'Iterated Cohen Extensions and Souslin's Problem', *Annals of Mathematics*, **94**, pp. 201–45.
Specker, E. P. [1953]: 'The Axiom of Choice in Quine's *New Foundations for Mathematical Logic*', *Proceedings of the National Academy of Sciences, U.S.A.*, **39**, pp. 972–5.
Stauffer, R. C. [1957]: 'Speculation and Experiment in the Background of Oersted's Discovery of Electromagnetism', *Isis*, **48**, pp. 51–7.
Stegmüller, W. [1957]: *Das Wahrheitsproblem und die Idee der Semantik*. Vienna: Springer.
Stove, D. [1960]: Review of Popper [1959], *Australasian Journal of Philosophy*, **38**, pp. 173–87.
Strawson, P. F. [1954]: 'Wittgenstein's *Philosophical Investigations*', *Mind*, **63**, pp. 70–94.
Suppes, P. [1957]: *Introduction to Logic*. New York: Van Nostrand.
Szabo, A. [1969]: *Anfänge der Griechischen Mathematik*. Budapest: Akademiai Kiadó.
Tarski, A. [1939]: 'On Undecidable Statements in Enlarged Systems of Logic and the Concept of Truth', *Journal of Symbolic Logic*, **4**, pp. 105–12.
Tarski, A. [1954]: 'Comments on Bernays: "Zur Beurteilung der Situation in der Beweistheoretischen Forschung"', *Revue Internationale de Philosophie*, **8**, pp. 17–21.
Tarski, A. [1956]: 'The Concept of Truth in Formalised Languages: Postscript', in J. H. Woodger (*ed.*): *Logic, Semantics and Metamathematics*, pp. 268–78. Oxford: Clarendon Press.
Tichý, P. [1974]: 'On Popper's Definitions of Verisimilitude', *British Journal for the Philosophy of Science*, **25**, pp. 155–60.
Toeplitz, O. [1963]: *The Calculus: A Genetic Approach*. Translated by L. Lange. University of Chicago Press.
Toulmin, S. [1950]: *The Place of Reason in Ethics*. Cambridge University Press.
Toulmin, S. [1953*a*]: *The Philosophy of Science: an Introduction*. London: Hutchinson University Library.
Toulmin, S. [1953*b*]: 'Critical Notice of *Logical Foundations of Probability* by R. Carnap', *Mind*, **62**, pp. 86–99.
Toulmin, S. [1957]: 'Logical Positivism and After, or Back to Aristotle', *Universities Quarterly*, **11**, pp. 335–47.

Toulmin, S. [1958]: *The Uses of Argument.* Cambridge University Press.

Toulmin, S. [1961]: *Foresight and Understanding.* London: Hutchinson.

Toulmin, S. [1966]: 'Review of *Aspects of Scientific Explanation and Other Essays in the Philosophy of Science,* by Carl G. Hempel', *Scientific American,* **214**, Number 2, pp. 129–33.

Toulmin, S. [1968]: 'Ludwig Wittgenstein', *Encounter,* **68**, number 1, January, pp. 58–71.

Toulmin, S. [1971]: 'From Logical Systems to Conceptual Populations', in R. C. Buck and R. S. Cohen (*eds.*): *P.S.A.*, 1970, *Boston Studies in the Philosophy of Science,* **8**, pp. 552–64. Dordrecht: Reidel.

Toulmin, S. [1972]: *Human Understanding, 1: General Introduction and Part 1.* Oxford University Press.

Toulmin, S. [1974]: 'Rationality and Scientific Discovery', in R. S. Cohen and K. F. Schaffner (*eds.*): *P.S.A.*, 1972, *Boston Studies in the Philosophy of Science,* **15**, pp. 387–406. Dordrecht: Reidel.

Turing, A. M. [1939]: 'Systems of Logic Based on Ordinals', *Proceedings of the London Mathematical Society,* **45**, pp. 161–228.

Urbach, P. [1974]: 'Progress and Degeneration in the I.Q. Debate', *British Journal for the Philosophy of Science,* **25**, pp. 99–135 and 235–59.

Waismann, F. [1936]: *Einführung in das Mathematische Denken.* Translated into English as *Introduction to Mathematical Thinking.* London: Hafner Publishing Company, 1951.

Wang, H. [1959]: 'Ordinal Numbers and Predicative Set-Theory', *Zeitschrift für Mathematik und Grundlagen der Mathematik,* **5**, pp. 216–39.

Warnock, M. [1960]: *Ethics Since 1900.* Oxford University Press.

Watkins, J. W. N. [1952]: 'Political Traditions and Political Theory: An Examination of Professor Oakeshott's Political Philosophy', *Philosophical Quarterly,* **2**, pp. 323–37.

Watkins, J. W. N. [1957]: 'Farewell to the Paradigm Case Argument', *Analysis,* **18**, pp. 25–33.

Watkins, J. W. N. [1958]: 'Confirmable and Influential Metaphysics', *Mind,* **67**, pp. 344–65.

Watkins, J. W. N. [1968*a*]: 'Non-Inductive Corroboration', in I. Lakatos (*ed.*): [1968*a*], pp. 61–6.

Watkins, J. W. N. [1968*b*]: 'Hume, Carnap and Popper', in I. Lakatos (*ed.*): [1968*a*], pp. 271–82.

Watkins, J. W. N. [1970]: 'Against Normal Science', in I. Lakatos and A. Musgrave (*eds.*): [1970], pp. 25–37.

Watson, W. H. [1967]: *Understanding Physics Today.* Cambridge University Press.

Wehr, M. R. and Richards, J. A. [1960]: *Physics of the Atom.* Addison-Wesley.

Weitz, M. [1944]: 'Analysis and the Unity of Russell's Philosophy', in P. A. Schilpp (*ed.*): [1944], pp. 55–122.

Weyl, H. [1928]: 'Diskussionsbemerkungen zu dem Zweiten Hilbertschen Vortrag über die Grundlagen der Mathematik', *Abhandlungen aus dem Mathematischen Seminar der Hamburgischen Universität,* **6**, pp. 86–8.

Weyl, H. [1949]: *Philosophy of Mathematics and Natural Science.* Princeton University Press.

Whewell, W. [1858]: *History of Scientific Ideas,* volume 1. (Part One of the third edition of *The Philosophy of the Inductive Sciences.*)

Whewell, W. [1860]: *On the Philosophy of Discovery.* London: Parker.

Whittaker, E. [1951]: *A History of the Theories of Aether and Electricity: The*

Classical Theories. Enlarged and revised edition. London and New York: Nelson and Sons.

Wisdom, J. O. [1952]: *Foundations of Inference in Natural Science*. London: Methuen.

Wisdom, J. O. [1959]: 'Esotericism', *Philosophy*, **34**, pp. 338–54.

Wittgenstein, L. [1951]: *Philosophical Investigations*. Edited by G. E. M. Anscombe and R. Rhees. Oxford: Basil Blackwell.

Wittgenstein, L. [1956]: *Remarks on the Foundations of Mathematics*. Edited by G. H. von Wright, R. Rhees and G. E. M. Anscombe. Translated by G. E. M. Anscombe. Oxford: Basil Blackwell.

Wittgenstein, L. [1966]: *Lectures and Conversations in Aesthetics, Psychology and Religious Belief*. Edited by C. Barrett. Oxford: Basil Blackwell.

Wittgenstein, L. [1969]: *On Certainty*. Edited by G. E. M. Anscombe and G. H. von Wright. Oxford: Basil Blackwell.

Worrall, J. [1976]: 'Thomas Young and the "Refutation" of Newtonian Optics', in C. Howson (*ed.*): *Method and Appraisal in the Physical Sciences*, pp. 102–79. Cambridge University Press.

Zahar, E. G. [1973]: 'Why did Einstein's Research Programme Supersede Lorentz's?', *British Journal for the Philosophy of Science*, **24**, pp. 95–123 and 223–62.

Zahar, E. G. [1977]: 'Did Mach's Positivism Influence the Rise of Modern Science?', *British Journal for the Philosophy of Science*, **28**, pp. 195–213.

Lakatos bibliography[1]

[1946*a*]: 'Citoyen és Munkasosztály', *Valosag*, **1**, pp. 77–88.

[1946*b*]: 'A Fizikalai Idealizmus Biralata', *Athenaeum*, **1**, pp. 28–33.

[1947*a*]: 'Huszadik Szarsad: Tarsadalomtudomanyi és politikoi szemle, Budapest', *Forum*, **1**, pp. 316–20.

[1947*b*]: 'Eötvos Collegium – Györffy Kollégium', *Valosag*, **2**, pp. 107–24.

[1947*c*]: Review of K. Jeges: *Megtanulom a Fizikat* in *Tarsadalmi Szemle*, **1**.

[1947*d*]: Review of J. Hersey: *Hirosima* in *Tarsadalmi Szemle*, **1**.

[1947*e*]: 'Vigolia, Szerkeszti Johasz Vilmos es Sik Sandor', *Forum*, **1**, pp. 733–6.

[1961]: *Essays in the Logic of Mathematical Discovery*. Unpublished PhD dissertation. Cambridge.

[1962]: 'Infinite Regress and Foundations of Mathematics', *Aristotelian Society Supplementary Volume*, **36**, pp. 155–84.

[1963]: Discussion of 'History of Science as an Academic Discipline' by A. C. Crombie and M. A. Hoskin, in A. C. Crombie (*ed.*): *Scientific Change*, pp. 781–5. London: Heinemann. Republished as chapter 13 of this volume.

[1963–4]: 'Proofs and Refutations', *British Journal for the Philosophy of Science*, **14**, pp. 1–25, 120–39, 221–43, 296, 342. Republished in revised form as part of Lakatos [1976*c*].

[1967*a*]: *Problems in the Philosophy of Mathematics*. Edited by Lakatos. Amsterdam: North Holland.

[1967*b*]: 'A Renaissance of Empiricism in the Recent Philosophy of Mathematics?' in I. Lakatos (*ed.*): [1967*a*], pp. 199–202. Republished in much expanded form as Lakatos [1976*b*].

[1967*c*]: *Dokatatelstva i Oprovershenia*. Russian translation of [1963–4] by I. N. Veselovski. Moscow: Publishing House of the Soviet Academy of Sciences.

[1968*a*]: *The Problem of Inductive Logic*. Edited by Lakatos. Amsterdam: North Holland.

[1968*b*]: 'Changes in the Problem of Inductive Logic', in I. Lakatos (*ed.*): [1968*a*], pp. 315–417. Republished as chapter 8 of this volume.

[1968*c*]: 'Criticism and the Methodology of Scientific Research Programmes', *Proceedings of the Aristotelian Society*, **69**, pp. 149–86.

[1968*d*]: 'A Letter to the Director of the London School of Economics', in C. B. Cox and A. E. Dyson (*eds.*): *Fight for Education, A Black Paper*, pp. 28–31. London: Critical Quarterly Society. Republished as chapter 12 of this volume.

[1969]: 'Sophisticated versus Naive Methodological Falsificationism', *Architectural Design*, **9**, pp. 482–3. Reprint of part of [1968*c*].

[1970*a*]: 'Falsification and the Methodology of Scientific Research Programmes', in Lakatos and A. Musgrave (*eds.*) [1970], pp. 91–196. Republished as chapter 1 of volume 1.

[1] References to 'volume 1' are to Lakatos [1977*a*]. We have included as many of Lakatos's Hungarian writings as we have been able to trace.

[1970*b*]: Discussion of 'Scepticism and the Study of History' by R. H. Popkin, in A. D. Breck and W. Yourgrau (*eds.*): *Physics, Logic and History*, pp. 220–3. New York: Plenum Press.

[1970*c*]: Discussion of 'Knowledge and Physical Reality' by A. Mercier, in A. D. Breck and W. Yourgrau (*eds.*): *Physics, Logic and History*, pp. 53–4. New York: Plenum Press.

[1971*a*]: 'Popper zum Abgrenzungs- und Induktionsproblem', in H. Lenk (*ed.*): *Neue Aspekte der Wissenschaftstheorie*, pp. 75–110. Braunschweig: Vieweg. German translation of [1974*c*] by H. F. Fischer.
volume 1.

[1971*b*]: 'History of Science and its Rational Reconstructions', in R. C. Buck and R. S. Cohen (*eds.*): *P.S.A.*, 1970, *Boston Studies in the Philosophy of Science*, **8**, pp. 91–135. Dordrecht: Reidel. Republished as chapter 2 of volume 1.

[1971*c*]: 'Replies to Critics', in R. C. Buck and R. S. Cohen (*eds.*): *P.S.A.* 1970, *Boston Studies in the Philosophy of Science*, **8**, pp. 174–82. Dordrecht: Reidel.

[1974*a*]: 'History of Science and its Rational Reconstructions', in Y. Elkana (*ed.*): *The Interaction Between Science and Philosophy*, pp. 195–241. Atlantic Highland; New Jersey: Humanities Press. Reprint of [1971*b*].

[1974*b*]: Discussion Remarks on Papers by Ne'eman, Yahil, Beckler, Sambursky, Elkana, Agassi, Mendelsohn, in Y. Elkana (*ed.*): *The Interaction Between Science and Philosophy*, pp. 41, 155–6, 159–60, 163, 165, 167, 280–3, 285–6, 288–9, 292, 294–6, 427–8, 430–1, 435. Atlantic Highlands, New Jersey: Humanities Press.

[1974*c*]: 'Popper on Demarcation and Induction', in P. A. Schilpp (*ed.*): *The Philosophy of Karl Popper*, pp. 241–73. La Salle: Open Court. Republished as chapter 3 of volume 1.

[1974*d*]: 'The Role of Crucial Experiments in Science', *Studies in the History and Philosophy of Science*, **4**, pp. 309–25.

[1974*e*]: 'Falsifikation und die Methodologie Wissenschaftlicher Forschungsprogramme', in I. Lakatos and A. Musgrave (*eds.*): *Kritizismus und Erkenntnisfortschritt*. German translation of [1970*a*] by A. Szabo.

[1974*f*]: 'Die Geschichte der Wissenschaft und Ihre Rationalen Reconstruktionen', in I. Lakatos and A. Musgrave (*eds.*): *Kritizismus und Erkenntnisfortschritt*. German translation of [1971*b*] by P. K. Feyerabend.

[1974*g*]: *Wetenschapsfilosofie en Wetenschapsgeschiedenis*. Boom: Mepple. Dutch translation of [1970*a*] by Karel van der Lenn.

[1974*h*]: 'Science and Pseudoscience', in G. Vesey (*ed.*): *Philosophy in the Open*. Open University Press. Republished as the introduction to volume 1.

[1976*a*]: 'Understanding Toulmin', *Minerva*, **14**, pp. 126–43. Republished as chapter 11 of this volume.

[1976*b*]: 'A Renaissance of Empiricism in the Recent Philosophy of Mathematics?', *British Journal for the Philosophy of Science*, **27**, pp. 201–23. Republished as chapter 2 of this volume.

[1976*c*]: *Proofs and Refutations: The Logic of Mathematical Discovery*. Edited by J. Worrall and E. G. Zahar. Cambridge University Press.

[1977*a*]: *The Methodology of Scientific Research Programmes: Philosophical Papers*, volume 1. Edited by J. Worrall and G. Currie. Cambridge University Press.

[1977*b*]: *Mathematics, Science and Epistemology: Philosophical Papers*, volume 2. Edited by J. Worrall and G. Currie. Cambridge University Press.

With other authors

[1968]: *Problems in the Philosophy of Science.* Edited by I. Lakatos and A. Musgrave. Amsterdam: North Holland.

[1970]: *Criticism and the Growth of Knowledge.* Edited by I. Lakatos and A. Musgrave. Cambridge University Press.

[1976]: 'Why Did Copernicus's Programme Supersede Ptolemy's?', by I. Lakatos and E. G. Zahar, in R. Westman (*ed.*): *The Copernican Achievement,* pp. 354–83. Los Angeles: University of California Press. Republished as chapter 5 of volume 1.

Index of names

(Indexes compiled by Allison Quick)

Subject Index